Ink Sandwiches, Electric Worms, and 37 Other Experiments for Saturday Science

Ink Sandwiches, Electric Worms,

and 37 Other Experiments for Saturday Science

Neil A. Downie

THE JOHNS HOPKINS UNIVERSITY PRESS
Baltimore and London

Printed in the United States of America on acid-free paper
9 8 7 6 5 4 3 2 1

The Johns Hopkins University Press
2715 North Charles Street
Baltimore, Maryland 21218-4363
www.press.jhu.edu

Library of Congress Cataloging-in-Publication Data

Downie, Neil A., 1955–
Ink sandwiches, electric worms, and 37 other experiments for Saturday science / by Neil A. Downie.
p. cm.
Includes bibliographical references and index.
ISBN 0-8018-7409-2 (hardcover : alk. paper) — ISBN 0-8018-7410-6 (pbk. : alk. paper)
1. Science—Experiments. 2. Science—Popular works. I. Title.
Q182.3.D68 2003
507′.8—dc21 2002154023

A catalog record for this book is available from the British Library.

Illustrations by Richard Williams. Figures that appear at each of the twelve part openers are selected U.S. patent drawings from the years 1928 to 2002.

Vertical thinking is a smooth progression, from one solid step to another solid step. . . . vertical thinking is high-probability thinking. Lateral thinking is low-probability thinking. . . . When low-probability thinking leads to an effective new idea there as a "Eureka moment." . . . The difference between lateral and vertical thinking is that with vertical thinking logic is in control of the mind, whereas with lateral thinking logic is at the service of the mind.

—Edward de Bono, *The Use of Lateral Thinking*

Contents

		Page	Difficulty	Cost
	Preface	xi		
CHAOTIC CLOCKS		**1**		
1	Chaotic Regularity Hammer randomly at a pendulum and get a remarkably accurate clock.	4		
2	Glacial Oscillations Ice melts and re-forms rapidly, then melts and re-forms slowly.	13		
3	Everlasting Hourglass Is it a clock or a concrete mixer?	20		
AERODYMAGIC		**31**		
4	Juggling Airstreams How many ping-pong balls can you balance on top of one another?	34		
5	Railroad Yacht Go sailing without seasickness pills.	41		
SOUNDS INTERESTING		**49**		
6	Musical Glugging Is this what Handel meant by water music?	51		
7	Pneumatic Drum Try out this percussion instrument for bagpipe players.	57		

Difficulty | Cost

8 Singing Contacts 63
Listen to those electrons.

JOLLY BOATING 71

9 Giant Putt-Putt Boat 73
Build a jet boat that's ideal for swamps.

10 Follow That Field! 79
Trace electric fields with a boat.

TRANSPORTS OF DELIGHT 91

11 Electric Worms 93
The magic of friction makes an unusual vehicle.

12 Vacuum Railroad 100
Travel by tube with help from a vacuum cleaner.

13 Naggobot, or Reverse Ice Vehicle 108
Jingle bells for the snowless make summer sleigh rides possible.

14 Boadicea's Autochariot 115
Keep the army of ancient Rome at bay with a whirling-arm vehicle.

15 Tubal Travelator 122
Construct a conveyance of convenience.

CENTRIPETAL FORCE AND CENTRIFUGAL PROJECTILES 131

16 Centripetal Chaos 134
Take a sideways look at the lava lamp.

17 The Rotapult 140
Watch out, Goliath!

EXOTIC AMPLIFICATION 147

18 Transformer Transistors 149
See the magic of magnetism with an amplifier that the Victorians could have made.

19 Electrolystor Amplifier 160
Control an AC output with bubbles from a DC control current.

			Difficulty	Cost
VIBRATIONS, ROTATIONS, AND CHANCE		**169**		
20	Waltzing Tube Observe curious rotation around two axes in this Victorian toy.	171		
21	Motor Dice Electrify your dice, and don't forget the nudge switches.	176		
MAVERICK MEASUREMENT		**183**		
22	Coffee-Cup Revolution Counter Track the approach of dinosaurs and measure rpm.	185		
23	Coulter's Bubbles Blow bubbles through your gas meter.	190		
24	Electronic Elastic Make a force sensor from a rubber band.	198		
25	Light Tunnels Measure angles with a bent tube.	204		
26	Reverse Electric Lamp Solar Tracker Put light into this bulb and get electricity out.	213		
27	Gravity Diode Electrons appear to know which way is up.	220		
28	AMIPLEX Go from an experiment to a graph in a millisecond.	226		
CURIOUS COMMUNICATIONS		**233**		
29	Servo Telegraph An output wheel in Anchorage follows every movement of an input wheel in Key West.	235		
30	Send Me a Bubblegram Make a telegraph from the orderly progress of bubbles in tubes.	245		
31	Pneumatic Morse Send messages with compressed air.	252		
32	Seven-Segment Telegraph Send messages with an ingenious scheme of wires and diodes.	259		
33	Six-Wire Telegraph Send reduced messages—"It's language, Jim, but not as we know it."	264		

			Difficulty	Cost
34	Moving Messages Rapid rotation deceives the eye in these prestidigitatory electric illusions.	276		
	UNUSUAL ACTUATORS	**285**		
35	Balloon Biceps Get muscles like Popeye's—from a balloon.	287		
36	Ink Sandwiches These sandwiches are not for eating but for controlling light.	292		
	ELECTROCHEMICAL MAGIC	**299**		
37	Red-Hot Batteries Need electric power? Plug in a flame!	301		
38	Unusually Cool Sunglasses Make shades from electrolytically etched aluminum foil.	307		
39	Wet Solar Cell Get power from sunshine and salt.	313		
	Hints and Tips	321		
	Suggested Further Reading	325		
	Index	329		

Preface

This book offers demonstrations of a number of unusual phenomena, with explanations and analyses of how and why they work. The demonstrations are novel in principle or detail, and most readers will not have seen them before or analyzed how they work. I hope that this novelty will refresh and stimulate the imaginations of readers and perhaps encourage what Edward de Bono calls "lateral thinking" in science and engineering.

The demonstrations come not only from my work as an industrial scientist but also from my search for new science projects to amuse and instruct the kids of the Saturday Activity Centre (SAC), a Saturday morning club in my hometown. One of the projects comes from some work I did for the Gatsby Technical Education Project. Doing something original for each session at the SAC made it more interesting for me and had the happy side effect of making the sessions more memorable for the children. It is, after all, human nature to remember surprises. When something is counterintuitive—when you first see the green flash as the sun goes down or when you feel a gyroscope wriggle like a live animal as you turn it over in your hands—it becomes etched in memory.

Those interested only in trying the experiments can do so. However, in sections titled The Science and the Math, I have expanded the mathematical analysis and explanations that go with each phenomenon for those who desire a deeper understanding. The level of mathematics required varies from elementary algebra and geometry to simple calculus. The physics theory centers on mechanics and waves, but it also touches on a variety of more complex subjects: hydrodynamics, electricity, electronics, and even chemistry. I hope that the university graduate, even the science specialist, will find the discussions stimulating. Nevertheless, the book has its roots in demonstrations for kids, and bright teenagers will understand most of the main text of the book.

I have tried to make the demonstrations reasonably foolproof, and I hope that they are reproducible—not too dependent on my own "magic touch." Nearly all are practical experiments you can try yourself without expensive apparatus, although they do vary in difficulty. In Hints and Tips at the back of the book, you will find material on practical matters common to a number of the projects, including a list of English and metric conversion factors.

The demonstrations are grouped in twelve parts so that they are easy to read and so that their common phenomena and analyses can be compared and contrasted. The material within most of the chapters is presented in the same order, which should make the book easier for the reader to navigate. Where necessary, though, the format has been varied a little to suit the individual subject under discussion. My more detailed theoretical analyses and explanations are clearly distinguished from the rest of the text and can thus be omitted if desired, at least on a first reading.

Science that has been forgotten in the mists of time holds a certain fascination for me. The history of science and technology is littered with false starts. Although they are interesting in themselves, a fraction of these false starts, though they led to dead ends in the past, will lead to useful technology in the future, as societal conditions change and "enabling technologies" arise. The introductions to the parts describe some of the cul-de-sacs of history that relate to the demonstrations.

New phenomena need new analytical understanding, and this is, I hope, one of the values of the book. It provides practice in applying the analytical principles of mathematical physics, practice that is made more relevant and less abstract because it is applied to real demonstrations and experiments. In addition, the novelty of the experiments means that there are no right answers to be found in standard books. Readers may therefore feel a little closer to the leading edge of science and be enthusiastic enough to carry the analyses in this book further on their own. For as Mark Twain once said:

> What is it that confers the noblest delight? What is that which swells a man's breast with pride above that which any other experience can bring to him? Discovery! To know that you are walking where none others have walked; that you are beholding what human eye has not see before; that you are breathing a virgin atmosphere. To give birth to an idea—an

> intellectual nugget, right under the dust of a field that many a brain-plow had gone over before. To be the first—that is the idea. To do something, say something, see something, before anybody else—these are the things that confer a pleasure. (*Innocents Abroad,* chap. 26)

The analyses and the simple mathematical models given could be extended for nearly all the experiments. I have made approximations and simplifying assumptions that the reader may want to modify or improve. The removal of some of these simplifications may well reveal more subtle points in the analyses. The theory of any real phenomenon is always, at the deepest level, more complex than can be fully analyzed with simple algebra. Even where an attempt at an exhaustive analysis has been made, some of the field is left unplowed.

I encourage readers to try the demonstrations for themselves. You will see effects in the real world that you might miss if you only read and analyze. I am a firm believer in doing both "the thinking and the doing," the practice and the theory. I have graded the projects with respect to degree of difficulty and cost. In addition to instructions for a basic version of the demonstration for each project, I include various hints on extending it. The further developments suggested obviously have a greater experimental content and will be more difficult, but they hold out the prospect of being correspondingly more rewarding.

Finally, please send your suggestions and feedback to me care of the publisher.

I am indebted to many people for their help and encouragement. My wife, Diane, and my daughters, Helen and Becky, not only have often been asked to "hold this while the glue sets" but also have made intellectual contributions.

Space limitations prevent me from listing everyone who helped in the preparation of this book, but a few people deserve special mention: Johns Hopkins University Press editor-in-chief Trevor Lipscombe for his stimulating discussion about the book and his wholehearted support; Peter Strupp and his team at Princeton Editorial Associates, who helped convert my tangled thoughts into something intelligible; Peter Madden, who helped with a couple of good projects; and Tim Rowett, toy collector and consultant extraordinaire, who has kept me abreast of the sometimes extraordinary things that are made to amuse.

My colleagues at Air Products & Chemicals have often been the inspiration, sometimes unwittingly, for projects. Thank you, guys! Thank you, ladies! I would also like to thank the kids (and their parents) at the Saturday Activity Centre in Guildford, where each Saturday we try out projects like the ones in this book.

I'd like to thank the many people who made suggestions. I could not use all of them, either because there was not enough space in the book or because I have not found time to carry out the necessary experiments. Perhaps these ideas will make it into a future book.

Ink Sandwiches, Electric Worms,

and 37 Other Experiments for Saturday Science

Chaotic Clocks

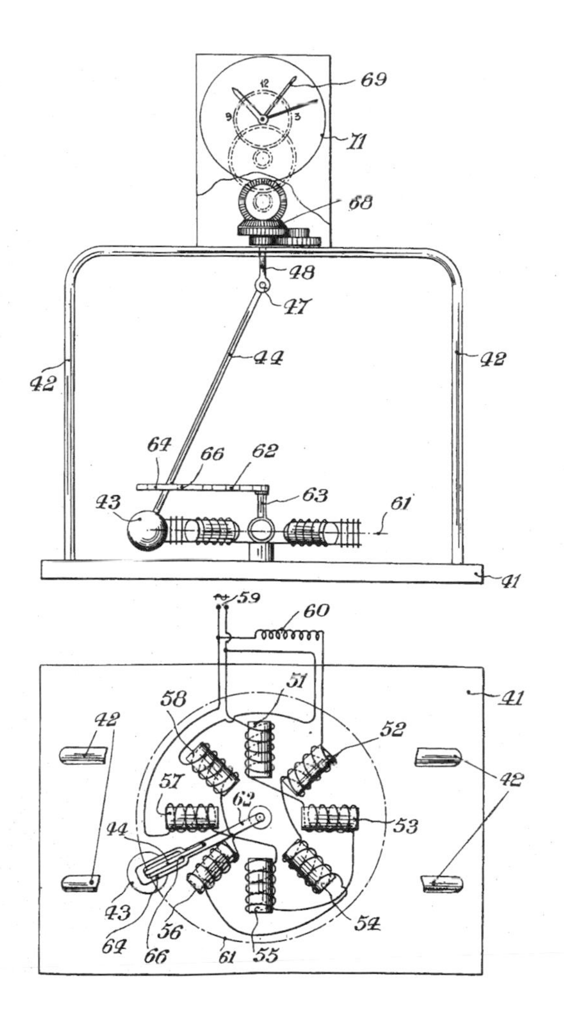

It turns out that an eerie type of chaos can lurk just behind a facade of order—and yet, deep inside the chaos lurks an even eerier type of order.

—Douglas R. Hofstadter, *Gödel, Escher, Bach: The Eternal Golden Braid*

Most clocks amount to an oscillator linked to a counting mechanism and a device for displaying the count. Look inside a modern wristwatch, and you will find a piezoelectric quartz oscillator "ticking" 32,000 times per second and an integrated circuit that counts the oscillations and passes the count on to the liquid crystal display. Look inside the elegant ormolu case of an eighteenth-century French carriage clock, and you will see something that has a completely different appearance, but whose functions are analogous. A delicate balance-wheel oscillator swings clockwise and then counterclockwise every second or two, its almost silent ticking activating a complex "going train" of cogs and gearwheels that convey a count of the swings to the hands on the ornately inscribed face.

The oscillators that power clocks are apparently steady in their operation. To and fro the balance wheel goes; the cogs slowly turn, with the smaller cogs visibly jerking around a precise angle, the larger moving almost imperceptibly but steadily, inexorably onward. Temperature changes and similar environmental changes have predictable effects, all carefully compensated by the cunningly designed parts of the mechanism, and overall the mechanism does not apparently deviate from complete regularity.

But look closely and carefully behind the orderly facade of a clock, and you will find random processes at work. These processes make clocks' operation intrinsically irregular—only slightly irregular, it is true, but nevertheless irregular. Designers try to minimize the influence of these processes. In old-fashioned clocks, the unpredictable effects of friction, for example, are reduced by the use of jeweled bearings, tiny pieces of drilled ruby. In modern electronics, random noise in oscillators—*jitter* is the technical term—is carefully measured, and strenuous efforts are made to minimize it. (I have myself been associated with such efforts; see the references.)

Here we consider three practical examples of clocks—periodic oscillators—that are actually powered by processes that have an obvious random component but seem to produce periodic results in two of the examples and random results in the third. The details of the chaotic impact clock, in which a rotating hammer strikes a pendulum, are complex, but the overall result is simple: the regular swinging of the pendulum. No one can precisely predict the flow of sand, yet that flow leads, at least some of the time, to the predictable oscillations of our everlasting hourglass. The freezing of ice might seem more predictable than the flow of sand, but the glacial oscillator we make using a freezing and thawing cycle produces an unpredictable output.

REFERENCES

The fascination factor of clocks and watches is so powerful that their early history is covered in many books. Here are a few titles; more are listed in Suggested Further Reading at the back of the book.

Britten, F. J. *Old Clocks and Watches & Their Makers: Being an Historical Descriptive Account of the Different Styles of Clocks and Watches of the Past, in England and Abroad, to Which Is Added a List of Ten Thousand Makers.* 2d ed. London: Batsford, 1904.

Downie, Neil A. "The Anti-Jitter Circuit." *Electronics World,* October 1999, 859–863. Describes electronics professor Mike Underhill's fundamental anti-jitter circuit concept and some of its variations.

Sobel, Dava. *Longitude: The True Story of a Lone Genius Who Solved the Greatest Scientific Problem of His Time.* 1st ed. London: Fourth Estate, 1996. Describes the extraordinary efforts of mechanical engineering that clockmaker John Harris exerted to produce a mechanical clock accurate enough (parts per million!) to measure the longitude of a ship at sea.

1 Chaotic Regularity

> Everything he heard about clock movements confirmed that they were precision instruments of exact design and construction.
>
> —J. G. Ballard, *Chronopolis*

Early clocks made with trains of gearwheels used mechanical regulators, just as today's mechanical clocks do. But medieval clocks relied on a weighted metal bar, the *foliot,* whose swinging to and fro regulated the speed of the gearwheels. The foliot was an inaccurate device that was inferior to a true oscillator as a regulator, and it relied on great skill on the part of the clockmaker to make it workable. Eventually, clockmakers realized the importance of finding a better regulator. If a regulator that strongly governed the speed of the mechanism could be found, an inexact instrument of indifferent construction could be made to run precisely. Galileo's observations about pendulums in 1641 made it clear that a pendulum would make a much better regulator. (Galileo famously observed the swinging of a church chandelier, checking its constancy of motion by reference to his own heartbeat.) However, it would be sixteen years before Christiaan Huygens and his associates began to make timepieces with pendulum regulators in the Netherlands, and thirty years before the accurate pendulum clocks that we know as long-case or grandfather clocks began to appear in London and elsewhere.

The key to the success of the pendulum is that its constancy of timing did not depend on the application of exactly the same amount of power on every stroke. Slight power variations were inevitable, particularly in portable clocks, which relied on a wound spring rather than a weight to drive them. These variations led to severe inaccuracy, even after the invention of ingenious devices such as fusée barrels and stackfreed mechanisms to control the spring's power. The use of pendulums in such clocks was a radical improvement because, as we will see, the frequency of a pendulum's swing varies very little with the amplitude of that swing. So although the pendulum in a newly wound clock will swing wide and that in a nearly exhausted one will swing only a fraction of its former amplitude, both pendulums will oscillate with nearly the same period.

In this project, we dispense with any attempt to maintain constant power to the pendulum and instead apply random impacts to keep it going. It is a curious fact that despite the randomness of the impacts on the pendulum, its timekeeping is surprisingly good.

What You Need

- ❑ Electric motor (I suggest a small 1- to 3-V type, but almost any would serve.)
- ❑ Single-cell battery (1.2 or 1.5 V)
- ❑ Battery holder
- ❑ Thread spool
- ❑ Light wooden dowel rod, 6 mm in diameter and 300–500 mm long
- ❑ Thin, round 15-mm nail (e.g., veneer pin or panel pin), as pivot
- ❑ Washers to fit over the nail
- ❑ Various pieces of wood for mounting the components
- ❑ Hot-melt glue
- ❑ Modeling clay (for pendulum bob)
- ❑ Resistance wire (e.g., Nichrome, 28 swg, 10 ohm/m)

What You Do

The key to this experiment is setting it up so that just as the pendulum returns to the vertically downward position, a blow from the thread spool causes it to swing out again. To accomplish this, mount the spool on the motor shaft with the shaft offset from the center of the spool, so that the spool rotates eccentrically and rapidly. I used a thread spool arranged to rotate with 5 mm or so of eccentricity on the motor, although there is obviously scope for variation here. The motor can be run on the battery alone or on the battery through a short

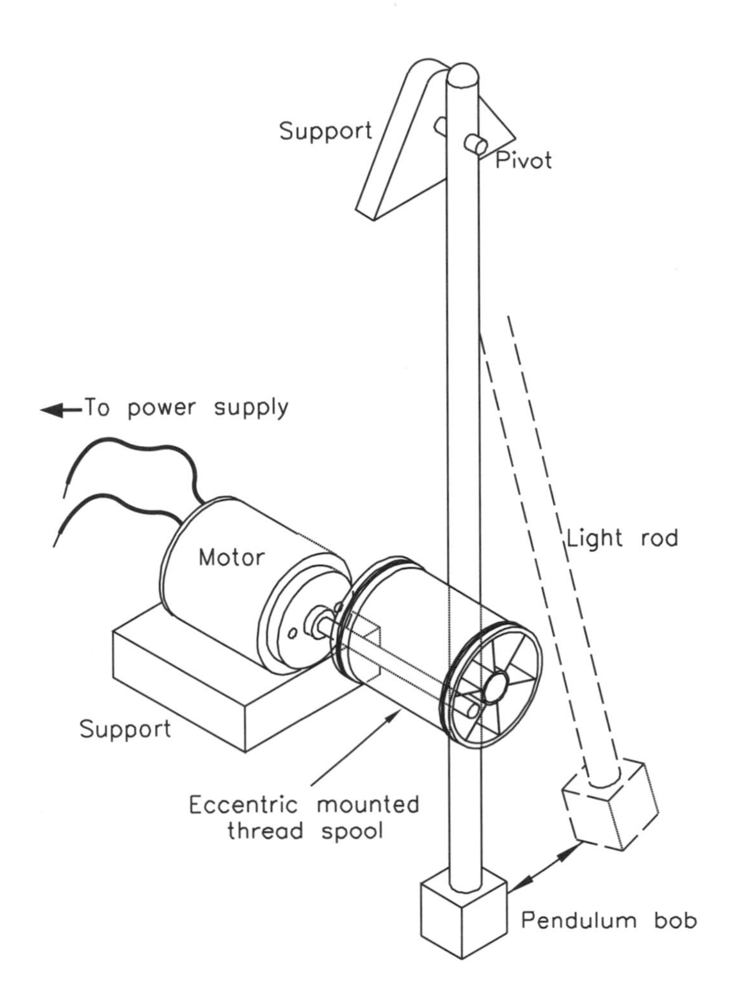

piece of resistance wire (try a 15-cm piece) to slow the motor slightly. The pendulum must swing freely, so predrill the hole in the rod to give a slightly loose fit, and space the rod out from its mounting using washers. When hanging vertically, the pendulum should only just be hit by the eccentric spool.

Now start the motor. After a moment, it will hit the pendulum hard enough to set it swinging a few tens of millimeters. When the pendulum approaches the vertically downward position on the return swing, the spool will hit it again. Because the motor turns many times during the pendulum's swing and because it does not run at a perfectly precise speed, the angle at which the eccentric spool hits the rod will be random from swing to swing. Therefore, the impact of the spool on the rod will vary, giving a random amplitude of swing.

At first, you may doubt the regularity of the pendulum. You will see that its amplitude varies enormously, and you may be fooled into thinking that the clock is running unevenly. Try closing your eyes, though. When you do, you will notice that the "clack" sounds that you hear when the eccentric hits the pendulum are remarkably periodic, sounding something like a metronome.

How It Works

The power of the pendulum concept in this clock, just as in Huygens's original pendulum clock, is that the amplitude of oscillation affects the period of oscillation only slightly. Whether the pendulum has been kicked out 10 cm by a lucky hit or has just barely been moved a millimeter or two, it will still take the same half second or so to swing out and back. You may find this surprising. Common sense would seem to dictate that a hard kick on the pendulum would make it take longer to swing back than a soft kick would. Here we have an example, one of many in science, where common sense is simply wrong. It's wrong because of the nature of the restoring force that pulls the pendulum back to the vertical position: this force is proportional to the angle from vertical. The farther the pendulum is kicked from vertical, the greater the restoring force. This greater force will give the pendulum a proportionally greater acceleration. The result is that, at least for fairly small angles to vertical, the extra distance given by the hard kick is neatly compensated by a correspondingly faster acceleration, and the time taken to return to vertical remains unchanged.

THE SCIENCE AND THE MATH

Recall that in a pendulum that is not deflected too far from vertical, the restoring force, F, is proportional to the horizontal displacement, x:

$$F = Mgx/L,$$

where M is the mass of the pendulum bob, g is the acceleration of gravity, and L is the pendulum length. Using Newton's second law of motion, we obtain

$$F = M(d^2x/dt^2) = -Mgx/L,$$

or

$$d^2x/dt^2 = -(g/L)x.$$

By setting $\omega^2 = g/L$, we arrive at

$$d^2x/dt^2 = -\omega^2 x.$$

This equation has the solution

$$x = x_0 \sin(\omega t),$$

where x_0 is the amplitude. This solution is periodic—it predicts that the pendulum will repeat its motion every $2\pi\omega$. No matter what value of x_0 is chosen, the period will be the same. In our clock, x_0 is determined randomly on each stroke, but this does not affect the period of the clock. Another feature of our clock is that the period, $\pi\omega$, is half that of the usual pendulum clock, $2\pi\omega$, because the pendulum flies out only to one side of vertical.

We can use a computer to make a more sophisticated model for the motion of the pendulum, accounting for the fact that the restoring force is not exactly proportional to the amplitude and accounting for the different impacts delivered by the eccentric on each stroke. This model requires that we determine how much the impact of the eccentric and, hence, the amplitude of swing vary. Taking a model in which the period of the pendulum, T, is given in terms of the amplitude by

$$T = \pi(1 + A^2/16)/\omega,$$

where A is the amplitude in radians, we can construct the computer model. This formula, but for the factor of two discussed above, is derived in many places (see, for example, Kibble, *Classical Mechanics,* 35–36). The formula gives a period that varies with amplitude, but only slightly for large swings and very slightly indeed for small swings:

Swing (degrees)	Period (relative to zero amplitude period)
0	1
5	1.0005
10	1.0019
20	1.0076
30	1.0171

Using a spreadsheet program, we can simulate the effect of the random impacts on the regularity of the pendulum. First, we assume that the eccentric gives random impacts that result in swings ranging from 5 to 30 degrees, and then we assign the pendulum swing periods according to the above formula and plot successive amplitudes and swing periods

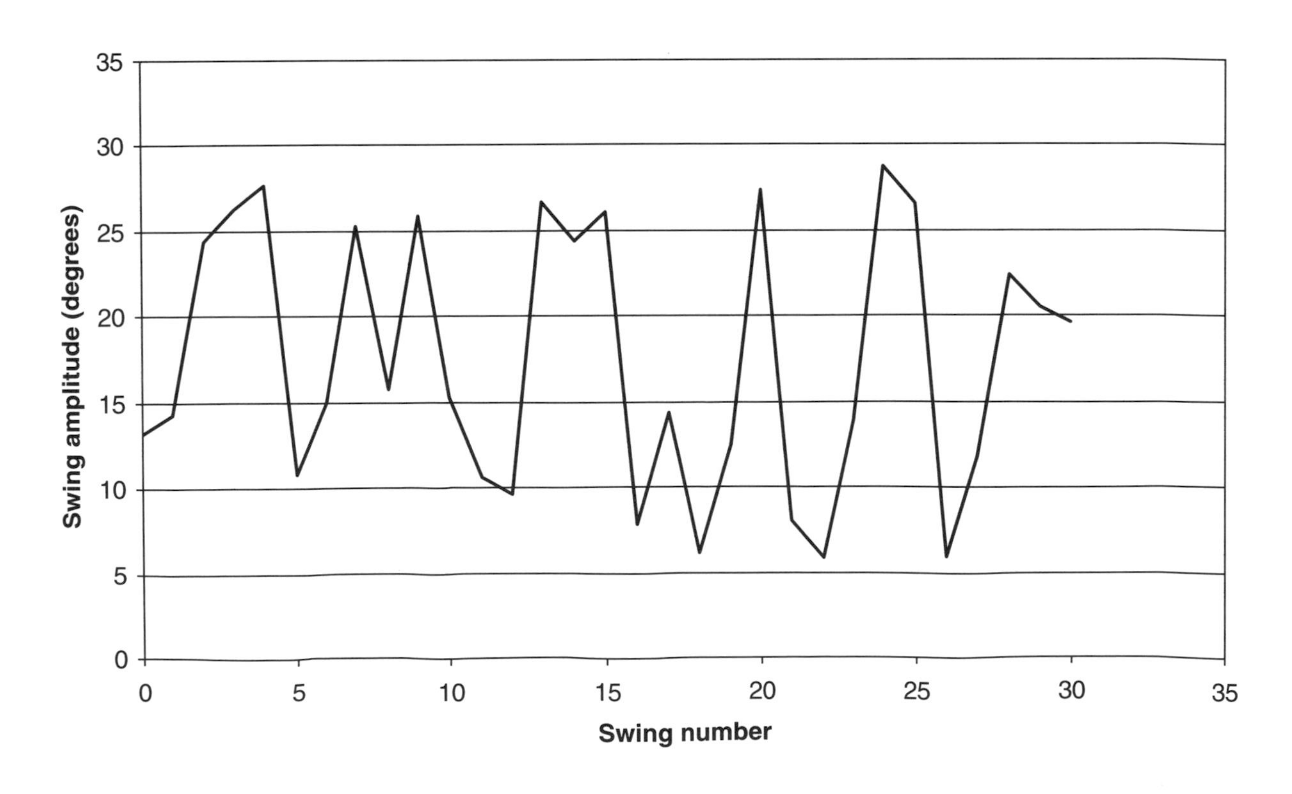

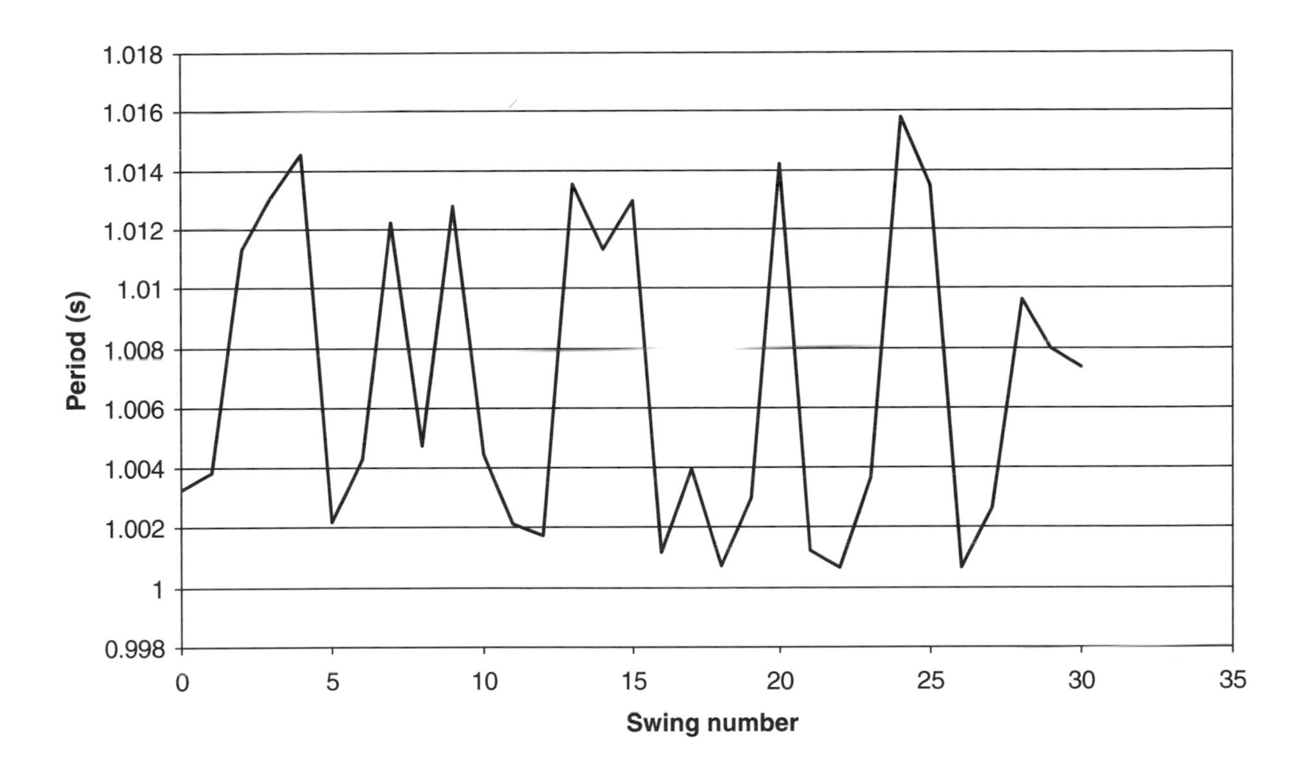

versus time. Although simplified by our assumptions, this simulation does seem to reproduce the behavior of the actual pendulum, at least qualitatively. The two plots also clearly indicate that the chaotic nature of the pendulum's motion is little disadvantage in terms of keeping good time: although individual periods vary from 0.005 to 1.7 percent in error, the average over a number of cycles is much less variable, ranging from about 0.06 percent to 0.07 percent for an average of 200 cycles. The percentage by which the clock is in error will decrease further as the number of cycles increases.

And Finally . . . Advanced Random Pendulums

If we are to use the periodicity of our pendulum for anything other than a metronome, then we need to connect it to mechanisms for counting and displaying the number of oscillations. Here, however, the variable amplitude of the swing is a real problem. The normal clockwork mechanisms of ratchets and escapement wheels rely on a fairly steady amplitude of swing. Perhaps an escapement wheel linked by a springy, mechanically compliant link would work. Set up to release the "going train" of the clock on the slightest swing, the springy link would simply absorb the extra motion of the pendulum on larger swings. Or perhaps a photoelectric device—such as an LED-photodiode pair found, for example, inside a computer pointing device (a mouse)—would be a better idea because it would impose no mechanical load on the pendulum. The impulses from the photodiode could be amplified by a simple transistor to operate a mechanical counter, or used to pull in a solenoid on a ratchet. Whatever mechanism we used would have to operate down to small swing values so as not to miss a swing.

Our spreadsheet analysis of the random motion of the pendulum could be made considerably more sophisticated. We could try to simulate the dynamics of the impact, that is, the variations in the speed-up of the motor and its eccentric during the beat of the pendulum. The motor speeds up more during a long beat than during a short beat. Simulating the dynamics might shed light on features that do not show up in the simple random model.

How do variations in motor speed affect the pendulum amplitude? One possibility is that the motor speed-up effect might tend to steady the amplitude of the pendulum. During a long beat of the pendulum, the motor reaches a higher speed, which might increase the probability that the next beat will also be a long beat, because the eccentric will most likely hit the pendulum harder. The opposite might occur with a short beat: surely a short beat will, on the average, be followed by another short beat. This effect will be very small for small pendulum swings because the effective duration for the speed-up of the motor varies by only milliseconds for swings of less than 20 degrees.

A second possibility is that the motor speed-up effect might actually reduce the regularity of the pendulum. A hard impact of the eccentric on the pendulum will slow the motor, and perhaps the large swing caused by the hard impact will be followed by a smaller swing because the motor will not have recovered its full speed for the next impact. This effect ought to predominate over the previous one because it depends on the amplitude rather than the period of the previous swing.

We can also use a spreadsheet program to model the motor speed-up effect. In the model, the new swing amplitude, A_n (in degrees), is given by

$$A_n = \varphi(2 - kA_{n-1})^2,$$

where φ is a random factor from 0 to 1 given by the computer's random number generator, k is a constant, normally between 0 and 1, and A_{n-1} is the amplitude of the previous swing. The amplitude correlation behavior is adjusted using k. If you make $k > 1$, the swings grow in amplitude with time, whereas when $k = 0$,

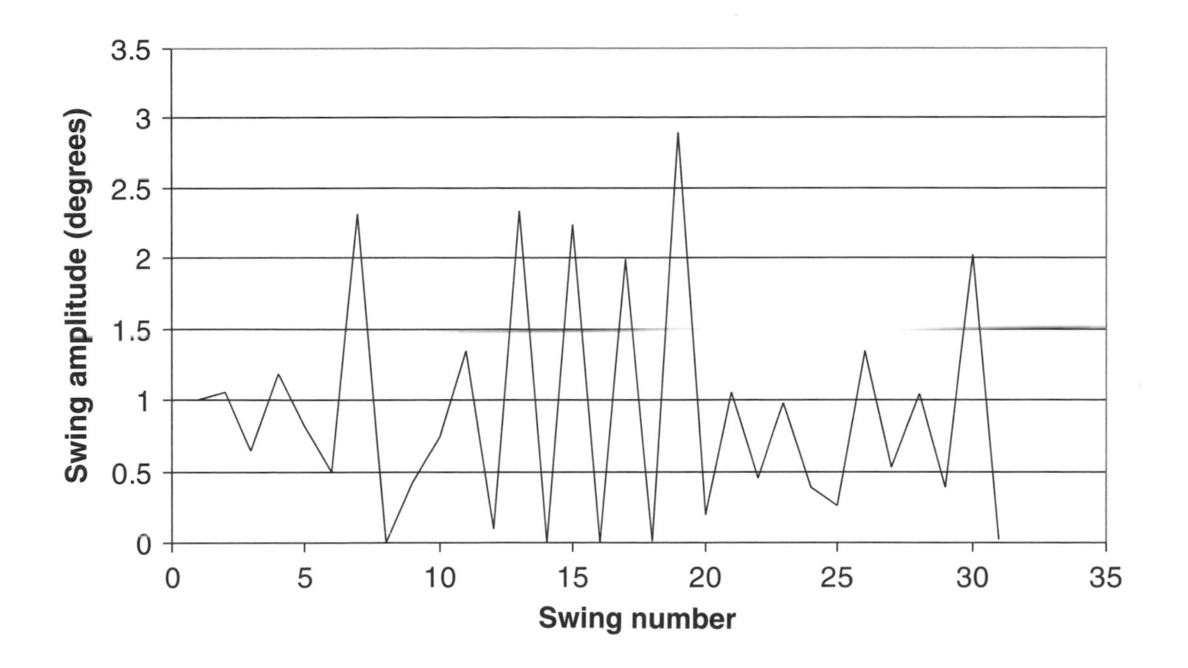

the swings are random with no correlation. When you plot this function, you'll see that there is a higher frequency in the simulation with increased *k:* an increased tendency for a large swing to be followed by a small one, which gives a more jagged set of peaks on the graph. But is this effect really seen in practice? And what is the best test for the correlation between swings?

Another source of inaccuracy in our pendulum occurs at small amplitudes of swing. Our pendulum undergoes half a swing between impacts of the eccentric: its period is half that of a conventional pendulum clock and, as we have seen, is nearly independent of how far the pendulum swings out. However, the swing amplitude is not the only variable. There is also variation in where the pendulum is in its swing when the eccentric hits it: sometimes the pendulum will be hit just before its half-swing has been completed, sometimes just afterward.

REFERENCES

Chaotic phenomena that are approximately periodic are described in many popular math books. See, for example,

Coveney, Peter V., and Roger Highsmith. *The Arrow of Time: A Voyage through Science to Solve Time's Greatest Mystery.* London: Flamingo, 1991.

Gleick, James. *Chaos*. London: Heinemann, 1988.

Kibble, T.W.B. *Classical Mechanics*. London: McGraw-Hill, 1973.

Stewart, Ian. *Does God Play Dice? The Mathematics of Chaos*. 2d ed. London: Penguin, 1997. See Chapter 13.

2 Glacial Oscillations

When water turns to ice does it remember one time it
was water?
When ice turns back into water does it remember it
was ice?

—Carl Sandburg, "Metamorphosis"

The chaotic impact clock in the previous chapter reveals how order can emerge from chaos. By contrast, the glacial oscillator shows how chaos can break out in what should be an orderly periodic oscillation. If you remove heat from water—that is, cool it down—it will eventually freeze. If you have a source of constant heat removal, say, a freezer, and a sensor that stops the heat removal when the water has frozen and then signals for the application of heat, you have the basis for an oscillator. Once the ice sensor has given the signal and heat is applied faster than it is removed, the ice will melt back to water, which will stop the heater and thus allow freezing. Freezing will in turn signal the ice sensor to apply heat again and so on.

The application of heat is straightforward: a simple electrical resistor, such as a radio resistor, will do. But detecting ice formation is more challenging. The curious but true fact that water conducts electricity whereas ice does not is the basis for our glacial oscillator. We arrange a simple inverting amplifier to switch current to a heating resistor when a small current is conducted through electrodes in the water. We complete the device by putting it in a freezer.

What You Need

- ❑ 2 transistors, almost any type will do (I used cheap BC184 devices.)
- ❑ Resistors, 3.3 kohm, 330 kohm, 100 ohm (this must be a high rating, above 0.5W)
- ❑ Small plastic soda bottle cap, or other 25-mm plastic container
- ❑ Salt
- ❑ Wires
- ❑ 12-V DC power supply (e.g., power supply for an electronic device like a radio, rated at more than 200 mA)
- ❑ Readout (multimeter, datalogger, or chart recorder)
- ❑ Matrix board for circuit (optional)
- ❑ Ribbon cable (optional)

What You Do

Make the assembly that goes in the bottle cap first. The 100-ohm resistor is the heater; choose one with a good rating, 0.5 W or more, or it may overheat if it accidentally comes out of the water/ice. Because one end of the heating resistor can serve as a counter electrode, you need only one electrode. This approach also guarantees that ice formation in the vicinity of the resistor contributes more to the operation of the oscillator than does ice formation elsewhere. The water should be laced with just a little salt—a pinch in a small wine glass should be sufficient.* Use a small amount of the slightly salty water, just 1 or 2 ml.

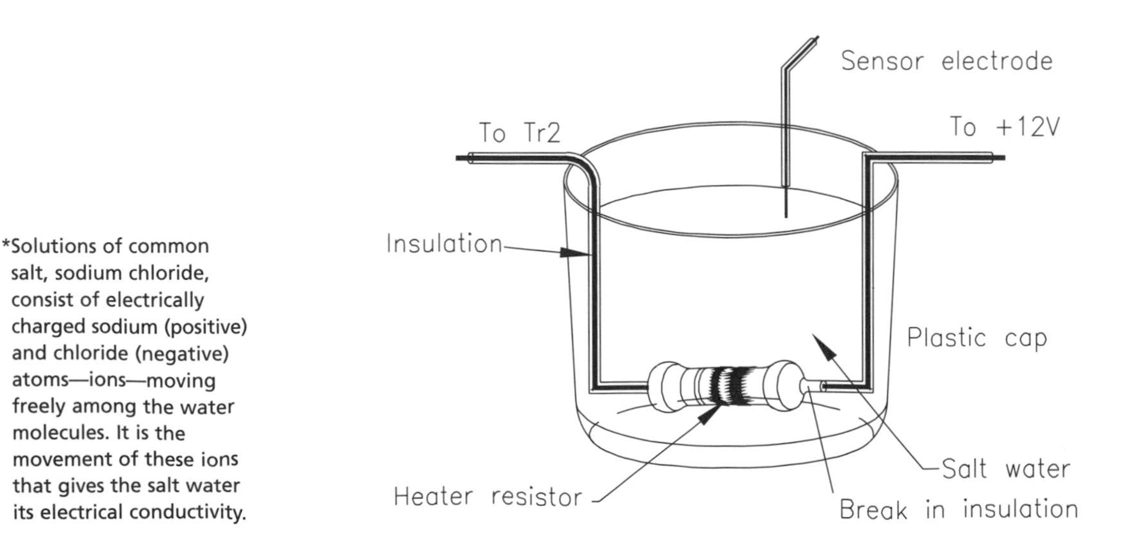

*Solutions of common salt, sodium chloride, consist of electrically charged sodium (positive) and chloride (negative) atoms—ions—moving freely among the water molecules. It is the movement of these ions that gives the salt water its electrical conductivity.

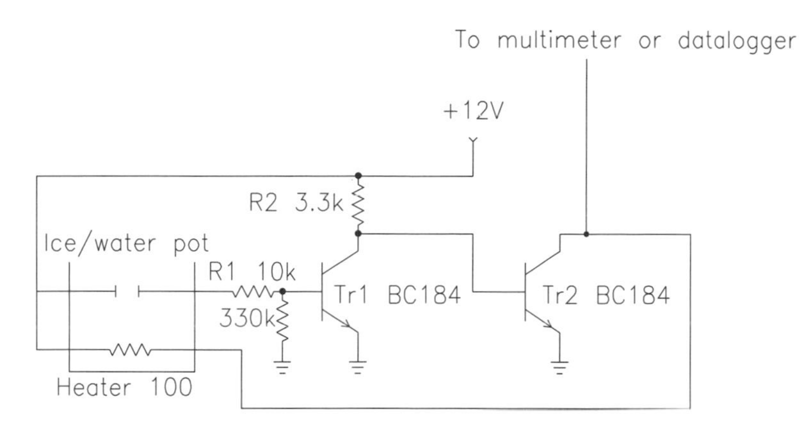

Next you need to build the circuit, which is so simple that you could just solder the components to one another, but you may find it helpful to lay them out on a piece of matrix board.

Now you need to attach the power supply and a multimeter to the circuit and then install the bottle cap with water, electrode, and heater in the freezer. Using thin connecting wires, you can simply lead the wires in on the surface of the freezer with the door open, and then when you close the door, the magnetic rubber sealing strip will close almost tight, with only a little loss of cold air. You can minimize air leaks by using a piece of ribbon cable to keep the wires flat against the freezer frame. Place the device a few inches into the freezer to minimize the effect of any air currents from the leak.

I monitored the circuit's oscillations by measuring the voltage across the heater resistor using the multimeter. The system appears capable of oscillating at frequencies varying from seconds to many minutes or even hours, depending on chance and on the precise geometry of the system. However, fixing the system geometry does not fix the frequency: surprisingly, there is still a large random component that allows the frequency of successive oscillations to vary by one or two orders of magnitude. Why is the oscillator capable of such enormous swings in frequency? I confess that I don't know precisely. But my analysis demonstrates at least one frequency at which operation should be seen and hints at why the system exhibits such a large random component.

You could just log timed voltage readouts from the clock directly onto graph paper, or perhaps just note the transition times. But connecting the clock to a recording or counting mechanism would be better: without such a mechanism, you will find it hard to understand how the clock is behaving. You could use one of the simple dataloggers or analog input boards that you can now buy to

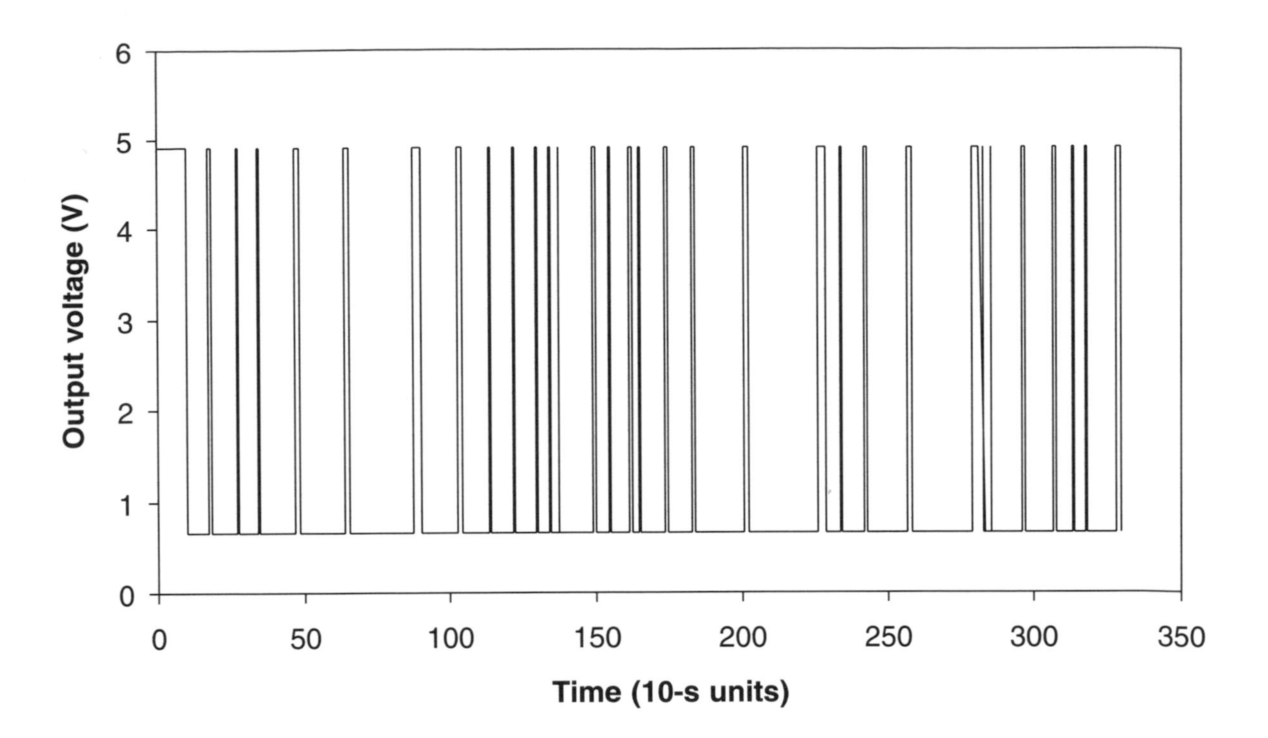

plug into your personal computer. Some of these fit into the printer port or some other port, so you don't have to delve into the computer's case (see Hints and Tips at the back of the book). The simplest of these, which will almost certainly be entirely suitable, costs only a few tens of dollars. A chart recorder is an alternative, if you can get your hands on an old one for a reasonable sum (they tend to be expensive new). Connect your recording device ground to ground on the power supply, and connect the recorder input to the sensor electrode. Let the recorder log data for a longish period, say, a couple of hours, and then examine the record of activity.

How It Works

In the completely frozen oscillator assembly, current flows down resistor R_2 and is amplified by Tr_2, which allows heating current to flow through the 100-ohm heating resistor. When the heat from the resistor melts the ice, the sensor electrode allows current to flow through limiting resistor R_1 into the base-emitter junction of transistor Tr_1. This small current causes a larger current to flow through Tr_1's load resistor, R_2, and the voltage on the output falls, which removes the standing current to the base of transistor Tr_2. Tr_2 then stops conducting, and the current to the resistor that provides the heat is stopped, allowing the ice to re-form.

THE SCIENCE AND THE MATH

The period of glacial oscillations, T_{glac}, can be approximated from the following equation:

$$T_{glac} = T_{melt} + [(\rho \Delta H_f L^3)/(dQ/dt)],$$

where T_{melt} is the time required to melt ice with the electric power supplied; L, the distance between the electrodes; ρ, the density of ice; ΔH_f, the latent heat of fusion of ice; and dQ/dt, the rate of heat conduction from the device by the freezer. You can roughly estimate dQ/dt by measuring the time, T_{freeze}, it takes your freezer to freeze cold water (water just a little above its freezing point) in the device. Inverting the T_{glac} equation (neglecting T_{melt}) and using the known volume of the water, V_w for L^3 and T_{freeze} for T_{glac} gives

$$dQ/dt = (\rho \Delta H_f V_w)/T_{freeze}.$$

T_{melt} can be estimated from a similar formula if you substitute the power of the heater resistor, V^2/R, for dQ/dt. Since T_{melt} will likely be small compared to the refreezing time, it can be ignored in an approximate calculation:

$$T_{melt} = (\rho \Delta H_f L^3)/(V^2/R).$$

This estimate assumes that the resistor heats up the ice rather rapidly (so that the heat flow from the freezer is the limiting factor) and that a substantial volume between the electrodes, roughly a cube of side L, must be melted before switching can occur.

With reasonable assumptions for dQ/dt, the equation for T_{glac} leads to quite long time constants for the oscillator. With a 100-ohm heater resistor running on 12 V (1.44 W), an assumed 0.1-W cooling power, and an ice volume of $(4\ \text{mm})^3$ (electrodes 4 mm apart), the time to melt, T_{melt}, is only15 s, but that to refreeze is 213 s, which gives nearly 4 minutes for the total cycle time.

Although my simple analysis indicates that the time constant for the clock should be long, in practice the clock often runs for a time at a much higher oscillation frequency. I am not certain why this should be so, but here are two possible explanations. Because the water contains salt, as it cools and begins to freeze, the first ice that forms may well contain a little less salt than the water.* In fact, with some solutions, small crystals of relatively pure ice are formed at first. This happens because, essentially, the intermolecular forces between water molecules are greater than those between water and the dissolved ions. As the cooling continues and the temperature decreases further, the concentration of salt in the ice continues to increase, until finally all the water is frozen into a mass of interlaced ice crystals of slightly different salt content. Only when

*The lowered freezing point of a solid containing dissolved material is, of course, a familiar phenomenon: we take advantage of it every winter, when we sprinkle salt on our roads and sidewalks to stop ice formation. We can understand the effect by considering the number of molecules at the melting interface: the more dissolved material there is, the fewer liquid molecules there are to form the solid and the lower the temperature must be before solid can form. If you asked someone to measure the minute mass of a single molecule, you might doubt their sanity if they began by dissolving things in water and measuring freezing points. But, unlikely as it may seem, you can measure molecular masses in this way. Freezing-point depression is exactly proportional to the relative *number* concentrations of liquid and solute molecules. The depression gives a measure of the number of molecules and, hence, a means for measuring molecular weight, if one knows the masses of solute and liquid used.

a complete ice barrier between the electrodes has formed does the conductivity seen by the circuit fall to zero.

During thawing, the heat from the resistor warms the ice, and the first crystals to melt are those with the highest salt content, because they have the lowest melting point. These also provide most of the electrical conductivity. In fact, if a series of spicular (spiky) crystals melt and form a path through the ice from one electrode to the other, then only a tiny percentage of the ice need melt to give a full conductivity signal to the electrodes—even though the bulk of the ice remains frozen. In this way, the period of oscillation could be one-hundredth the period predicted by the simple formula above. I suspect that this is what we are seeing in some of the experiments.

Perhaps we should replace the volume in our simple formula with something on the order of $R_p^2 \times L$, where L is the length and R_p the average diameter of a roughly cylindrical pathway between the electrodes, which might be only a millimeter or so, about the size of the ice crystals that form in a typical domestic freezer. If we use 1 mm for R_p and make the same assumptions described above, we get a T_{glac} value of $(T_{melt} + 13)$ s, as opposed to the 4 minutes we arrived at earlier.

Another phenomenon that may be causing random effects is nucleation. In a liquid below its melting point, solids form when molecules that are diffusing—jiggling about randomly—find a piece of crystal to attach themselves to. As solids continue to form in a liquid melt that already contains a lot of solid, the growth of the crystals reaches a steady rate determined by the nature of the compound that is freezing and by the rate of heat removal from the melt. The molecules of liquid are now close enough to a piece of crystal to diffuse there rapidly and find a site on the crystal surface to attach themselves to. If a molecule finds a steplike site, it will have two of its fellow molecules as neighbors; or, even better, at a corner in a step, it will have three neighbors. With three neighbors, the molecule will be bonded to the surface and unable to escape, despite thermal agitation.

However, in a substantially liquid melt, many of the liquid molecules are more than a typical diffusion length from a crystal of solid. Although they will meet other liquid molecules and form clusters, they will typically not have three neighbors; and these clusters will be easily broken up by thermally driven collisions.

This phenomenon results in the nucleation effect often seen in crystallization. A liquid state will persist to several degrees below a material's freezing point until a seed crystal is added, whereupon the liquid will rapidly solidify. This phenomenon is most easily seen in melts of salts like sodium thiosulfate, which can lie "dormant" at 30 or 40 degrees below the freezing point, waiting for a seed crystal to start the solidification process. Once the seed crystal is added, solidification proceeds in seconds. The solidification is accompanied by a remarkable release of heat. There have been attempts to use this heat to make rechargeable chemical heat packs for warming your hands or feet, for example. (The recharge process is achieved simply by melting the solid.)

And Finally . . . Advanced Glacial Oscillators

There is clearly much scope for investigating the frequency shifts of glacial oscillators as well as their "normal" periodic behavior. If the frequency shifts are related to the thawing of ice crystals of different salt content, then the use of low salt concentrations may reduce the effect, and different salts should show the effects to a different extent. In addition to common table salt (sodium chloride), we could try other completely ionized neutral salts, acid salts, alkaline salts such as washing soda (sodium carbonate), or incompletely ionized materials such as vinegar.

The use of an electronic circuit with greater sensitivity (which would permit the use of less added salt) might reduce the frequency shifts, whereas the use of strong salt solutions and an insensitive amplifier might increase the shifts to the extent that you would see high-frequency modes of the glacial oscillator.

REFERENCES

Although nucleation phenomena are described in physics textbooks, they are not, in my opinion, given the prominence that they deserve. Nucleation is after all a subtle and interesting process that is important in many different areas of science and industry, and it is still not fully understood.

Downie, Neil A. *Industrial Gases*, p. 445. London: Chapman and Hall, 1997. A brief look at an analogous process, nucleation in boiling.

Kashchiev, Dimo. *Nucleation: Basic Theory with Applications*. London: Butterworth Heinemann, 2000. A detailed specialist monograph.

Moore, Walter J. *Physical Chemistry*. 5th ed. London: Longman, 1972. Chapter 6 is a good introduction to the subject of freezing (and boiling), with a little on nucleation.

3 Everlasting Hourglass

Lives of great men all remind us
We can make our lives sublime,
And, departing, leave behind us
Footprints on the sands of time.

—Henry Wadsworth Longfellow, "A Psalm of Life"

The hourglass has been a familiar object for many centuries, so much so that it has become a classic symbol of the passing of time. In the form of the "wait icon," a virtual hourglass is used by millions of computers to inform the user that the computer is busy at the moment. The real hourglass, sometimes referred to as a sand-glass, is of considerable antiquity, and there are written records of hourglasses, and paintings and frescoes depicting them, dating as far back as the 1300s. By the end of that century, the hourglass had become a ubiquitous household item because, despite its apparent simplicity, the hourglass is a remarkably effective piece of technology. I have a crude egg timer that seems to be accurate to fewer than a couple of seconds in 3 minutes. That is equivalent to 16 minutes in a whole day, which is not bad for a gadget made for less than a dollar using five-hundred-year-old technology.

The famous water clocks of the Greeks, although based on a similar principle (water flowing out of the bottom of a container), were not nearly as accurate. Sand flow is complex—when you look into an hourglass, you might be surprised to see that the sand doesn't just line up and flow from the cone in a streamlined pattern narrowing down through the cone—but it is more predictable than liquid flow.

There have from time to time been attempts to improve on the hourglass. As recently as 2001, for example, Richard K. Coleman filed a patent (U.S. Patent no. 6,260,996) describing an hourglass with a large bypass duct between the chambers; in the duct there is a valve that allows you to reset the hourglass back to zero halfway through its time period if you want.

The real problem with the hourglass, however, is that it must be turned. An hourglass watcher has to check it periodically until the last few minutes, when the watcher must monitor it continuously and turn it over the instant the sand runs out. But reliable help is so difficult to come by these days. How do you keep accurate time with an hourglass without an hourglass watcher?

The mechanical device in this project is one possible solution. In the everlasting hourglass, sand is contained in a large, cheese-shaped wheel or drum that turns continuously under the power of a falling weight, so no hourglass watcher is needed. The sand flows from one wedge-shaped compartment to another as the drum turns.

What You Need

- ❑ A large cylinder (15–30 cm in diameter and a few centimeters high) with at least one transparent face (I used a fully transparent candy box for maximum visibility.)
- ❑ Plastic sector dividers, with holes (or mesh dividers, perhaps made from a sieve or a flour sifter)
- ❑ Circles of plastic foam to fill the center of the cylinder (optional)
- ❑ Sand, fine, dry, and free of lumps (I dug some out of a hillside nearby, ran it through a flour sifter, and dried it thoroughly in the oven; but beach sand is probably best.)
- ❑ Metal rod for the driveshaft (e.g., 6 mm in diameter and 30 cm long)
- ❑ Bearings for rod
- ❑ Weight bucket (e.g., an open-top box filled with 1–2 kg of fine gravel)
- ❑ String
- ❑ Thread spool
- ❑ Various pieces of wood (to mount bearings so that the drum can rotate freely and so that the weight bucket can fall 1 m or so)
- ❑ Modeling clay (30 g or so for balancing the drum)

What You Do

Assemble your drum with sand inside, taping the joint carefully so that you won't lose too much sand. I held the dividers in with hot-melt glue, sealing any gaps in the edges with sticky tape. Your sector dividers must have the same number of similar-size holes to meter the sand out. Mount the drum as close as you can to centrally on the rod, canceling out any residual balance error by attaching small pieces of clay to the perimeter of the drum. Make sure that the bearings allow the rod to turn smoothly and that the thread spool, which must take quite a bit of torque, is firmly glued to the rod. You can save a bit of wood by arranging for the spool to hang over the edge of a table, allowing the drive weight to

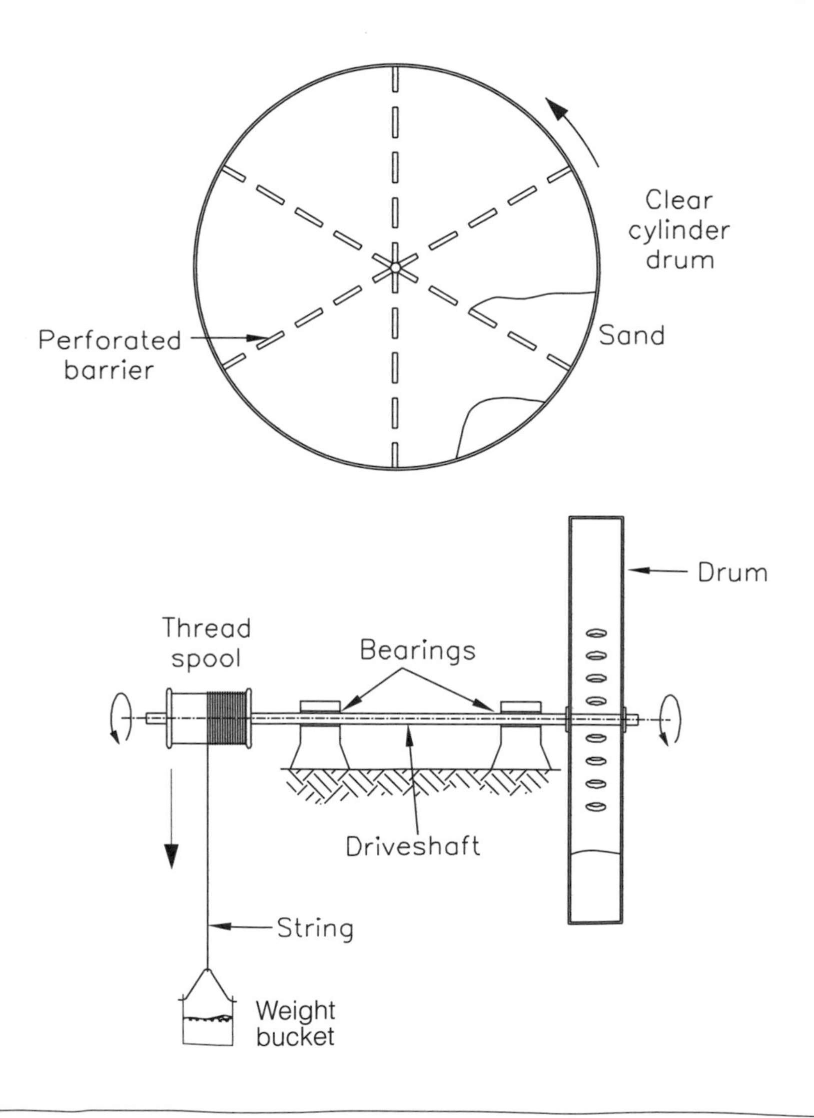

descend to the floor. To adjust the falling weight fairly precisely, you can simply add or remove some gravel. Experiment with different amounts of sand in the drum. Filling it about 30 percent full will probably work best, but almost any amount between 10 percent and 90 percent full might be workable.

You may find that the sectors in the drum do not quite drain completely, owing to uneven flow patterns and "bridging" in the sand. Sector dividers made from sieve material are more reliable than plastic dividers with holes: dividers with many holes in parallel are less prone to jamming up. The type of sand used also affects its tendency to bridge. Beach sand has more-rounded grains and runs more smoothly than the more-angular grains of inland sandpits. Antique hourglasses often used grains of crushed marble that had been boiled in vinegar to etch away the sharp corners and leave more-rounded grains.

The accumulation of undrained sand in the middle of the drum and around the walls of the sectors reduces the amount of sand available in the active counterbalancing sector and leads to runaway when the sand in that sector becomes insufficient to balance the drive weight. You can reduce this effect by designing the sectors so that the sand cannot run into the center of the drum—that is, so that each sector forms a little sand-dispensing hopper, a triangular shape with a downward point. A piece of circular foam in the middle will help achieve this. An alternative is to put partial dividers inside the drum. I found that a drum with the following specifications worked quite well:

Drum diameter	200 mm
Thickness	32 mm
Length of vanes, edge to middle	40 mm
Diameter of hole in vane near edge	5 mm
Number of sectors	8 (45 degrees each)
Drive drum diameter	25 mm
Drive weight	~500 g
Rotation time	~50 s (when correctly adjusted)

Finding the correct drive weight to make the device run can be difficult. If the weight is too light, the drum will get stuck; if it's too heavy, the drum will begin to spin around at an increasing speed. Centrifugal force will push the sand to the perimeter, and the weight will plummet to the floor in a second or two. The change in the center of gravity caused by the sand flow is relatively small,

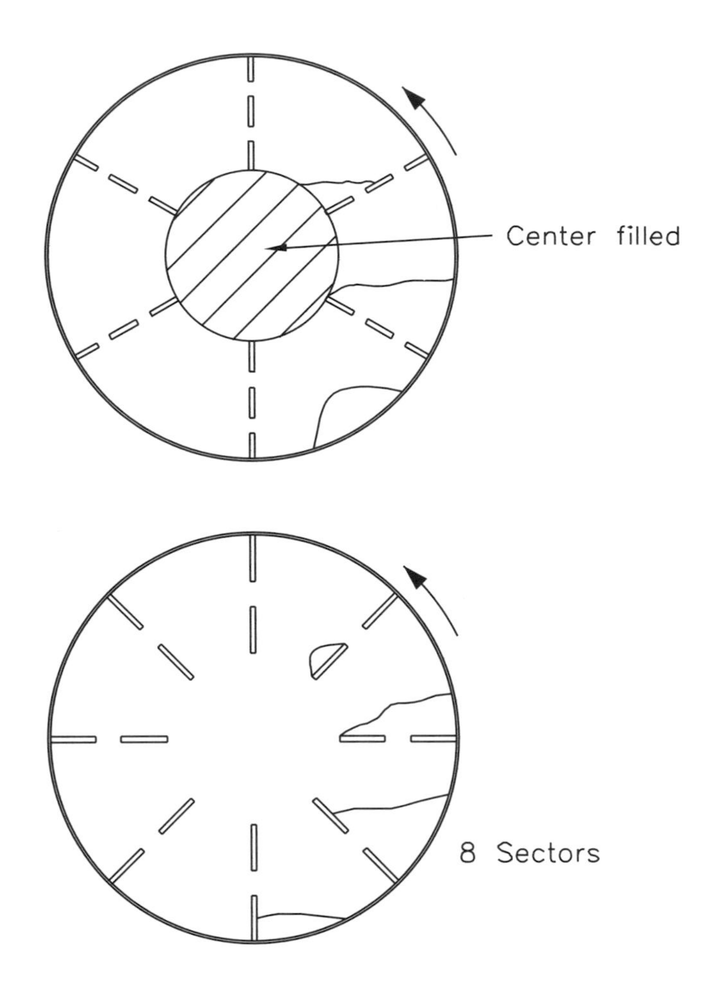

so the margin for error is also small. The drum must be carefully made with good bearings and counterbalanced with small pieces of modeling clay or similar material if it is not well balanced.

As the drum turns from one active sector to the next, the sudden lurch can lead to runaway. You can solve this problem with a speed-sensitive brake or damping device. A simple fix is to use a high-viscosity lubricant on the axles. One pack of a two-pack slow-setting epoxy resin adhesive is a possibility (the whitish adhesive stuff is better than the yellowish hardener stuff). However, something as simple as molasses from your kitchen will also work, at least until it dries out and hardens.

An air brake is a more advanced project that might have the same effect. A large pulley is fixed to the sand drum and connected to a small pulley with a rubber band. With paddle blades affixed to the small pulley and with the speed-up given by the transmission ratio of the pulleys, this brake suppresses the lurching motion. With the air brake or the viscous brake, you can use a larger drive weight, one that will be better at preventing jamming. If you use these anti-jamming and anti-runaway measures, they must not be allowed to determine the wheel's speed: the sand flow is supposed to be the regulating phenomenon. The drum is certainly a temperamental beast, however; and even with these precautions, you may have difficulty getting it to run smoothly. You have been warned!

How It Works

If the everlasting hourglass is to operate correctly, the maximum clockwise torque from the sand in the drum must exceed the anticlockwise torque from the drive weight. In addition, the drive weight has to raise the sand high enough so that it can flow. The sand torque is given by the weight of the sand retained in the sectors on the draining side of the drum, multiplied by the sand's mean horizontal distance from a vertical line through the axle. The drive-weight torque is given by the weight multiplied by the radius of the thread spool. When the sand torque is slightly larger than the drive-weight torque, the drum will turn until the horizontal distance of the sand from the axle is reduced. Then, as the sand begins to drain into a lower sector, closer to the vertical line through the axle, the torque from the sand decreases, and the drum will turn a little, raising the sand upward, until balance is restored. In this way, the drum will rotate, the many tiny movements making it appear to move continuously.

The sand in the everlasting hourglass flows just as sand does in an ordinary hourglass, but the sand does not flow in the same way that water does. Water and air are made of tiny particles (atoms and molecules), and sand is similarly made of grains. So why doesn't sand flow in the same way as water? The main reason is that water molecules are in a state of continuous agitation at a microscopic level, whereas sand grains are not.* The same thermal energy that agitates the water molecules is incapable of moving the much larger sand grains appreciably. Thus whereas water will run down slopes of just a few degrees at high speed, sand will not flow at all until the slope exceeds 30 degrees or so. This is why the drive weight must provide enough torque to raise the sand in one sector high up: a small drive weight is not sufficient.

In principle, our sand clock, or any hourglass, should keep better time than a Greek water clock. The water clocks had several drawbacks. First, they ran

*Under special circumstances, sand can be made to behave like water. In fluidized beds, for example, a blast of air up through the sand can lift and agitate the sand grains so that their behavior resembles in many ways the behavior of water molecules.

faster when full because the higher pressure at the outlet forced the water out faster. Hourglasses and our sand clock do not suffer from this defect. The flow of sand from the funnel exit of a hopper is almost completely steady, irrespective of the height of the sand column above the exit. Second, water clocks varied with temperature because the viscosity of water varies with temperature. Our sand clock does not experience this viscosity effect: it flows in exactly the same way no matter what the temperature.

THE SCIENCE AND THE MATH

Analyzing sand flow is extremely complex, but we can start off by thinking about why Greek water clocks worked so poorly. How can we model the problem of differential pressure? If we use a long, thin capillary tube with laminar flow as a model for the water outlet of such clocks, the flow rate is proportional to the height of the water column, H:

$$dH/dt = -aH,$$

where a is a constant. Integration gives an exponential variation of water height:

$$H = H_i \exp(-t/T),$$

where H_i is the initial water height and T the time constant ($1/a$) for its sinking. With this model, the water never completely runs out!

If we use a model in which the water outlet has a pressure drop ΔP versus flow more typical of an orifice,* then

$$\Delta P \propto \tfrac{1}{2}\rho(V^2 - V_i^2),$$

*You can think about the pressure change at an orifice in a pipe as being similar to a Bernoulli restriction, or venturi, in a pipe. Bernoulli's law tells us that

$$P + \tfrac{1}{2}(\rho V^2) = P_i + \tfrac{1}{2}(\rho V_i^2),$$

where V and P are the velocity and pressure after the restriction, and V_i and P_i are the velocity and pressure before the restriction. This equation allows us to determine the change in pressure at the orifice:

$$\Delta P = P - P_i = \rho(V^2 - V_i^2)/2.$$

where ρ is the density, V_i is the initial velocity of the water (before ejection from the exit), and V is the velocity after ejection. Because V_i is negligible compared with V and because the pressure in a liquid at depth H is given by ρgH, we arrive at

$$V^2 \propto \Delta P \propto H.$$

So $V \propto \sqrt{H}$. Taking the velocity, V, of the liquid as proportional to the rate of loss of liquid height in the tank, dH/dt, gives us

$$V \propto dH/dt \propto \sqrt{H},$$

so

$$dH/dt = -b\sqrt{H},$$

where b is the constant of proportionality. Integration gives

$$H = H_0 - (\sqrt{H_0} - bt/2)^2,$$

where H_0 is another constant. The plot of this function compares typical values, showing the water level in the orifice version of the clock diving more rapidly than the level in the laminar (capillary) version.

As I mentioned, the second thing wrong with the Greek water clocks was their reliance on the constancy of the viscosity of water. In fact, the viscosity of water, like that of most liquids, falls with temperature. (Motor oil at –20 °C is like molasses in winter, which makes cold engine starting difficult; but the

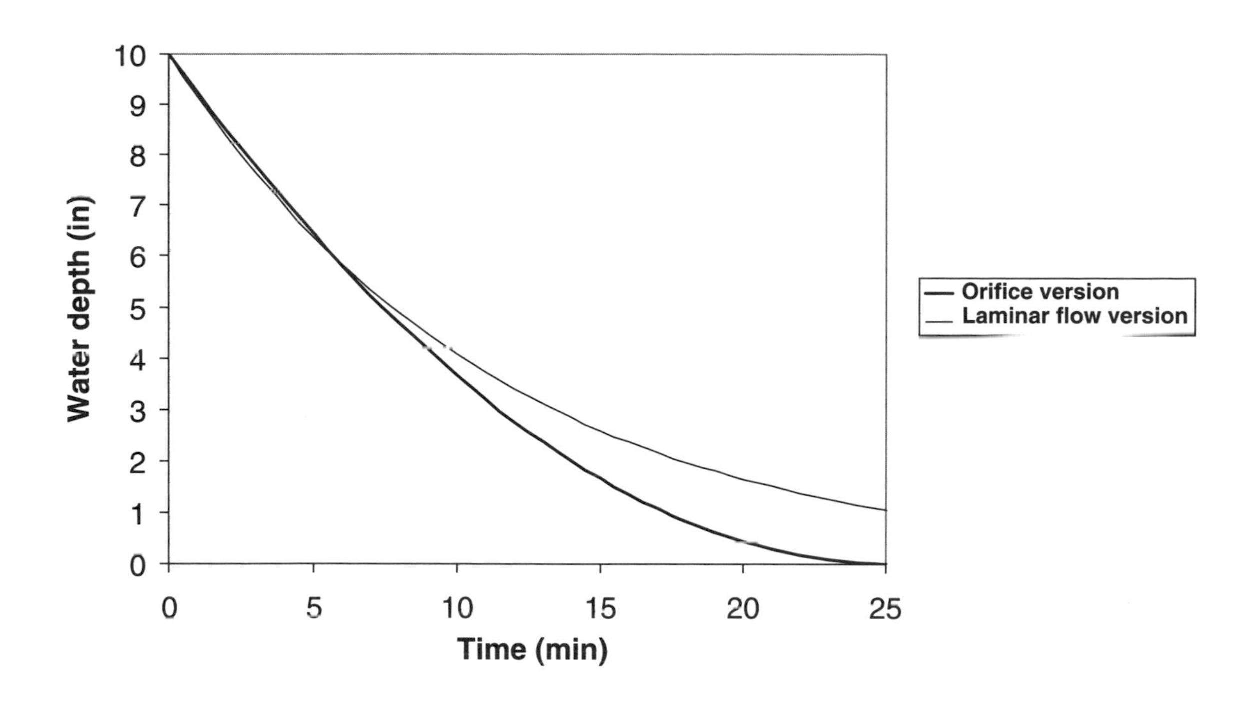

oil is watery at 150 °C and higher, which reduces its effectiveness at preventing engine wear.)

Now let's use what we've learned about water clocks to analyze the flow of sand in our everlasting hourglass. Some useful rules of thumb and simple math have evolved to help engineers design sand-handling machinery, and certain phenomena have become recognized.

First, it is well known that a huge column of sand has much less effect than a huge column of water: the effective pressure at the exit of an hourglass does not vary much with the height of the column of sand above it. Try pouring sand out of an hourglass-size "hopper" cone onto a scale and observing how the weight increases with time. You will find that a full hopper pours at a constant rate, provided that the depth of the bed is at least four times the hopper diameter. In huge industrial hoppers, sand actually forms bridges—arches of sand that hold up the sand above them, reducing the pressure below the arch to zero.

Second, the flow pattern of sand from a vessel through an orifice in a hopper is quite different from liquid flow. Above the orifice, a column of sand a few times the diameter of the orifice flows first: the bulk of the sand at the sides of the hopper does not move. With no incoming sand to the top layers, a conical depression begins to form in the middle of the sand in the hopper, and sand rolls down the sides of this depression. In this way, the sand at the walls of the hopper near the bottom flows through the orifice last, long after the sand in the middle at the top.

The flow of sand is affected by many other parameters that are difficult to examine in detail but are nonetheless vital for studying flow in industrial processes. Particle size is an obvious parameter of interest, but not in the way you might think. Batches of sand that have the same average particle size but different ranges of particle sizes will behave differently: batches with a narrow range of particle sizes will have a greater tendency to form solid blockages, whereas sand with a wide spectrum of size packs down much more densely than monosize sand. (To understand this, think of an array of spheres and the gaps between them that could be filled by smaller spheres.)

The presence of moisture is clearly undesirable because of the increased possibility of blockage: moist sand grains will stick together to some extent. Try building a sandcastle on the beach without a little moisture and you'll appreciate this fact. Some electronic goods supplied from countries with a humid tropical atmosphere include a tiny silica gel sack to absorb moisture. Maybe the addition of one of these to the everlasting hourglass would be useful if high humidity was a problem.

And Finally . . . Everlasting Electric Hourglasses

Have you got your drum running smoothly? If so, why not measure how it works? Rotation time, steadiness of rotation (my cheese drums all ran rather unevenly), and speed versus drive weight are obvious measurements to take. Is your drum affected by changes in ambient temperature or humidity? If your everlasting hourglass doesn't run smoothly, the sources listed in the references, which describe some of the fascinating properties of sand, may help you understand why.

You may find that an electric drive—a motor and transmission—may be more useful than the weight drive for these measurements. You need to ensure, however, that the electric drive is not capable of producing more torque than the sand in the drum does: otherwise the drum will not govern the speed. I don't think you will find the drum nearly as accurate as a genuine hourglass, but remember, the hourglass was under development for at least five hundred years, so you shouldn't be apologetic about this.

REFERENCES

The flow of solids under the influence of gravity and wind has been extensively studied, although reports of this work tend to be published in isolated papers at extended intervals in a variety of scientific journals. Perhaps the easiest references to find are those dealing with the chemical-engineering aspects of handling coal and crushed mineral ores.

Bagnold, R. A. *Physics of Blown Sand and Desert Dunes.* London: Methuen, 1941.

Coleman, Richard K., Jr. "Hourglass with Bypass Duct." U.S. Patent no. 6,260,996, July 17, 2001.

James, C. A., T. A. Yates, M.E.R. Walford, and D. F. Gibbs. "The Hourglass." *Physics Education* 28 (1993): 117.

Richards, J. C., ed. *The Storage and Recovery of Particulate Solids: A Report of a Working Party of the Institution of Chemical Engineers*. London: Institution of Chemical Engineers, 1966.

Veje, C. T., and P. Dimon. "Power Spectra of Flow in an Hourglass." *Physical Review E* 56 (1997): 4376.

Aerodymagic

Blow, winds, and crack your cheeks! Rage!
Blow!
You cataracts and huriccanoes.

—Shakespeare, *King Lear*

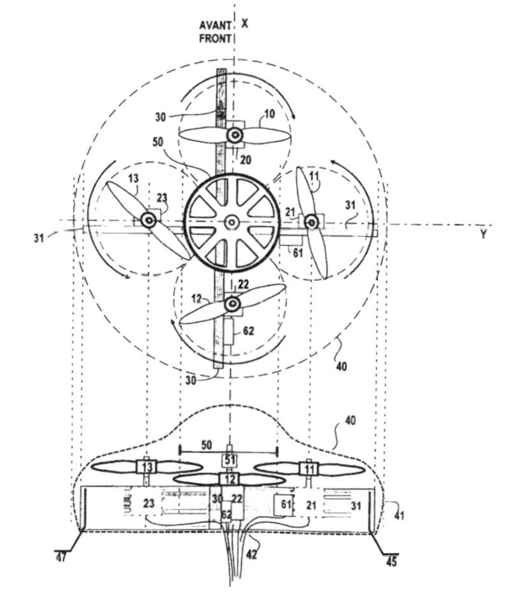

Although air is constantly flowing all around us, our scientific understanding of its flow has been slow to develop and is still incomplete. Perhaps air's invisibility and intangibility are to blame. Although in stormy weather the effects of moving air are all too evident, it is only when air drives particulates—dust motes or leaves, for example—that we can visualize its motion with any accuracy.

Once the fundamentals of the motion of air began to be understood, pioneers like Daniel Bernoulli (working from 1730 on), Siméon-Denis Poisson (from 1810 on), and George Stokes (from 1845 on) founded the study of aerodynamics. The blossoming of the calculus techniques developed by Leibniz and Newton in the 1790s allowed for the mathematical analysis of flowing air. However, although the math was relatively simple—partial differential equations that could easily fit on the back of an envelope*—it was not simple enough for practical use. Although the equations of aerodynamics are relatively simple to write down, they are generally quite impossible to solve. Simple situations of incompressible laminar flow can be calculated analytically, but almost every other form of flow—for example, those involving compressibility or turbulence—can be calculated only with computer simulations.

Because of the limitations of computers, experiments with models in wind tunnels and other rigs are still vital to understanding and prediction in aerodynamics. However, building a wind tunnel the diameter of a 747 jet would cost more than a fleet of the airplanes themselves. Fortunately, wind tunnels of a conveniently small size can accurately mimic flow at much larger scales. We can understand this by considering a dimensionless parameter called the Reynolds number, R_N:

$$R_N = \rho VL/\mu,$$

where ρ is the density of the fluid, V is its speed, μ is its viscosity, and L is a typical linear dimension. Fluid systems that have the same Reynolds number exhibit the same behavior. Thus, for example, by setting up a wind tunnel with a high ρ value (using cryogenically cooled air at 90 K, or –183 °C, for example) and a low μ value (the viscosity of a gas decreases with temperature), you can use a small airliner model to predict the flow around a large airliner of similar design. The real thing can be six times as large as the model in linear dimensions, which is two hundred times as large in volume.

*Consider, for example, the Navier-Stokes equation for irrotational flow:

$$\rho(\partial \boldsymbol{U}/\partial t) + \rho \boldsymbol{U}\nabla \boldsymbol{U} = -\nabla P + \mu\nabla^2 \boldsymbol{U} + \boldsymbol{F},$$

where ρ is the fluid density, $\boldsymbol{U}$ the fluid velocity, P the pressure, and $\boldsymbol{F}$ the "body force" (e.g., gravity) on a fluid element. These vector partial differential equations do not take into account the compressibility of the air. However, even with this simplification, it is rather remarkable that a one-line equation can describe—at least in principle—so much of real fluid flow.

One of the phenomena that we consider in this part depends upon scale, upon the details of the air flow forming in front of and re-forming behind obstacles. The other phenomenon, which involves a kind of yacht, would probably work at any scale: after all, model yachts just a few inches long seem to work in pretty much the same way as giant America's Cup boats.

4 Juggling Airstreams

Can Apples Fall Upwards?

—Fabrizio Pinto, grant title

If you saw an apple falling upward, you would probably be surprised, but you might not be surprised to see a ball being supported by a stream of air issuing from a nozzle. Since at least the mid–nineteenth century, this principle has been used in games in which a ball is propelled upward into a collecting bucket. If you go shooting rifles at a fairground, you may find that one of the targets is a ping-pong ball suspended on an air jet. The phenomenon seems simple: the force of the air blasting on the ball is bound to push it upward. On reflection, however, you will begin to see that it is actually rather surprising.

Because the ball is streamlined, the flow around the ball should re-form smoothly on the other side of the ball, leaving viscous drag as the only force supporting the weight of the ball. However, air is low in viscosity, so this force is so small that unless the ball were made of something little more substantial than spider's web, it would fall. One of King Lear's "raging hurricanoes" would be needed to support even a ping-pong ball. There is another mystery: why does the ball show such remarkable stability, hovering in the middle of the airstream? You might expect the system to be just as unstable as a pencil balanced on its point. If the ball moved slightly off to one side, say, the left, the force from the air impinging upon it would no longer be directly below the center of mass, so surely the ball should be propelled off to the left, with the impinging airstream deflected slightly to the right.

So why does the ball stay up? First, it is supported mostly by the drag force produced by the formation of eddies behind it—that is, by pressure drag as opposed to the viscous drag that derives from the viscosity of the air. The drag force produced is much larger than the viscous force, so the airstream velocity needed is reasonably slow. Second, the ball is stabilized by the rotation induced when it goes off-center (the Magnus effect).* These happy circumstances occur only when quite a few conditions are satisfied, however. The airstream velocity must lie between certain limits; the airstream must be not too wide; and the ball must be roughly spherical. Try hovering a ping-pong ball above a wide nozzle—a leaf blower with a cone to widen the airstream, for example—and you will have trouble. If you glue a paper cone on the back of the ball to make it more streamlined, it won't even hover above the blast from the most powerful hair dryer. Other shapes such as cubes or cylinders also seem to be useless, and very tiny spheres probably won't work, because flow around them is not turbulent and the drag force on them may be too tiny to support their weight.

Have I now convinced you that supporting a single ball in an airstream, although possible, is really quite improbable? If so, then you won't need to be convinced that suspending four or five balls in a column above one another in a jet of air is really quite preposterous. But if it didn't work, it wouldn't be in this book, would it? Well, it nearly doesn't work. If you just pile a lot of ping-pong balls into the airstream willy-nilly, it certainly won't. But it can be done. Read on.

*The Magnus effect is produced when a symmetrical spherical or cylindrical body rotates around its axis. When the spinning body flows through an airstream, the thin layer of air near the body is affected by the body's spin. The part of the body that is spinning with the airstream experiences lower pressure relative to the airstream, whereas the part that is spinning against the stream experiences higher pressure; and the sum of the forces produces lift.

In our case, when the ping-pong ball moves off-center, it begins to rotate because of the differential drag of the airstream, and the rotation generates a Magnus effect. But is the Magnus effect sufficient to stabilize the hovering ball?

What You Need

- ❑ Hair dryer with a "cool" setting
- ❑ Card-stock cone (to concentrate the airstream)
- ❑ Ping-pong balls
- ❑ Cotton thread
- ❑ Balloons (Use the smallest ones you can find, 15 cm or less; the really small ones, 10 cm, sold as water balloons are ideal.)

What You Do

Make a paper cone for the hair dryer from the card stock. When taped to the dryer, the cone should reduce the diameter of the air outlet by approximately a third or a half and should speed up the air enough to make the ping-pong balls hover. Don't overdo it and make the exit from the cone too small—the hair dryer will overheat if you do. I found that an exit with a diameter of 3 cm was about right for the average hair dryer. Fix the hair dryer with its cone in a vertical position.

Now you need a selection of ping-pong balls with different drag factors. One way of increasing the drag factor is to carefully attach three or four diametrical rings of cotton thread around a ball. Despite the apparent insignificance of this modification, the resulting ping-pong ball will hover 10 cm or more higher than a plain ball. The extra drag is in fact substantial.

Another method of making a high-drag ball is to expand it slightly. Warm a ball in boiling or nearly boiling water for a few seconds. Use a pair of spoons, a

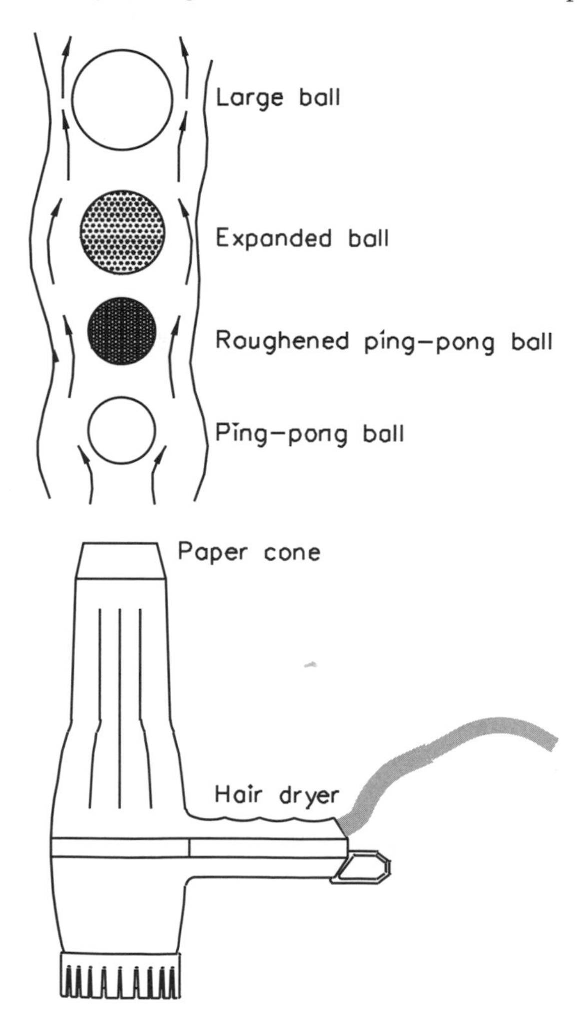

cage of wires like a wire eggbeater, or some other device to suspend the ball in the water without scalding yourself and without crushing or denting the ball. The ball will expand slightly, although not evenly, and will be 4–6 mm bigger in its largest diameter and will be shaped like a slightly squashed sphere (an oblate spheroid), like an M&M's candy. Such a ball will also hover higher than a standard ball, again by a surprisingly large margin.

You need to ensure that the drag factor of each ball after the first is sufficiently high so that all the balls are separated by at least a ball diameter. Once you have a selection of appropriately modified balls, you will be able to support two or three balls in a column above the air jet.

Lower the ping-pong balls, virgin or high-drag versions, into the airstream, and measure the height at which they hover. If you mark stripes or sectors on the balls, you will be able to see whether, and around what axis, they rotate. The higher-drag balls will hover higher: they have a greater force on them, so they can fly higher in the airstream, where the stream has widened and slowed slightly relative to the speed just past the nozzle. To extend the principle further, you need much lighter, larger (higher-drag) balls. I used balloons, but I inflated them only very slightly. Once you have inflated them to 10–15 cm in diameter, tie the neck off close to the balloons' surface. Then trim off the excess neck as close as you can, and maybe use a small piece of tape to hold the remainder of the knotted neck flush with the surface. Armed with these additional balls, you should be able to balance at least five balls in the air jet.

How It Works

The key to getting the airstream to juggle multiple balls is to ensure that they differ in size, drag factor, or weight. The airstream weakens as it travels upward from the nozzle of the hair dryer, and each position in the airstream above the nozzle is characterized by its own particular airstream speed, high near the nozzle and low far from the nozzle. A ball will hover only when its weight is proportional to the product of its drag factor and its cross section and the square of the airstream speed (approximately). A ball with a small size (area), a large weight, or a small drag factor will hover lower than a ball with a larger size, a smaller weight, or a larger drag factor. Two balls with the same size, weight, and drag factor will hover at the same place. If you test two identical balls, you will find that they hover only briefly one above the other before they collide and one or both are ejected. The two balls are trying to occupy the same position in the airstream.

In addition, the airstream must more or less reconstruct its linear form after impinging on the first ball, so that there is a good jet to support the second ball. For this, you need balls that are fairly smooth and don't break up the airstream too much. (They have to generate some turbulence, though, or they won't hover in the first place.) Spherical or nearly spherical balls are therefore preferred.

THE SCIENCE AND THE MATH

When the ping-pong balls are hovering over the airstream, their weight is exactly balanced by the aerodynamic drag force of the air flowing past in the upward direction. In dealing with fluid flows, we should always check the Reynolds number of the system to determine whether viscous forces or inertial forces will dominate. If the inertial forces are large, the Reynolds number will be large, and the flow will be turbulent; whereas at Reynolds numbers of less than a few hundred, the flow will likely be smoothly laminar.

If we use the 18-mm radius of a ping-pong ball for L and a characteristic value of 2 m/s for the airstream speed, the equation for the Reynolds number,

$$R_N = \rho VL/\mu,$$

gives a value of about 2,600, which indicates that the inertial forces are large and that the flow is turbulent. (The density of air is about 1.2 kg/m^3, and its viscosity is about 18 μPa s at room temperature on an average day at sea level.) In this situation, we should use Newton's approximation, which assumes a turbulent flow, for the drag force:

$$D = \tfrac{1}{2}\rho C_D A V^2,$$

where D is the drag force on the ping-pong ball, C_D is the drag coefficient (approximately 0.44 for a rigid sphere), and A is the maximum cross-sectional area of the sphere (πR^2, where R is its radius). The drag force is the upward force on the ball. The downward force, F, is the force of gravity: $F = Mg$, where g is the acceleration due to gravity. At equilibrium, with the ball hovering, the upward drag force must be equal to the downward gravitational force, so

$$Mg = \tfrac{1}{2}\rho C_D A V^2.$$

If we substitute πR^2 for A and use the terminal velocity, V_0, for V, then we arrive at

$$Mg = \tfrac{1}{2}\rho C_D \pi R^2 V_0^2.$$

Solving for V_0 gives

$$V_0 = (1/R)\sqrt{[2Mg/(\pi\rho C_D)]},$$

which indicates that the terminal velocity increases as the radius of the ball decreases. This should not be too surprising, as we are talking about a ball whose weight is constant (2.5 g). Simply by changing the radii of the balls, we can make them hover at different heights. If we had two balls with the same weight but different radii, we would expect the ball with the smaller radius to fall more quickly through the air. Our formula gives a V_0 of 4 m/s.

Although the flow field—that is, the values and directions of air velocity at different positions above the hair dryer—is complex, we may be able to model it by making a few simplifying assumptions. If we assume that the jet of air spreads out and slows down without any overall loss of upward momentum and that the air velocity is the same, and vertical, across horizontal slices, then we can say that the momentum, P, of a cylindrical volume of air (with a radius R_a and a length 1 s of airstream long) is equal to the mass of that volume, M, times the air velocity, V:

$$P = MV = (\rho\pi VR_a^2)V = \rho\pi R_a^2V^2.$$

So if we can determine how the radius of the airstream changes with its height above the nozzle, we can estimate how the velocity varies with height. Suppose the radius simply increases linearly:

$$R_a = R_{a0} + kH,$$

where R_{a0} is the starting radius of the airstream, H is the height above the nozzle, and k is a constant. If, for example, k had a value of 0.1, then a starting radius of 20 mm would widen to 70 mm at 500 mm above the nozzle and to 120 mm at 1,000 mm. Because we've assumed that momentum is conserved,

$$2\pi(R_{a0} + kH)^2\rho V^2 = 2\pi R_{a0}{}^2\rho V_{st}{}^2,$$

where V_{st} is the starting velocity. Solving for V gives

$$V = V_{st}[R_{a0}/(R_{a0} + kH)].$$

And with our assumptions about the radius of the airstream at different heights, a starting velocity V_{st} of 4 m/s would decrease to 1.1 m/s at 500 mm and to 0.7 m/s at 1,000 mm.

Using this field of flow speeds, we can simply engineer the mass, radius, or drag of the balls to make them float at different heights. The equilibrium height, H_0, of a spherical ball of mass M and radius R is given by

$$H_0 = \{V_{st}R\sqrt{[\pi\rho C_D/(2Mg)]} - 1\}R_{a0}/k.$$

From this equation you can see how a carefully arranged set of balls with different radii can float nicely above one another, with their hovering heights proportional to their radii. Substantial differences in drag factor, C_D, will also separate the balls, as will substantial weight differences.

And Finally . . . Taller Towers of Hovering Balls

It is remarkable that the stream of air from the hair dryer, although it clearly spreads and slows as it rises, seems to be largely re-formed behind each successive ball: the turbulent wake from the ball below does not disturb the airstream impinging on the one above very much. However, to achieve the tallest tower of hovering balls, some mechanism to realign or reinforce the airstream in some way might be helpful because the stream becomes more and more chaotically turbulent as it passes over more and more balls. Could a carefully designed nozzle with a high-speed inner core and a low-speed outer annulus (or vice versa) result in a more spectacular arrangement of balls? By raising the temperature of the air (that is, by turning the hair dryer to a higher setting), you can get a somewhat higher velocity. Does the added heat have any other beneficial effects? Does, for example, the lower density of the hot air allow a buoyancy effect to help the airstream work on the hovering balls better?

REFERENCE

Douglas, J. F., J. M. Gasiorek, and J. A. Swaffield. *Fluid Mechanics*. London: Pitman, 1979. Engineering texts give many basic formulas and useful tips on flow properties, but this book gives a better overview.

5 Railroad Yacht

The pessimist complains about the wind;
The optimist expects it to change;
The realist adjusts the sails.

—William Arthur Ward

Railroads have been linked to engine power, particularly steam engine power, for more than 150 years. This link is so strong that we sometimes forget that early railroads did not always use "locomotive" engines, that is, engines that ran down the track with the train. This arrangement, which now seems obvious, was not so obvious, or so practical, in the pioneering days of railroads. Other systems were available, though. Vacuum railroads, for example, used a stationary engine to provide a vacuum for propulsion, and others, such as the streetcars of San Francisco, used a cable hauling system running across pulleys set into slots in the ground, again powered by a stationary engine. Even older systems, especially in mines, involved human power or ponies. Special breeds of small but strong horses called pit ponies were used in coal mines. In 1828, there was even one attempt, described in *Eurekaaargh!* by Adam Hart-Davis, to haul a tricycle with a kite for propulsion. Here now is the ultimate "green machine," combining the low propulsive-energy requirements of the railroad with the ultimate in zero-emission power-generation technology—the sail. But your engine driver will need to follow William Arthur Ward's advice: he or she must be a realist and adjust the sails.

What You Need

For the artificial-wind generator

- ❑ *N* (say, 8) propellers (e.g., 100 mm or 150 mm)
- ❑ *N* small motors (e.g., 1.5–3 V)
- ❑ Auto battery charger, 12 V

For the railroad yacht

- ❑ Model railroad track
- ❑ Model railroad car
- ❑ Balsa wood for mast, boom, and countersail boom
- ❑ Countersail
- ❑ Sail
- ❑ Clay for ballast and other weights

What You Do

Assemble the railroad yacht along the lines suggested in the diagram. The sail and countersail can be made from paper and supported by balsa wood mast and booms. A clay weight in the car will help to keep it from tipping, and a little additional clay can be added as a counterweight underneath the countersail to tip the car toward the airstream from the propellers.

Mount a propeller on each of the motors, and then mount a row of the resulting fan assemblies along the side of track, as suggested by the one fan shown in side view in the drawing. Arrange them so that the fan centers are a little less than two propeller diameters apart. Now wire the fans up in parallel and series so that they go at a good speed without overheating. If, for example, they run happily on 3 V and you have eight of them, then wire two series chains of four motors to the 12-V battery charger.

When the wind from the fans blows, the railroad yacht, like any sailing vessel, is subject to a sail torque that tends to overturn it. To offset this overturning torque, our yacht uses a counterweight, as well as a countersail; and the offset torque, like the sail torque itself, is proportional to the wind strength. Without the counterweight and the countersail, the narrow gauge of the track and the top-heaviness of the yacht will not allow you to use much sail. With them, however, much more sail can be put up, and much higher speeds reached—despite the

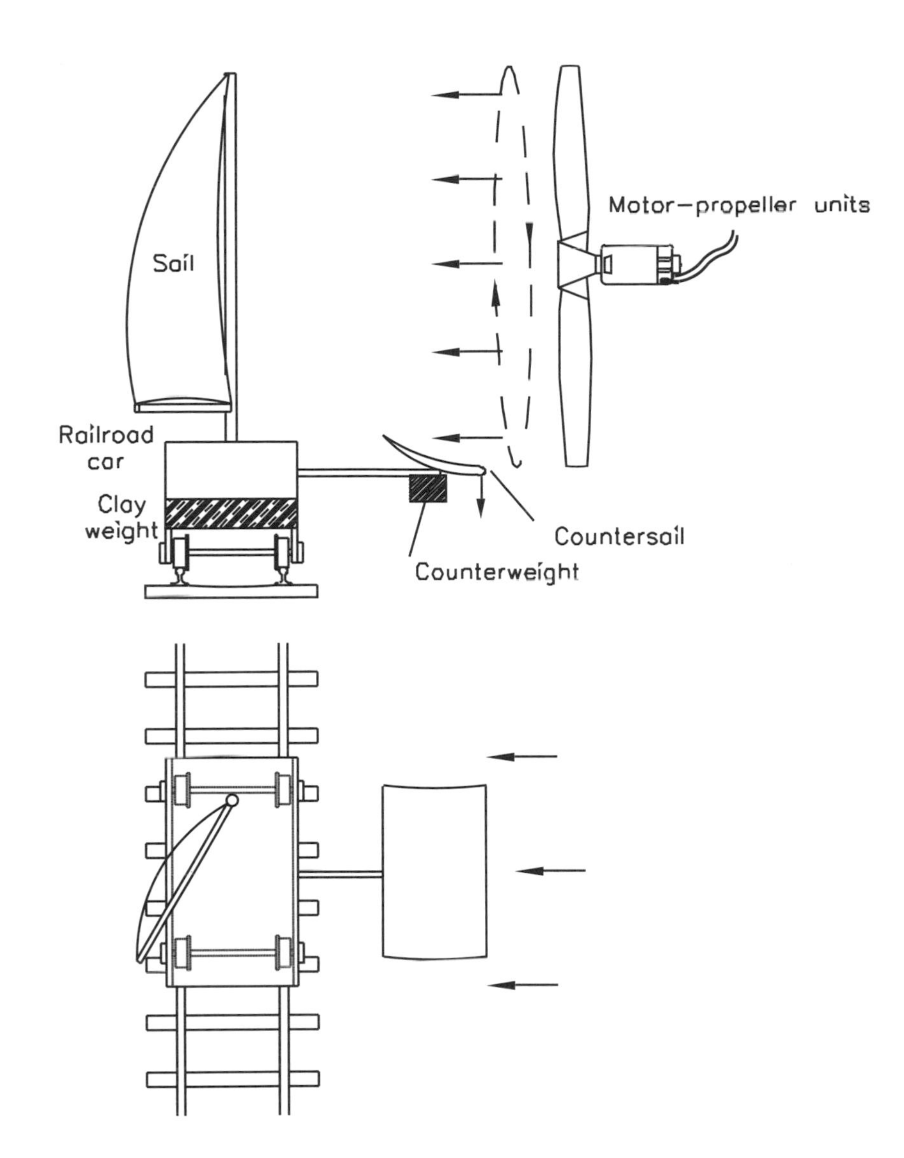

extra mass and drag of the added devices. The counterweight helps to lower the center of gravity, too, by adding ballast to the body of the yacht.

The positioning of the counterweight and countersail is quite critical. The counterweight should be increased until the yacht is just short of overturning into the wind. The countersail is set so that when wind comes, the yacht shows little or no overturning tendency. You can find the best arrangement by temporarily removing the ballast and applying wind.

The railroad yacht, like a sand yacht, should be capable of traveling surprisingly fast—several times faster than the wind speed in ideal conditions. With care, you can probably determine the yacht speed over a measured 2-m track with a stopwatch. An electronic timer that could be switched on and off by the yacht's passing would be more accurate and would work even on a short track. Similarly, if you have a datalogger board for your PC, then you could use it to detect the passing of the yacht with a photocell of some kind.

Measuring the wind speed from the fans is more difficult. A rotating-vane anemometer is really far too big. A thermistor or hot-wire anemometer is a possibility, but it is perhaps easier to measure the rpm of the fans and multiply that number by their pitch, which will give you at least an idea of the idealized airstream speed. The pitch is the distance that the propeller would go if it screwed itself forward through an imaginary solid gel material. The pitch is given by the product of π times the propeller diameter times the tangent of the tip angle to the plane of rotation. Model airplane propellers for glowplug internal-combustion engines typically have a pitch of 75–100 mm on a 150-mm diameter, whereas propellers driven by twisted rubber bands have a much bigger pitch, 200 mm or more, on a 150-mm diameter.

How It Works

The mainsail of the railroad yacht works in the same way that most other yacht sails do: by deflecting the wind coming from the side off in a backward direction. Newton's third law of motion, which holds that every action force has an equal and opposite reaction force, ensures that the yacht itself will go off in the opposite direction, that is, forward. Some yacht sails—such as spinnakers, oversize lightweight sails—are designed for use when a yacht is running before the wind. However, as any sailor will tell you, you sail the fastest not when the wind is behind you but when it comes from the side. The countersail of our railroad yacht works in an analogous way, of course, but provides a downward force to oppose the overturning torque from the mainsail.

There is an element of self-correction in the mainsail of a yacht: if the wind blows too hard, the yacht will tend to heel over, which reduces the force on the sail, and an equilibrium will be established in this heeled over position. In the railroad yacht, we don't really have that luxury. If the wind exceeds a certain value, the vehicle will simply tip over. The countersail causes an even worse problem: if the wind blows too hard, the vehicle can tip slightly, and this tipping motion will cause the countersail to exert more force and tip the vehicle even more, a process that will swiftly cause a capsize. However, the fact that a stronger airstream will have an equal effect on the mainsail and the countersail helps considerably to prevent a capsize, provided the two sails are sized correctly.

THE SCIENCE AND THE MATH

You can easily calculate the maximum counterweight that can be used in your railroad yacht by summing the turning moment about the windward-side wheel. With the yacht pivoting around the windward-side wheel, the turning moments, T, will be

$$T_{\text{clockwise}} = LM_c g,$$

where L is the boom length, M_c the mass of the counterweight, and g the acceleration due to gravity, and

$$T_{\text{anticlockwise}} = RMg/2,$$

where R is the distance between the rails. When the car is about to tip over, these moments will be equal:

$$LM_c g = RMg/2.$$

Therefore,

$$M_c = RM/2L$$

gives the maximum usable counterweight mass.

Although the countersail calculations are somewhat similar, there is the additional difficulty that the force produced by the countersail depends on the wind speed. In an ideal world, both the countersail and the mainsail would produce forces that rose by the same proportion with an increase in wind speed. Let's assume (as is roughly correct) that both the drag, D, on the mainsail and the negative lift, F_l, from the countersail are proportional to the square of the wind speed, V; to the sail areas A_s (mainsail) and A_c (countersail) projected in the flow direction; and to the drag coefficients C_{ds} (mainsail) and C_{dc} (countersail):

$$D = {}^1\!/\!_2 \rho C_{ds} A_s V^2$$

and

$$F_l = {}^1\!/\!_2 \rho C_{dc} A_c V^2.$$

Then the yacht will suffer an overturning moment, T_o, due to the drag acting at height $H/2$ (half way up the sail) on the order of

$$T_o = aA_s V^2 H/2$$

and a correcting moment, T_c, from the countersail on its boom (length L) of

$$T_c = bA_c V^2 L,$$

where a and b are constants. There is a further correcting torque, T_w, due to the weight of the truck on

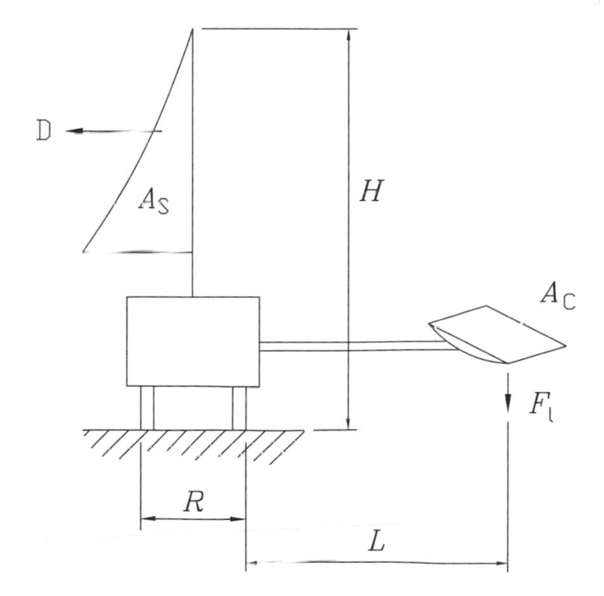

the rails. With rails spaced distance R apart and a truck of mass M, the maximum T_w is given by

$$T_w = MgR/2.$$

For the yacht to stay upright, T_o must be less than the sum of T_c and T_w. So at the tipping point,

$$aA_sV^2H/2 = bA_cV^2L + MgR/2.$$

By equating the maximum yacht speed and the maximum pressure on the sails (and hence the maximum T_o), we can maximize yacht speed simply by maximizing the sum of T_c and T_w. A heavier railroad car is obviously an advantage, but we can also gain advantage by increasing the area of the countersail and the length of its boom.

And Finally . . . Stabilizing Your Yacht and Sailing in Circles

There are several possible methods for increasing the stability of your yacht. The mast could be sprung in its mounting, so that the sail, like that of a windsurfer in a gusting wind, would blow from upright to lie with the wind somewhat, reducing the overturning moment during a gust. There is another option for stability that is uniquely available to a railroad system: a hook on the underside of the top of the rail—but maybe that would be cheating. Also, using a hook would require that the track bed be firmly anchored down, which would make railroad construction more difficult.

You could try setting up a complete oval track. Unless you vary the wind direction, you will need to provide for tacking by means of a sail that either flops to and fro on its own or can be forced to do so with a control. The special servo motor winch units often sold by model shops for trimming the sails of model yachts might be suitable and would probably cost only twenty or thirty dollars. You could avoid the issue of sail trimming by keep the wind at right angles to the track all the way round. This could be achieved by directing all the fans inward toward a volcano-shaped hill or even a duct—like a power-station cooling tower perhaps—in the middle of the oval. You might also be able to use a ceiling fan pointed downward over the middle of the track with a volcano-shaped hill deflector giving an outward wind. Maybe with a twin-track layout you could have races.

What about a regular (water-borne) yacht with a flying outrigger to help stabilize it against tipping? Could the outrigger actually be made big enough? Its area would need to be a sizeable fraction of the mainsail area. Would it be possible to control the flying outrigger well enough? Could the flying outrigger dodge the waves in heavy seas? Would wind drag on the outrigger negate the advantages it gives in terms of reduced water drag and increased mainsail area?

REFERENCES

Claughton, Andrew, John Wellicome, and Ajit Shenoi, eds. *Sailing Yacht Design: Theory.* London: Longman, 1999. One of the many books on sail and yacht design.

Hart-Davis, Adam. *Eurekaaargh! A Spectacular Collection of Inventions That Nearly Worked.* London: Michael O'Mara, 1999.

Sounds Interesting

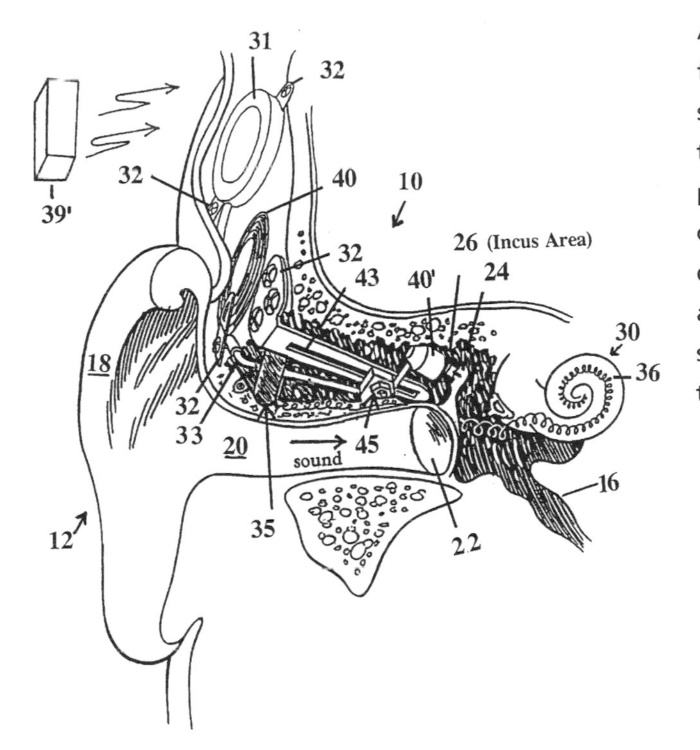

As the harmony and discord of sounds proceed from the proportion of the aereal vibrations, so may the harmony of some colours . . . and the discord of others . . . proceed from the proportions of the aethereal. And possibly color may be distinguished into its principal degrees, Red, Orange, Green, Blew, Indigo and deep Violet on the same ground, that sound within an eight is graduated into tones.

—Sir Isaac Newton, letter to the Royal Society

Hearing seems to be the most "mathematical" of the senses, as Sir Isaac Newton tacitly acknowledged in a 1675 letter to the Royal Society in which he described the phenomenon of color with the terminology of musical sound. People who are good at math are often musically gifted, and music is fundamentally mathematical, as authors like Ian Johnston in *Measured Tones* have noted. Pleasing harmonies between different pitches are heard only when they relate to one another in whole number ratios.

The human ear is mathematical in both form and function. The inner ear, the cochlea, has a spiral shape and is lined with sensitive—possibly actively driven—hairs (see Thomas Duke's "The Power of Hearing"). The cochlea performs a Fourier analysis on incoming sounds just as a spectrum analyzer does, resolving the rapid, jagged waveforms into simple patterns of notes and harmonics on the hairs. The cochlea can also compare the phase between two sounds to determine their directions, just as the antenna arrays of a radio telescope do. In the projects in this part, we see yet more math in musical sounds.

REFERENCES

Duke, Thomas. "The Power of Hearing." *Physics World,* May 2002, 29–35.

Johnston, Ian H. *Measured Tones: The Interplay of Physics and Music.* Bristol, U.K.: Institute of Physics Publishing, 1989.

6 Musical Glugging

In came a fiddler with a music-book, and went up to the lofty desk, and made an orchestra of it—and tuned like fifty stomach aches. In came Mrs Fezziwig, one vast substantial smile.

—Charles Dickens, *A Christmas Carol*

Can you use bubbles to make music like Handel's *Water Music*, or will the sound be more like the fifty stomach aches Charles Dickens described in *A Christmas Carol?* What might a musical stomach ache sound like? Perhaps a sort of gurgling sound would be reasonable. But can you make musical gurgling? When you pour a liquid from a bottle, the liquid rushes out through the neck of the bottle while air rushes back in, and the resulting air bubbles often make a fairly musical glugging sound. (A wine aficionado might declare that the finer the vintage, the finer the music.) The pitch of the glugging varies rapidly as the liquid level changes, and the sound quits altogether when the liquid level goes below a third or so.

A interesting bit of theory explains the frequency of this momentary but delightful sound. In this project, you can use different methods to prolong the glugging sound so that you can appreciate it more fully, perhaps even long enough to make "water music."

What You Need

Either

- ❑ Large container, at least 2-l size
- ❑ Air bleed valve to fit the container (or make your own; see instructions)
- ❑ Water
- ❑ Flexible tubing

Or

- ❑ Aquarium air pump
- ❑ Flexible tubing
- ❑ Tubes of different sizes to fit the ends of the flexible tubing
- ❑ Water

What You Do

Although you can simply tip water out of a wine bottle to hear musical glugging, the effect is only momentary. One way to extend the glugging time is to find a large container and fit it with a small neck. The pitch of the glugging still changes quite rapidly, however. Instrumenting the container with a pressure gauge will give you a clue as to why this happens: the static pressure in the bottle changes. This pressure change does not in itself change the frequency of the glugging

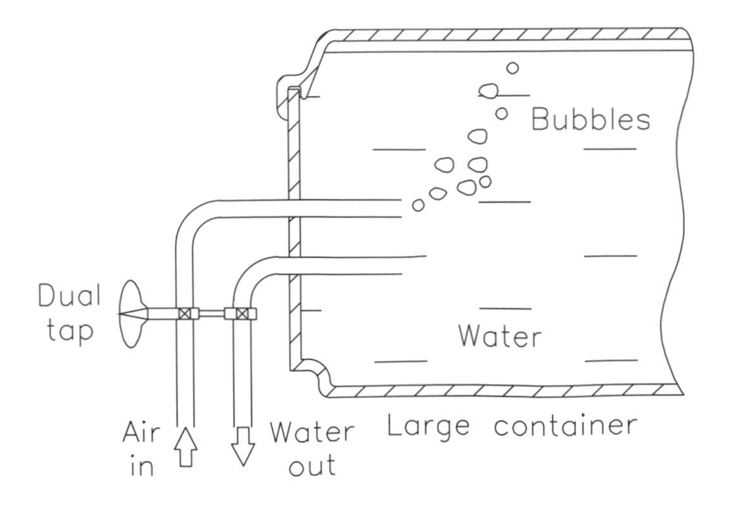

much, but the fall in static pressure changes the size of the bubbles being produced, as well as the rate at which they are produced. The smaller bubbles yield higher frequencies, which explains why the pitch of the glugging rises as you pour out the liquid.

You can better control of the size of the bubbles if you provide separate paths for the water going out and the air coming in. Installing an air pipe will control the bubble size much better than just relying on the air space at the top of the neck of a bottle. However, the pipe should exit underneath the water in the container, otherwise glugging won't happen at all. Choosing a large container also helps because it gives you more time to test out the bubble frequency and see what is going on.

With the largish container suggested, you will find that the pitch of the glugging changes only slowly with time. The container should ideally be fitted with a proper air bleed valve and water outlet nozzle. If you look around, you may find a tap or faucet that does exactly what you need. There are "no-drip" taps for barrels and other containers that have a small plastic assembly with an air inlet, a water outlet, and a lever that cuts off both flow paths. Containers with this kind of faucet are sold for putting in a refrigerator to dispense cool water, for example. But if you can't find one of these, just glue two tubes into the neck of your container, the lower, larger pipe for water, and the upper, smaller pipe for air. The water outlet could be 8–10 mm in inner diameter, and the inlet 3–4 mm in inner diameter, for example. The air inlet should end a short way, say, 5 cm, inside the container.

Another simple technique is to abandon the container altogether and pump air at a suitable rate into a tube of a suitable diameter. Changing the tube diameter at the same volumetric flow rate, which is what I am suggesting here, changes the pitch of the bubbles. You can use a simple aquarium oscillating-diaphragm pump. These pumps are inexpensive and readily available, but they do have certain disadvantages. They produce a pulsatile air flow—pulsating at the domestic electricity frequency of 50 or 60 Hz—rather than a smooth flow. A reservoir vessel will smooth out these pulses.

How It Works

Glugging occurs when newly formed bubbles compress and expand rhythmically. This rhythmic vibration is just loud enough to escape the liquid and just the right frequency for us to hear it. Smaller bubbles compress and expand more rapidly than larger ones, and the frequency of oscillation is inversely proportional to bubble size.

The pump setup simply blows bubbles into the water. In the container setup, the glugging occurs when the pressure in the container falls below the ambient air pressure. The water runs out but air is not getting back in, so the pressure in the air space falls as the air expands. The lower pressure means that ambient air can blow bubbles from the air bleed nozzle inside the neck of the container. Whatever their source, the bubbles will oscillate a little after they are formed, creating the musical sound.

THE SCIENCE AND THE MATH

We can estimate the oscillation frequency of a bubble as follows. Suppose a bubble of volume V_0 ($4/3\pi r^3$) expands to volume $V_0 + \Delta V$, undergoing a corresponding change in its radius, Δr. The mass, M, of the surrounding liquid that is displaced by the expansion is related to the product of the bubble volume and the density of the liquid. The force, F, required to achieve this expansion is given by

$$F = M(d^2\Delta r/dt^2) = V_0\rho(d^2\Delta r/dt^2).$$

This force derives from a change in pressure inside the bubble, $F = A\Delta P$, where ΔP is the change in pressure caused by the change in volume, ΔV, and A is the surface area of the bubble. If we assume that the bubble is spherical ($A = 4\pi r^2$) and that we have approximately isothermal "ideal gas" conditions, we can use the ideal gas law,

$$PV = nRT,$$

where n is the number of moles of gas, R the gas constant, and T the absolute temperature in Kelvin, to determine the pressure change:

$$(P_0 + \Delta P)(V_0 + \Delta V) = nRT,$$

where P_0 is the ambient pressure. Solving for small pressure changes, ΔP, from P_0 and small volume changes, ΔV, from V_0 gives

$$P_0 = nRT/V_0$$

and

$$P_0 + \Delta P = nRT/(V_0 + \Delta V).$$

So

$$\Delta P = -(nRT/V_0)(\Delta V/V_0).$$

Because $F = A\Delta P$, we arrive at

$$V_0\rho(d^2\Delta r/dt^2) = -A(nRT/V_0)(A\Delta r)/V_0$$

and

$$d^2\Delta r/dt^2 = -(A/V_0)^2(P_0/\rho)\Delta r.$$

This equation is rather like the equation for simple harmonic motion for a mass M on a spring of rate K:

$$d^2x/dt^2 = -(K/M)x,$$

where x is the displacement of the spring. Integration gives $x = \sin(2\pi ft)$, which has frequency $f = (1/2\pi)\sqrt{(K/M)}$. Because A/V_0 is just $3/r$ and because $(K/M) = (A/V_0)^2(P_0/\rho)$, our bubble frequency, f, is expected to be approximately

$$f = [3/(2\pi r)]\sqrt{(P_0/\rho)}.$$

In practice, our assumption of isothermal conditions is likely to be incorrect. At the fast bubble oscillations we are talking about, heat does not have time to diffuse from the middle of the bubble to the edges. In other words, the gas expansion is adiabatic, so that the ideal gas equation does not apply. The ratio γ of the specific heats at constant pressure and constant volume for the gas might enter the equation to compensate for this, as often happens when adiabatic processes occur in gases. In addition, our simplistic assumption of the mass of liquid that is moved by the bubble expansion is not quite right. In fact, Marcel Minnaert first estimated the oscillation frequency of a bubble to be

$$f = [\sqrt{3\gamma/(2\pi r)}]\sqrt{(P_0/\rho)}.$$

For monatomic gases such as argon ($\gamma = 5/3$), our simple formula is in fact remarkably accurate: Minnaert's equation gives $\sqrt{3\gamma/2\pi}$ ($= 2.9/2\pi$) for the constant multiplier, as opposed to our value of $3/2\pi$ for the equivalent numerical factor in our formula. However, air contains only 1 percent argon, being mainly nitrogen and oxygen, and has a γ value of 7/5, which makes our result less spectacularly accurate.

And now for some example numbers: for a 5-mm air bubble near the surface of water (at atmospheric pressure), the frequency should be 770 Hz, a musical tone roughly equivalent to the A# above middle C.

Our analysis so far deals with the sound emitted by oscillating bubbles, but it does not deal specifically with the glugging sound that occurs when liquid is poured from a container. As you pour liquid from a container, the pressure in the air space above the liquid (the *ullage* space) falls below the ambient air pressure because air is not getting back in. If we start with an initial volume V_i at an initial pressure P_i, Boyle's law tells us that

$$P_i V_i = P_f V_f,$$

where P_f and V_f are the pressure and volume after some liquid has been poured out. This means that if $V_f > V_i$, then $P_f < P_i$. If, for example, we started with a 100-ml volume at the top of the bottle and poured out 10 ml of liquid (thus increasing the ullage volume to 110 ml), then the pressure should drop from 1,000 mbara (atmospheric pressure) to 909 mbara, a drop of 91 mbar. I measured the pressure in the ullage space in a 2-l container as it glugged. Depending upon the angle of pouring and how much water was left, the pressure varied from about −3 mbar to −15 mbar.

The pressure in the water is higher at the level of the air bleed nozzle; the pressure, P, varies with water depth, H, according to the following equation:

$$P = \rho g H,$$

where ρ is the water's density. For the kind of water depths we are talking about in our experimental arrangements, we can take a value of about 10 cm. Now 1 cm of water exerts about 1 mbar of pressure. So we might expect to need pressure on the order of a few millibars to push bubbles of air back into the bottle.

Our negative Boyle's law pressure needs to exceed the hydrostatic pressure in order for glugging to take place—a minimum of a few millibars with the container nearly horizontal or 15 mbar or so with the container nearly vertical. Once glugging starts, the air coming in reduces the negative pressure until an equilibrium is reached with air coming in at roughly the same rate as water is leaving through the main spout, with a fairly steady pressure in the ullage space, and with the pressure in the bottle never exceeding 10–20 mbar.

And Finally . . .
Continuous Glugging

To utilize a continuous supply of water, you will need to provide just the level of vacuum that I measured above. You need to control the vacuum above the water in order to get controlled glugging. In that case, maybe you can set up a glugging bottle with a truly continuous capability. I started to do this, but I soon realized how much simpler the setup described above was. However, if you really want specific and precise information about bottle glugging—maybe you are designing a bottle for a whiskey company, one that must have a really satisfying glug—then maybe a continuous supply is the only way.

REFERENCE

Minnaert, M. "On Musical Air Bubbles and the Sound of Running Water." *Philosophical Magazine* 16 (1933): 235–248.

7 Pneumatic Drum

I understand the inventor of the bagpipes was inspired when he saw a man carrying an indignant, asthmatic pig under his arm. Unfortunately, the manmade sound never equaled the purity of the sound achieved by the pig.

—Alfred Hitchcock

A drummer can vary the frequency, the musical pitch, of a drum by changing the tension on the strings attached to the drumhead. However, this process usually takes a second or two and is not normally done during a performance. An exception is the African talking drum, or dondo, whose strings can be almost instantly tightened by squeezing the instrument or pressing on a tensioning pedal.

Pneumatic pressure can also be used to tune a drum, as, for example, in U.S. Patent no. 5,392,681, which describes how a pneumatically inflated annular bladder can be used to increase the tension in the drumhead membrane by pulling it tighter over the opening in the drum body. The pneumatic pressure is applied not to the sound-producing part of the membrane but to an annular portion outside the central sound-producing part. This kind of pressure-tuned drum is complex, however, and a percussionist cannot adjust it quickly enough to play a tune on a single drum.

Why not just apply the air pressure to the inside of the drum? Surely it would be simpler to make a drum body sealed, except for a pipe connection, and then vary the air pressure inside the body to give different notes? In this project, we use this approach to create a strange mongrel instrument, a cross between a set of bagpipes and a tunable musical drum. It's simple and it works, at least roughly.

What You Need

- ❑ A metal or plastic cylinder for the drum body (e.g., a large cookie tin or a 10-l plastic wastepaper bin)
- ❑ Membrane (e.g., shrink-fit plastic for covering model airplanes like Mylar; not necessary with the cookie tin)
- ❑ Plastic tubing, about 1 m long, 6-mm internal diameter
- ❑ Hot-melt glue or Krazy Glue
- ❑ Soft drumstick
- ❑ U-tube manometer (optional)

What You Do

First you need to fix the tubing into the drum. A good way to do this is to punch a hole just big enough to thread the tubing through and then "belly out" one end of the tubing by heating it in boiling water and widening it with a cone-shaped piece of wood or metal. When you insert the tubing into the hole in the drum, the bellied-out section will prevent the tubing from pulling out. Nevertheless, you should glue the tubing in with hot-melt glue or Krazy Glue to make it doubly secure.

If you use a large cookie tin or some other sheet metal cylinder as your drum, you won't need a drum membrane: the sheet metal is it. Seal up the drum, either by sealing the lid on or by gluing on a wood bottom. If you are using an open-ended container such as a wastepaper bin, you will need to mount a membrane on the end of the drum and then glue or tape it as appropriate. The membrane must be stretched with an even and fairly tight tension in its surface. Mylar and many other plastics can be shrunk by means of the judicious application of heat. For Mylar, which is heat resistant, you can use an iron switched to low; and for other plastics, you can use the radiant heat from the burner on a cooking stove (hold the plastic at least 300 mm from the burner). Plastics that shrink easily can even be shrunk with the heat from a really hot hair dryer.

Now trying blowing or sucking on the tube as you drum on the drumhead. You vary the pitch of the note produced by the drum by varying the pressure in the drum body either up or down relative to atmospheric pressure. Creating an

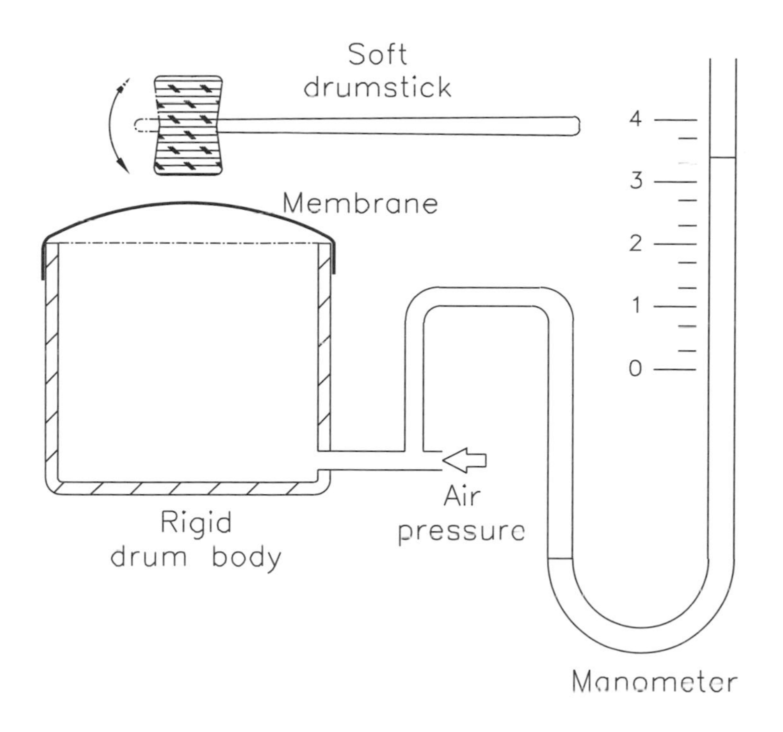

air-pressure differential between the sides of the membrane bellies it out (or in), stretching it and increasing the tension in it. It does not matter whether the differential pressure is positive or negative, that is, whether the membrane is convex or concave; you get a frequency shift to higher notes in either case.

Before sounding the drum, you could use a simple U-tube manometer or other pressure-indicating device to judge when the correct pressure has been applied. The manometer could be calibrated for different musical notes. This pneumatic drum allows for almost instantaneous changes of pitch, and you may find that with practice you can play a tune.

How It Works

Most of us are comfortable with the idea that the frequency of the note from a stretched string on a guitar or violin goes up as the tension in the string is increased. When the string is pulled away from its straight rest position, the tension pulls it back toward its rest position. The greater the tension, the greater the restoring force, and the faster the acceleration of the string, just as we expect from Newton's second law of motion: acceleration is proportional to the force producing it. The faster acceleration gives rise to the higher frequency we observe. The situation is similar with the drum membrane, except that we have a surface rather than a string. The restoring force increases with tension in the membrane, so the frequency increases too. Applying air pressure to the membrane increases the tension in it and so changes the pitch at which the drum sounds.

Another way to think about the drum is to consider the speed of the waves traveling across the membrane. You can think about the membrane oscillations as superpositions of traveling waves: if you add a left-going sine wave to an identical right-going sine wave, for example, you get a line that just goes up and down in a sine wave pattern without moving—something like a cross section of the simplest motion of the membrane. With increased tension, the membrane will oscillate at a higher frequency because the speed of the waves on the membrane surface increases with tension.

THE SCIENCE AND THE MATH

Increasing the tension in a membrane of finite size increases the speed of transverse vibrations and the frequency of stationary-wave oscillations in that membrane. In a drum, this frequency increase manifests itself as an increase in the pitch of the drum. The speed, C, of transverse waves through an idealized membrane is given by

$$C = \sqrt{(T/S)},$$

where T is the surface tension in the membrane and S the mass per unit area of the membrane. (The theory behind this equation is given in many acoustics textbooks, including William Elmore and Mark Heald's *Physics of Waves*.) The corresponding frequency, F, is given approximately by

$$F = R\sqrt{(S/T)},$$

where R is a characteristic length, here we take the radius of the drumhead, for the fundamental note.

You will find that when the drum is highly pressurized, the sound seems much duller and it stops more quickly. Scientists and engineers usually express this phenomenon in terms of the Q *factor*—the oscillator quality, Q, which is usually expressed as the stored energy divided by the energy lost per cycle and is proportional to the number of oscillations completed by the system. Bells and glockenspiels have high Q factors—their notes ring for the better part of a minute; whereas a typical drum has a low Q factor—its notes, though loud, die away in a fraction of a second. In our device, Q drops rapidly as the pneumatic pressure increases.

Just why Q should decrease as the pressure rises is not clear from the simple theory of the device. Oscillators like the drumhead here have many possible modes of oscillation, all of which contain energy. Some of the modes occur at frequencies that are similar to one another, and energy can be transferred between these modes. Perhaps the energy fed from the main "bowl" mode, which is normally lost to other modes and then regained, is not regained in the case of the curved membrane surface, where the other modes are damped by the curvature of the surface. However, the energy of the drum must be dissipated somewhere. Because the drumhead membrane itself cannot be absorbing more energy as the pressure increases and because the drum is apparently not emitting more acoustic energy, then more energy must be transferred into loss processes such as the flexing of the membrane at the edge of the mounting, or into nonadiabatic gas compression and expansion.

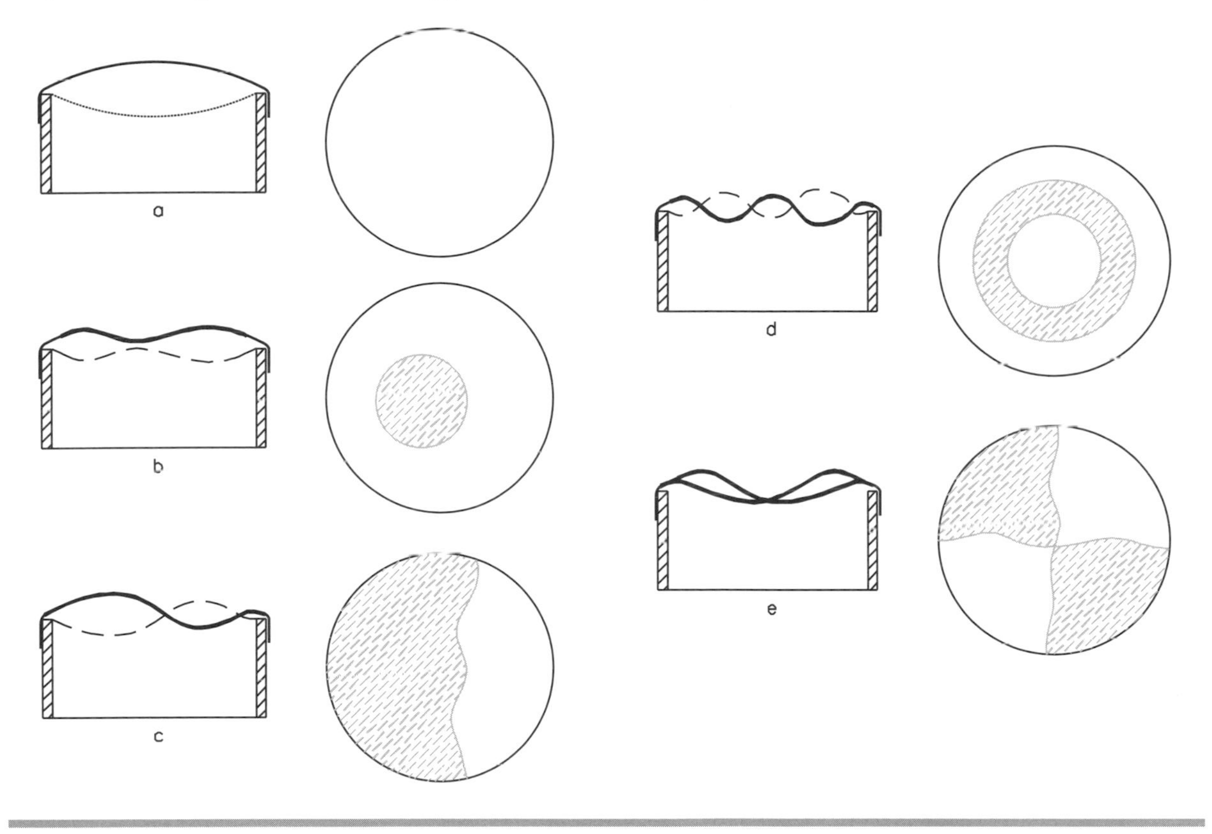

And Finally . . . Playing a Scale

It is difficult to accurately tune the pneumatic drum, except by long practice. (It is no different from a violin in this respect.) However, using valves to pre-set pressures quickly would allow a musical scale to be played on the drum. The valves, which could be activated by a computer or a piano-style keyboard, could connect a set of air reservoirs to the drum body, the reservoirs being at pre-set pressures corresponding to the desired pitches. When sounded with a drumstick, the drum would play a note at the pitch pre-set on the air reservoir connected via one of the valves.

In another advanced version, a high-speed, electronically controlled pressure regulator could be used to regulate the pressure inside the drum. This regulator could in turn be controlled by a computer, an electronic keyboard, or some other source of electrical signals—perhaps in that *lingua franca* of modern music, the MIDI format—that correspond to musical notes.

REFERENCES

Elmore, William C., and Mark A. Heald. *Physics of Waves*. New York: McGraw-Hill, 1969.

Hall, Walter L. "Drum Tuning Device." U.S. Patent no. 5,392,681, February 28, 1995.

8 Singing Contacts

'Tis strange, but true; for truth is always strange,—
Stranger than fiction.

—Lord Byron, "Don Juan"

The building blocks of matter—atoms, and their component parts, the nucleus and electrons—cannot ordinarily be sensed by humans. Lord Rutherford, in the early days of nuclear research, could "see" individual α particles—the nuclei of helium atoms—only by means of the tiny flashes that his dark-adapted eye could see when the particles hit a zinc sulfide screen. But these α particles possess enormous velocity and hence enormous energy, millions of electron volts. Electrons at similarly huge energies are moving at nearly the speed of light, and they too can be individually perceived as flashes on a zinc sulfide scintillation screen. But surely we cannot sense ordinary slow-moving electrons, those that have just a few electron volts of energy? Well, strangely, maybe we can, with our ears. It's no fiction.

What You Need

- ❑ An audio oscillator or a powerful radio set (e.g., one channel of a stereo hi-fi)
- ❑ Low-value resistor (e.g., 8 ohms) or a coil of resistance wire of 5, 10, or 20 ohms
- ❑ Wire, thin multistrand insulated connecting wire, such as the "hook-up" wire used inside electronic equipment

- ❑ 2 contact surfaces, one of which needs to be a sheet of material of at least a few square centimeters
- ❑ Insulating tape

What You Don't Need

- ❑ A loudspeaker!

What You Do

Your audio source must be able to withstand being short-circuited out completely while you are adjusting the contacts and at some points during operation. If you are in any doubt, put a resistor or a coil of resistance wire (5, 10, or

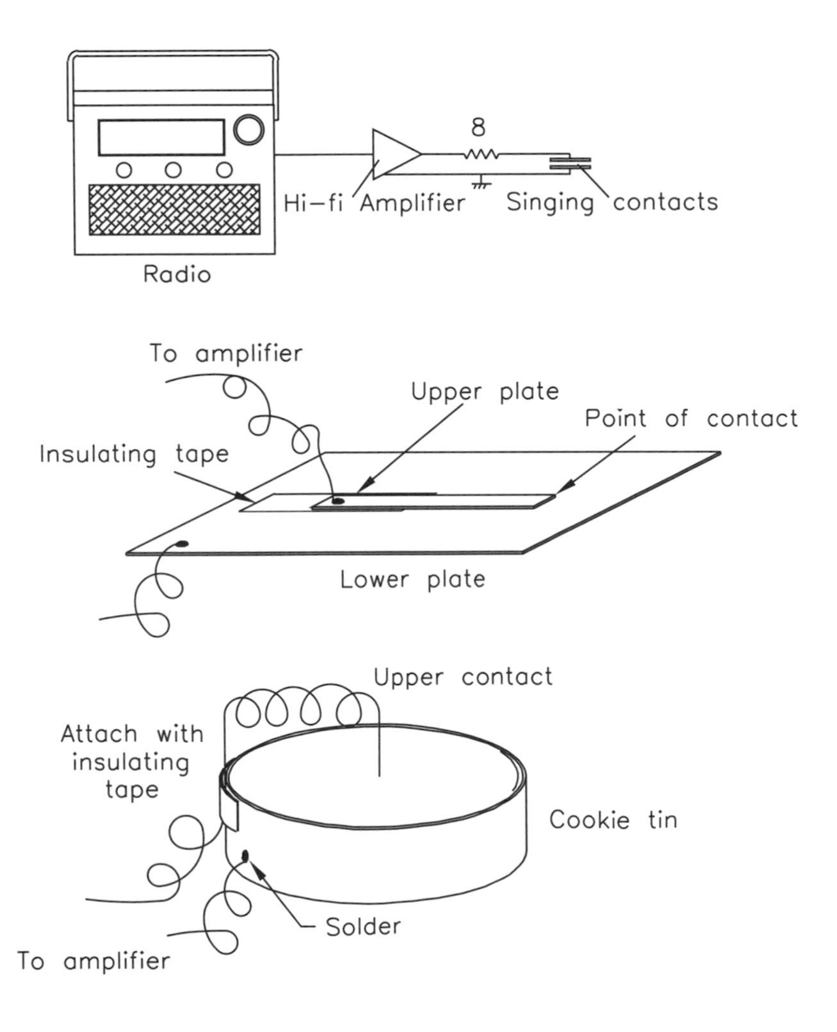

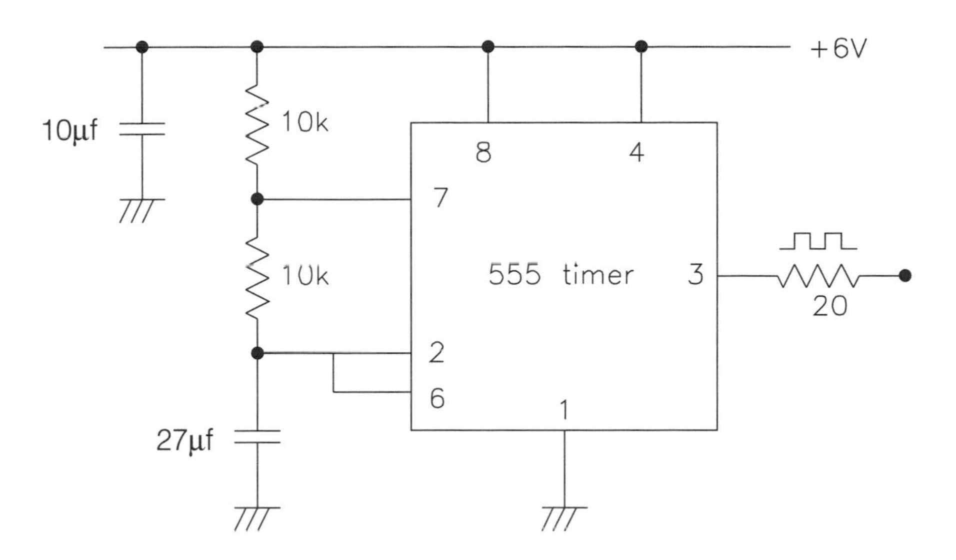

20 ohms) in one of the wire connections to limit the maximum current that can flow when the contacts are shorted together. Run two pieces of wire from the audio source and solder one wire to each of the two contacts. I tried a piece of tin-plated steel from a thick-walled can about 2 cm by 4 cm for one of the contacts because the tinplate was convenient for soldering the wires to. You can easily construct a larger "sounding board" by using the bottom of an empty cookie tin for one contact and a wire or a smaller piece tinplate for the other.

You now need to arrange for one contact to rest on the other with a small amount of force. Until a few years ago, most households possessed an old (or even still working) hi-fi turntable, or gramophone, which of course has a nicely engineered counterbalanced pickup arm that would work well for demonstrating singing contacts. The necessary adjustments could be made by varying the counterbalance position. However, such equipment went to the garbage in most households some time ago, so we need to come up with another way to adjust the contact force. If you are using a cookie tin and a wire as your contacts, the easiest way to do this is to coil the connecting wire around a pencil to give it more flexibility, then anchor the wire to the side of the cookie tin with insulating tape, and rest the end of the wire on the static contact. You can adjust the contact force by bending the wire. If you are using two flat contacts, you can put a sliver of an insulating tape (or some other insulator, such as plastic or paper) under one end of the upper plate and rest the other end on the lower plate.

Whatever way you do it, you must delicately adjust the force between the contacts until, magically, sound bursts forth from the contact area. It won't be very loud: even a 10-W hi-fi won't produce more than the volume you might get from a telephone earpiece or a pair of headphones. You may just be able to measure the sound produced if you place an industrial noise meter pretty close to the device.

How It Works

What is producing this sound? Electrostatic force is one possibility. The monitor for one of my personal computers has a loose contact, and I can sometimes make the faulty monitor "sing" at a high frequency. The sound from the monitor is not particularly loud, but—because it is at a very high pitch, 10 kHz or so—it must be at a relatively high power. (Human hearing is inefficient at high frequencies.) So maybe singing contacts are louder at higher frequencies. Sound emission from electrostatic devices increases in efficiency with frequency, which is why such devices are often used for the high-frequency units, the tweeters, in hi-fi systems.

Another possible cause of the sound might be the force due to a spark expanding the air in between the contacts. If you observe the contacts in the dark while they are emitting sound, do they glow like a low-energy spark?

Curious phenomena associated with contacts have been noted for years. Guglielmo Marconi used one of these phenomena—known as the "coherer" effect—in his early radio telegraphy systems to detect a tiny high-frequency AC current. When the current passed through a small bed of high-resistance metal filings, their resistance suddenly decreased. Once the current had been detected, the filings were jiggled to reset them back to their initial high-resistance state. Early telephone-circuit engineers also noticed curious contact phenomena (see Herbert and Procter's *Telephony,* 1:636). They noted, for example, that noise caused by sudden changes in contact resistance could be prevented by passing continuous small DC currents through the contacts.

THE SCIENCE AND THE MATH

Let's first consider the output power of the singing contacts. The arrangement is not very efficient. At 50-mA maximum current and 5 V (about 100 mW average power, at a duty cycle of roughly 50 percent), for example, my 2-cm by 4-cm tinplate singing contacts couldn't be heard distinctly from any farther than 3 or 4 m away in a room with sound-absorbing soft furnishings.

If we take the sound level of normal conversation at a power of 100 mW at 1 m as 50 dBA, then the power of a barely audible sound (0 dBA) would be 10^{-6} W at 1 m and 10^{-5} W at 3 m, assuming an inverse square relationship between distance and power. This estimate indicates an output power from the singing contacts of around 10^{-4} W. With a 100-mW input and a 0.1-mW output, we arrive at an efficiency of only 0.1 percent. (The sound meters used in industry for monitoring noise measure down to 30 dBA, which is why it is difficult to use such a meter to pick up the sound from the singing contacts.)

We haven't answered the question of how the singing contacts actually work. Could it be electrostatic attractive forces? The electrostatic force, F, between two plates of area A (say, 3 mm^2) very closely spaced (say, 0.1 mm apart) is given approximately by

$$F = EQ,$$

where E is the electric field and Q is the electric charge stored on the plates. Because

$$E = V/S$$

and

$$Q = CV = (\varepsilon A/S)V,$$

where V is the voltage on the plates (say, 10 V), C is the electrical capacitance of the plates, S is the distance between them, and ε is the dielectric constant of air (approximately 8.8×10^{-12}), the electrostatic force is given by

$$F = (V/S)(\varepsilon A/S)V = (\varepsilon AV^2)/S^2.$$

By making the spacing, S, small (as in our experiment), we can make the force appreciable.

Plugging the numbers into this equation reveals that the weight force on the plates is tiny, 1 μN. When this force is exerted over the distance between the plates as they vibrate 0.05 mm each way at a frequency, f, of 1,000 times per second, the work done, and hence the power produced as vibration (which approximates the radiated sound that we can hear), might be

$$P = fFS = 1{,}000 \times 10^{-6} \times 10^{-4} = 0.1\ \mu\text{W}.$$

If our crude system radiated all this power as sound, that sound would not be loud, but it would likely be audible. If a standard radio receiver is capable of emitting 70 dBA using a 1-W output speaker, we might expect to hear easily but faintly 10 dBA, which is around 1 μW. So the electrostatic theory does yield an audible sound intensity. As I noted above, power from electrostatic devices increases with frequency, and our simple theory agrees, saying that the power produced is proportional to frequency. This theory should not seem too outlandish: after all, there are hi-fi speaker units that operate on the electrostatic principle, albeit at large areas and high voltages. But is the electrostatic force theory correct?

Could the sound instead be the result of a magnetic field, produced by the flowing current, that pulls the plates to and fro, with the contact simply offering a small spacing between the wires and perhaps an intermittent contact, perhaps enhancing the effect. We can use Ampère's law to estimate the forces produced by the flowing current. With a current, I, of 100 mA, the two plates will feel an attractive force, F, given by

$$F = \mu_0 I^2 L/(2\pi S),$$

where μ_0 is the magnetic permeability ($4\pi 10^{-7}$), L the length of the plates (3 mm), and S the distance between the plates (0.1 mm). Plugging in the numbers gives us a force of 0.06 μN, which is not too far from the electrostatic result. This result should not appear too surprising either: if you replace the cone coil in a moving-coil loudspeaker with a single wire, and the magnet by the magnetic field of the other wire, then you can see how a conventional loudspeaker has similarities.

Could air expansion in a spark be the explanation for the singing contacts? How much power could air expansion generate? Suppose a 0.1-mm cube of air is heated by our tiny spark:

Heat capacity of air volume

$$= C_p(0.1 \times 10^{-2})^3/22.4 \sim 10^{-9} \text{ J/K},$$

where C_p is the specific heat capacity of the air, 29 J/K·mol, and 22.4 is the number of liters in a mole. To convert this into a continuous input power, P_{in}, we just multiply by the heat rise, ΔT, and the frequency of operation, f:

$$P_{in} = C_p(V/22.4)f\Delta T,$$

which gives 50 μW for $f = 1{,}000$ Hz and $\Delta T = 40$ °C. This tiny volume of gas will expand when heated, and it will compress the surrounding air, creating sound waves. Once again we can use the ideal gas equation,

$$PV = NRt,$$

and

$$P(V + \Delta V) = nR(T + \Delta T),$$

where P is the pressure, V the volume, ΔV the increase in volume, n the number of moles, R the gas constant, and T the temperature. It follows from these equations that

$$P\Delta V = nR\Delta T$$

or

$$P\Delta V = (\Delta T/T)PV.$$

And $P\Delta V$ is of course the work done by expanding the gas against the surrounding atmosphere. The acoustic power, P_{out}, generated is thus given by

$$P_{out} = (\Delta T/T)PVf,$$

which gives about 13μW for $f = 1{,}000$ Hz and $\Delta T = 40$ °C, a value that corresponds at least roughly with our observations. But are there really microsparks between the contacts? Could you see these sparks if you put the equipment in a darkroom? And how efficient is the heating process, that is, the heating of the gas and the transfer of the energy of thermal expansion into the energy of sound waves?

And Finally . . . Improving the Reproduction Quality

You could try to improve the "reproduction quality" of the singing contacts by adding a DC current to the AC current used in the experiment as described above. The use of AC current leads to a doubling of the fundamental frequency of any signal to the contacts if the singing contact effect is symmetrical—that is, independent of polarity—which it probably is.

To understand this doubling effect, imagine a 500-Hz AC current as a symmetrical sine curve distributed about the x-axis. If the contact effect is proportional to both the positive and the negative peaks, then you could redraw the curve so that the negative peaks were inverted. The resulting waveform for the singing contact effect would have a fundamental frequency of 1,000 Hz, that is, double the actual frequency of the current. (This is the reason that most small

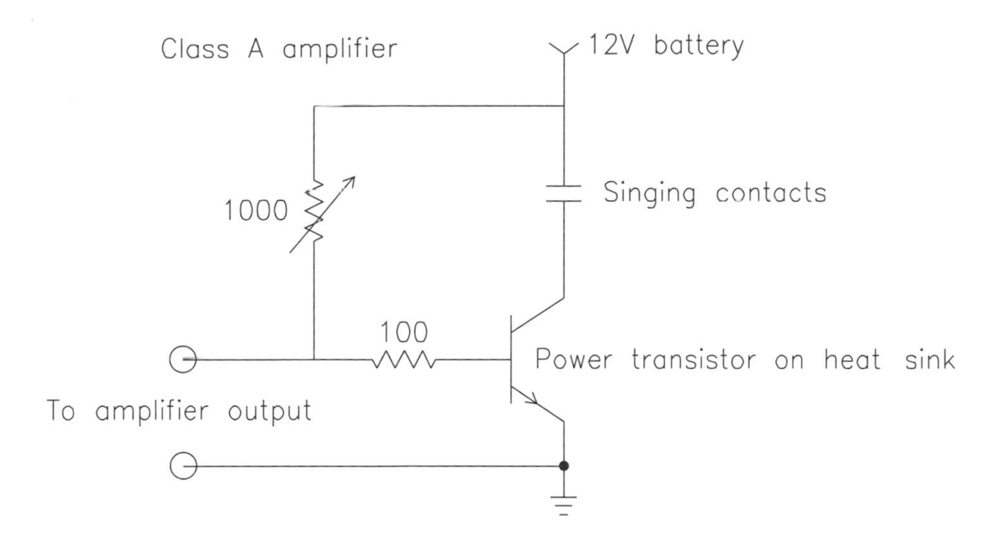

line transformers emit 120-Hz rather than 60-Hz notes when operating: they contract slightly when current passes through, but the contraction is the same no matter what the direction of the current.)

By adding to the AC current a DC current with an amplitude that is at least half that of the AC peaks, you can achieve a unidirectional current through the singing contacts. A unidirectional current will not lead to the doubling effect, and thus the distortion due to the doubling effect is removed. You could use a single-transistor class-A amplifier to add the DC current, or you could achieve a similar effect just by including a battery in series with the radio output in the circuit. You should be able to hear the pitch of the singing contact halve in frequency when you do this. By connecting a microphone to an oscilloscope and using the mike to listen to the singing contact, you could test the device more fully and see whether the added DC current gives a more linear response.

And speaking of reproduction quality, could you actually put a voice or broadcast signal through the singing contact? If so, what would reproduce the best—talk shows, musicals, grand opera?

REFERENCE

Herbert, T. E., and W. S. Procter. *Telephony.* London: Pitman, 1934.

Jolly Boating

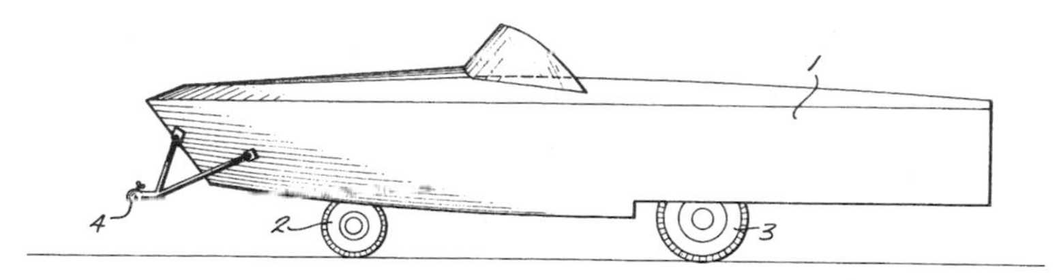

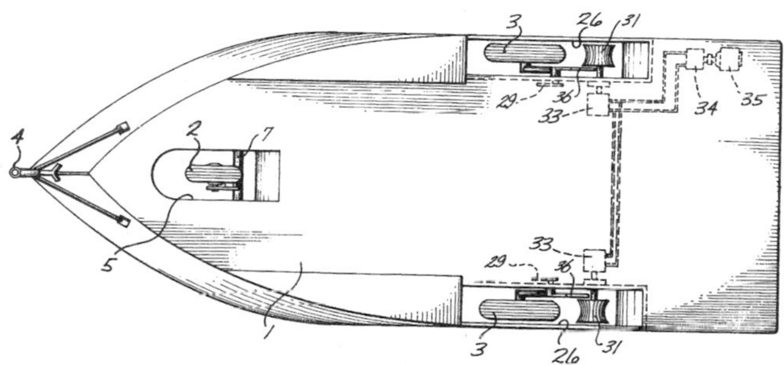

Jolly boating weather,
And a hay harvest breeze,
Blade on the feather,
Shade off the trees
Swing, swing together
With your body between your knees.

—William Cory, "The Eton Boating Song"

Marine vehicles seem to provide more opportunity for innovation than either airplanes or automobiles. The biggest ships, at 300,000 tons, are one hundred to one thousand times as big as the biggest land or air vehicles, for example, and the huge range of sizes allows a range of technologies to be used, appropriate to different scales. Varying from surface skimmers through ships to submarines, marine craft can be propelled by different forms of energy, from human muscles to nuclear steam turbines. Paddle wheels and oars can be used for propulsion, as well as screw propellers that correspond to the fans and propellers seen in aircraft. However, these are not the only possibilities for applying motive power to the task of propulsion through the sea. Dennis Normile has described how several Japanese ships have employed superconducting magnets that use the sea itself as the moving conductor in a kind of linear electric motor. In this part, we look at another possibility for propelling marine craft and also at an exotic control system that would not be possible for land or airborne vehicles.

REFERENCE

Normile, Dennis. "Superconductivity Goes to Sea." *Popular Science,* November 1992, 80–85.

9 Giant Putt-Putt Boat

Believe me, my young friend, there is nothing—absolutely nothing—half so much worth doing as simply messing about in boats.

—Kenneth Grahame, *The Wind in the Willows*

For many years, people have been making toy boats that have no mechanism at all. You can find old versions in junk shops and antique shops, although they are still being made new in Japan and Hong Kong. A lighted candle inside them heats a small metal chamber that is connected to the stern via two eductor pipes that are 2–3 mm in diameter. Magically, within a minute or two after the candle is lit, the boat propels little jets of water from its stern pipes and swims off across the water with a putt-putt sound.

If you have one of these putt-putt boats, you know that they are sometimes reluctant to start up. You can give your boat a boost by lighting the candle, launching the boat, and then squirting water from a syringe into one of the stern pipes. What makes the putt-putt boat go, or not, as the case may be? Hot air is expelled from the metal chamber as it heats, and when the heat decreases a little—because the candle has flickered, say—a small amount of water is drawn back up the pipe into the chamber. How does the squirt of water from the syringe help to get this process under way? When the water hits the hot metal, it evaporates, and the resulting steam drives a pulse of water out the back and a pulse of

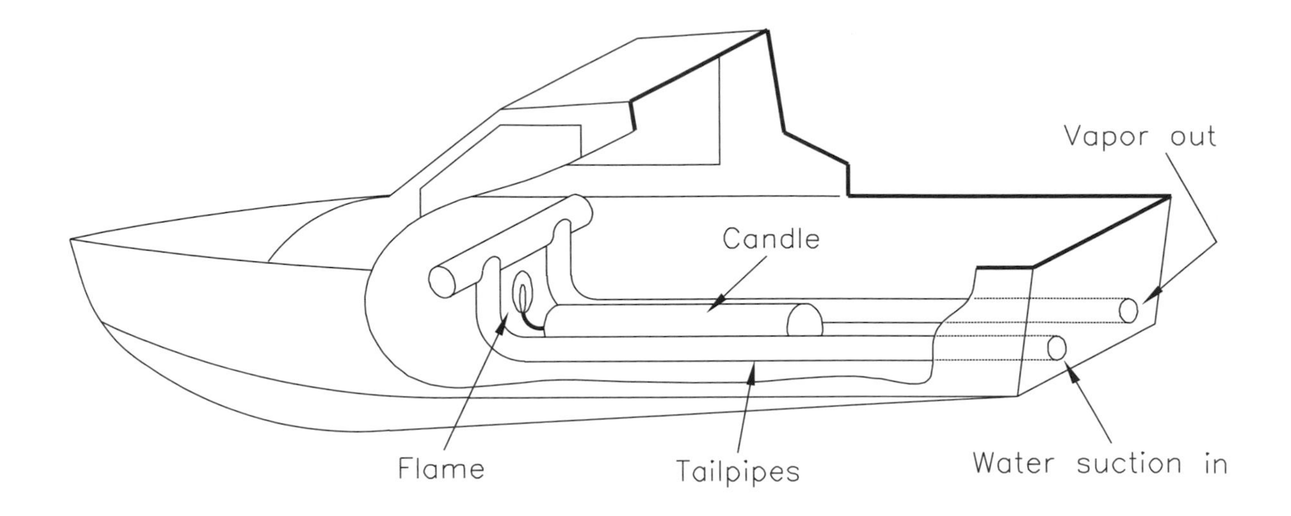

steam up into the chamber. As this steam condenses in the chamber, more water is drawn through the pipes, and the cycle repeats itself, with a typical putt-putt boat pulsing water more than once per second.

One might expect the pulsing water simply to push the putt-putt boat to and fro, but it actually goes forward quite efficiently. Why? Because the recharge of water is pulled in more or less isotropically (from all directions), whereas the water is pushed out in only one direction (backwards). The result is net propulsion. Of course, the boat would be more efficient still—nearly twice as efficient—if the pipes pulled water in from the front rather than isotropically.

In this project, we will make a large version of the putt-putt propulsion system so that its workings are easier to investigate. The steam-driven putt-putt system does not scale up easily, so we must shift to an electrically driven piston-and-cylinder-assembly as the source of the pulsing water jet.

What You Need

- ❑ Tire pump (I used the kind designed to run off 12-V DC current from an automobile cigarette-lighter socket. Inside the pump are a motor, a crank, a piston, a cylinder, and valves.)
- ❑ Rechargeable battery (e.g., 7.2 V, or 8.4 V NiCd pack used for radio-controlled model cars)
- ❑ Boat hull, 40–50 cm long and 10–15 cm wide (I used a plastic tool box with the drawers removed.)
- ❑ Short (5 cm) length of tubing, 1–2 cm in diameter, for the nozzle

What You Do

To shape my tool-box boat hull, I flattened the bow end underneath with the aid of a burner for softening the plastic and a trowel for pushing it into a more suitable shape—something like the bow of a military landing craft—but no doubt the more shapely boat hull of the kind used for model boats would work better.

Remove the valve head of the tire pump so that what is left is a piston that is driven up and down the cylinder by the electric motor via a crank. Using the

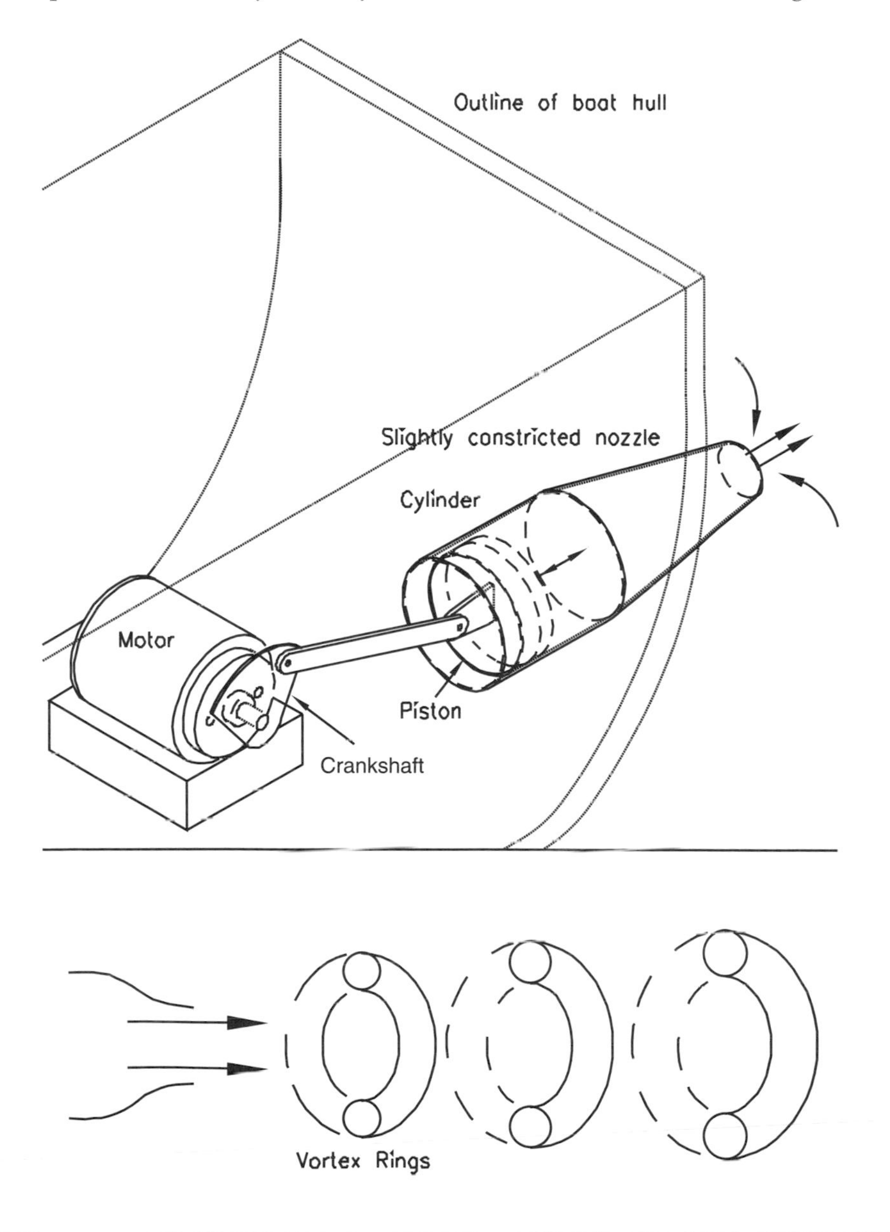

short tubing, connect the cylinder to the rear transom of the boat hull and form a projecting nozzle. The tubing should be fixed strongly to the cylinder: I glued mine to a flange that I fastened to the cylinder with small bolts. To constrict the nozzle, draw down the end of the tubing—that is, heat it up and stretch it—so that its diameter is somewhat smaller than that of the cylinder. Arrange the weight (principally the battery) inside the boat so that the boat floats on the level with the tube correctly immersed in the water.

When you switch on the motor, water will be sucked into the cylinder as the piston draws back toward the bow, and water will be squirted out of the cylinder as the piston pushes out toward the stern. The boat should spit out intermittent jets of water several times per second and be propelled forward powerfully. You will see patterns in the disturbed water behind the boat, and you may be able to make out hints that the nozzle is emitting a set of short-lived vortex rings. Once you have got the boat functioning, you can try the effect of different nozzles. Short, long, narrow, and wide are all worth trying, although I could not make divergent nozzles work. The boat will work even in water heavily contaminated with floating objects, which normally create problems for propeller craft. (I tried the effect of wood chips in the bath and found that blocking the propulsion system was almost impossible.)

How It Works

In the classic thought experiment of physics—a *gedanken* experiment, as Einstein would have said—a small team of imaginary gunners powers a railroad car by firing cannonballs backwards. When fired, each ball carries a momentum, call it $-P$. Newton's third law of motion—any force exerted (here on the cannonball) must be opposed by an equal and opposite reaction force—requires that the car with its team be given an equal and opposite momentum, $+P$. By firing off successive shots, our intrepid gunners can make their way down the track, accumulating an increment of speed each time they fire. If you have ever seen a cannon fired, however, you know that firing more than a couple shots a minute is difficult, so this propulsion method would not work particularly well.

The putt-putt boat is a much more practical scheme for, in effect, firing off cannonballs several times per second. Each time the piston pushes a slug of water backwards, the boat is pushed forward, just as the gunners' railroad car is pushed. In addition to our rate-of-fire advantage, however, we also have the advantage that, unlike the gunners, we don't need to carry our supply of ammunition with us—our slugs of water come from the water around the boat. The more slugs of water we can fire off behind the boat, and the bigger and faster we can make each slug, the faster we can go: the putt-putt's propulsive force is proportional to the product of these three factors.

THE SCIENCE AND THE MATH

The putt-putt boat is inefficient in at least two ways. First, in a moving boat, the mechanism provides thrust for less than 50 percent of the time: thrust is generated only when the jet is moving faster than the stream of water underneath the boat. Second, as I mentioned, thrust is reduced because the water is drawn in from all directions instead of from the front.

On reflection, these efficiency losses might not be so bad. How do they compare with the losses experience by propellers, which are continuous in action? Small propellers waste energy because they produce a small, high-speed jet of water; larger propellers produce a larger, lower-speed jet, which is more efficient, but they waste power because of the drag on the large blades. As if that were not enough, ultra-high-speed boats are often described as having a single-blade propeller because even if they have a propeller with two or more blades, only one blade can be in the water at a given time. The other blade or blades project above the water surface, because the larger more efficient propeller cannot run completely submerged in any practical design.

Although the arrangement in which the axially directed water is sucked partly from behind is clearly less desirable than one in which water is sucked in from the front, the rest of the flow pattern on the suck stroke is not much different than that seen with a propeller device. You could eliminate the intake from behind, perhaps by means of a forward-facing valved vent, but this setup would be complex.

Why is the small jet produced by a small propeller less efficient than a large jet, you ask? Consider this. The minimum energy required to accelerate a mass M_i to velocity V_i, is the kinetic energy, E, of that slug of fluid:

$$E = \tfrac{1}{2}M_iV_i^2.$$

But if a larger slug of fluid, mass M_j, is accelerated to a lower velocity, V_j, with the same energy, then the following equation applies:

$$\tfrac{1}{2}M_jV_j^2 = E = \tfrac{1}{2}M_iV_i^2.$$

Now think about the "impulse," P, that the slug of fluid gives to the boat. When you apply a force F for a time t to the boat, the resulting impulse equals Ft (impulse has the dimensions of mass times velocity, just as momentum does). Applying an impulse, P, to the boat will increase its momentum proportionately. But $P_i = M_iV_i$ and $P_j = M_jV_j$, so

$$P_j = P_i\sqrt{(M_j/M_i)}$$

and

$$P_j = P_i(V_i/V_j),$$

which clearly shows the propulsive advantage of a slower, more massive jet of water.

So why, in the case of the putt-putt boat, does a smaller jet work better? The first bit of reasoning you can apply is a reductio ad absurdum: when V_j is very small, P_j is big, but if V_j is smaller than the boat's forward speed, then no acceleration will occur. With the boat initially stationary, you might achieve a faster acceleration with a small V_j, but that acceleration will stop when the boat gets to speed V_j. A larger V_j will decrease acceleration, but the ultimate speed will be higher.

The second bit of reasoning you might like to apply is hinted at in our earlier assumptions: "if a larger slug of fluid . . . is accelerated . . . with the same energy." This statement does not in fact apply to our situation: with a nozzle that is not too large, we are limited not by the amount of energy that can be put into the nozzle but by the volume of water that a stroke of the piston provides to the nozzle. With the same mass M of fluid being ejected at each stroke, the impulse provided to the boat (neglecting the forward motion of the boat for now) is given by

$$P = MV_j.$$

When V_j is small, then P will be small; when V_j is large, P will be large. Therefore, a small nozzle giving a big V_j is best.

And Finally . . . Putt-Putts Go to Sea

You could try putt-putt propulsion on a even larger scale. With a putt-putt outboard motor, there is no propeller to get tangled, break, or wear out, and there is no torque effect. As I mentioned, the propulsion method, although it has a lower efficiency, is more or less contamination proof. Maybe the Marines could use it in amphibious landings. Or maybe those who harvest *fruits de mer* from the shallows of the sea—mussel or oyster cultivators, for example—would find it useful?

REFERENCES

For more on putt-putt boats, see the following sources.

Finnie, I., and R. L. Curl. "Physics in a Toy Boat." *American Journal of Physics* 31 (1963): 289.

Walker, Jearl. *The Flying Circus of Physics*. New York: Wiley, 1975.

10 Follow That Field!

The Sixth Sense . . . the ability of animals to detect the electrical currents flowing from other organisms.

—Mark W. Denny, *Air and Water*

It is commonly assumed that because humans can't detect electric fields, other creatures cannot either. Not so! Some of the scariest-looking denizens of the deep, the hammerhead shark, for example, can detect electric fields. Mark Denny, in his book *Air and Water,* describes how these extraordinary animals find at least some of their meals by sensing the tiny electric fields emitted by prey. I am not sure how easy it would be to emulate the extraordinary sensitivity of natural electric-field sensing, but this project gives us the flavor of the phenomenon. A toy boat is equipped with a propulsion system and a rudder or some other steering mechanism. The rudder is governed not by remote control or by compass bearing (like a yacht autopilot) but by "invisible railroad lines": electric field lines in the water in which it sails. For this quite complicated project, you need some confidence with electronics to build the oscillator and the servo motor. In addition, the mechanical drive, or "rotarudder," arrangement will require some effort. However, once it is all done and working, the operation is straightforward.

What You Need

Either

- ❑ Servo motor
- ❑ Servo driver circuit board
- ❑ Rudder
- ❑ Electric model boat

Or

- ❑ Electric motor
- ❑ Cylinder for rotarudder (see text)
- ❑ Motor drive circuit board
- ❑ Electric model boat

Or

- ❑ Twin-screw model boat
- ❑ Twin propeller drive circuit board

And

- ❑ Amplifier
- ❑ Oscillator
- ❑ Electrodes
- ❑ Balsa wood for electrode probes
- ❑ Wire
- ❑ Pond or swimming pool

What You Do

You must first assemble and test the circuits needed and ensure that your oscillator gives the necessary waveform. Because the oscillator works at audio frequencies, you can test it by connecting a set of headphones to output and ground via a 100-ohm resistor and checking for an output. The simple amplifier must boost the signals from its inputs and then average the output. If you connect the amplifier inputs via very large resistors, say, 1 Mohm, to the oscillator output

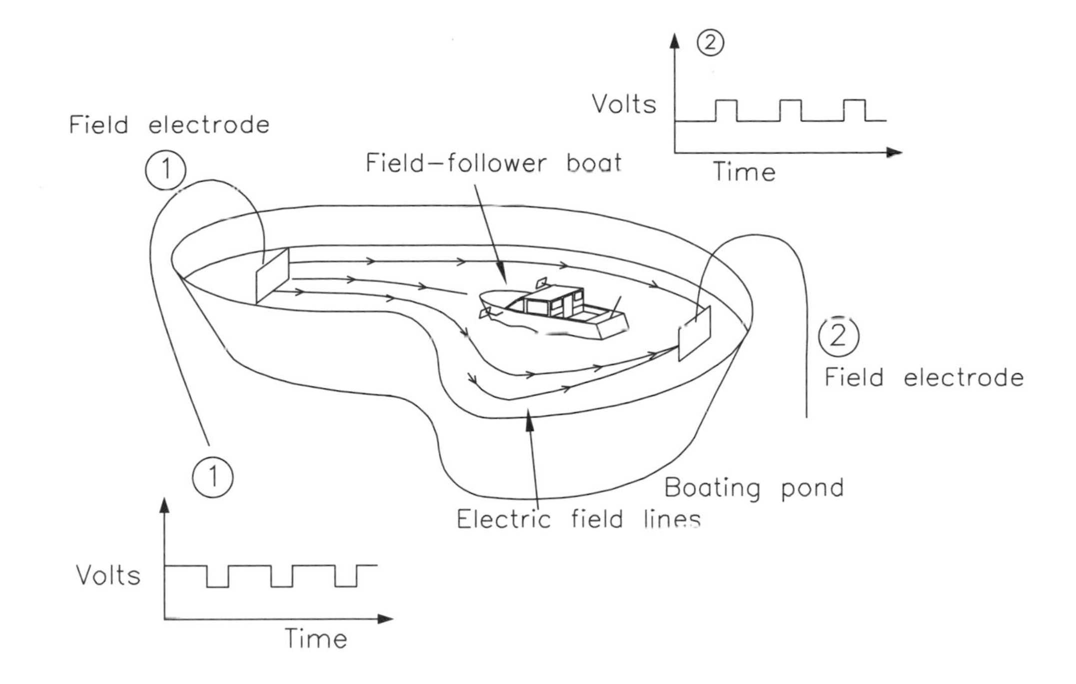

and ground, you should be able to see the output of the averaging circuit, which goes up and down as you swap the inputs over from 2 V to 4 V with a 6-V power supply. The servo motor rudder, if that is what you are using, must follow these voltage changes and swing from hard left to hard right as you swap the inputs. Similarly, if you use a rotarudder or twin screws, you will need to check that swapping the inputs produces the appropriate response. The boat's circuits should be mounted inside the boat so that they can't get wet. Try enclosing them in a tough plastic bag and sealing the openings around the wires with tape.

The oscillator needs to be equipped with two connecting wires that are long enough to go halfway around the pond you intend to use. The electrodes can be simple rods, say, 600 mm long, bent in the middle at right angles—put a brick or stone on the end of these so they don't fall in the pond, and then dip the other end in the water.

The boat will also need propulsion: the simplest possible motor arrangement driving a propeller via a shaft fitted in a long tube will serve.

The field-follower boat works best if you have a pond that is a reasonable size. I did my testing in the bath, which was not big enough to give the device a fair run, although it was convenient while I was modifying the boat to work properly. However, too big a body of water might also cause problems. Something like a private swimming pool or largish garden pond is best.

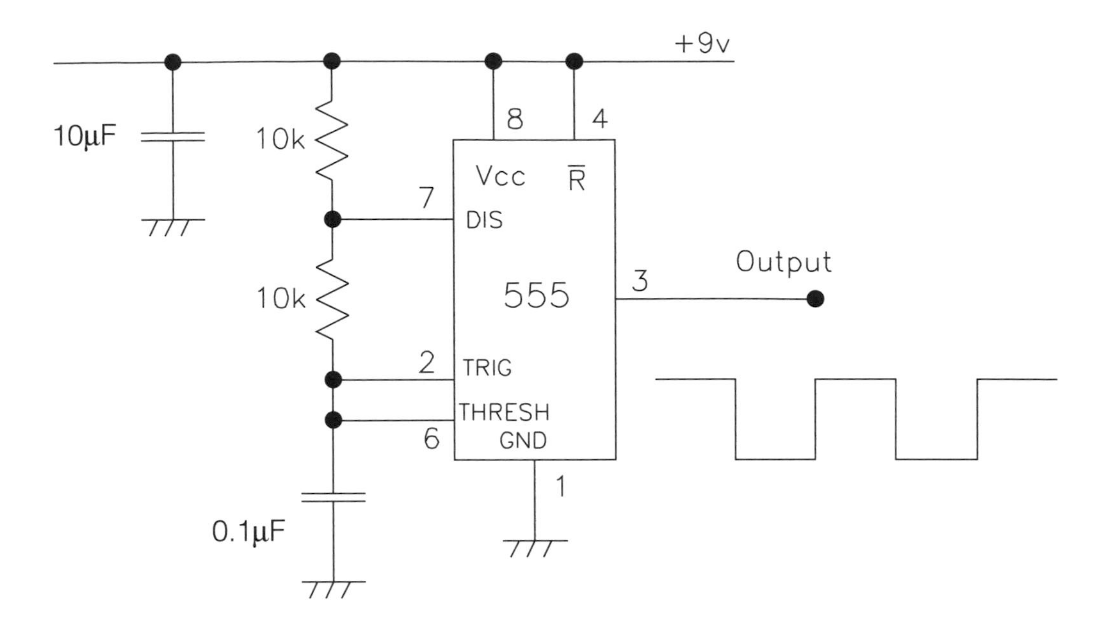
+9v
10μF
10k
10k
0.1μF
8
4
Vcc
R̄
7
DIS
555
3
Output
2
TRIG
THRESH
6
GND
1

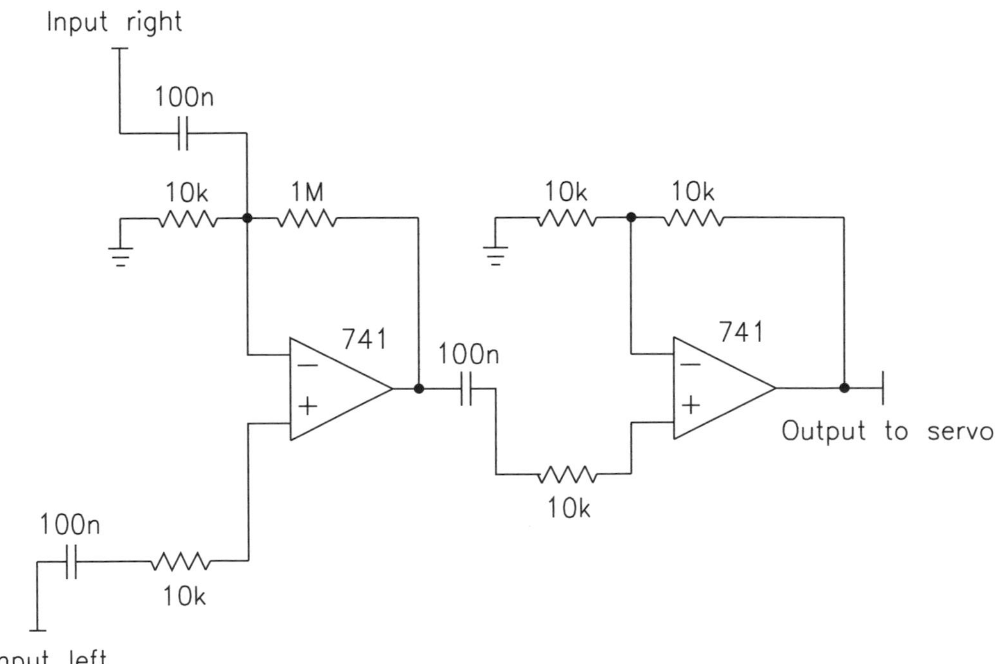
Input right
100n
10k
1M
741
100n
10k
10k
741
10k
Output to servo
100n
10k

Input left

How It Works

The field-follower boat works by probing the water around it for the electric field imposed on the pond by an oscillator connected to electrodes in the water. The electric field between two small electrodes a long way apart near the surface of a body of water consists of lines of force that pass through each electrode and curve out into the volume of water in a circular path. The pattern is more complicated if the electrodes are not near the surface. You could put several electrodes in a row at each end of the pool to get approximately parallel electric field lines, like the lanes of a swimming pool.

The "dipole" pair of sensor electrodes on the boat will cause it to turn until it is cruising along the field lines. If the voltage on the left-hand electrode is higher than that on the right, then the boat is turned too far to the right, and the rudder will turn it slightly to the left. When there is a higher voltage on the right-hand electrode, the rudder will turn the boat to the right. The result is that the boat tends to follow the field lines. If the probes see the same voltage, then the boat is sailing along a field line. You will find that when you release the boat approximately along a field line, it will simply carry on along the line, with some twitching from side to side. If the boat is released at 90 degrees to a field line, then it might not find the field lines and might thrash about with the servo motor revving and reversing.

We use an oscillator rather than a DC voltage to set up the field lines in the water. The oscillator creates an asymmetrical pulse waveform voltage: 30 percent on and 70 percent off. This pulse waveform is used in exactly the same way as a DC voltage. (The reason for using the pulse waveform is explained in the science and math section.) The amplifier circuit amplifies the small voltages received by

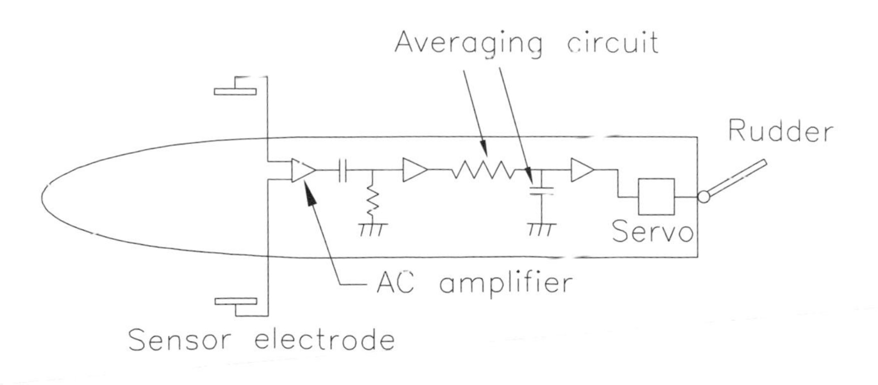

the sensor electrodes to the level at which the voltages on the circuit can be used to govern the rudder via the servo motor.

The job of a servo, or "slave," motor is to copy with its output, a small control input given to it. In this case, servo's job is to move the rudder, which is connected to the motor by a transmission, to the left or the right, depending upon the polarity of the signal picked up. However, we need some kind of a feedback system that must, at minimum, ensure that the motor stops when the rudder is fully left or fully right. I chose to make the servo mimic the voltage on its input, giving a proportional response, with feedback from a potentiometer on the rudder.

The function of the servo driver is simply to drive the motor proportionally in the direction indicated by the voltage from the amplifier. The 100-kohm potentiometer in the servo is the feedback device (see "Light Tunnels" for an explanation). The circuit makes the motor shaft move the potentiometer wiper until its voltage is equal to the input voltage, whereupon the motor is switched off, leaving the servo motor output shaft at a position proportional to the input. The two transistors are "emitter followers" to boost the output of the operational amplifier (op amp). I used a "solar motor" with a gearbox, and just enough batteries—eight 1.2 V—to make the servo work well (more might have overheated the small output transistors used). The 1-Mohm resistor provides negative feedback around the op amp, defining its gain factor, which governs how accurately the rudder servo will track its voltage input. The potentiometer will not go through its full travel as shown, swinging only about 45 degrees either way, but this is enough to move the boat rudder. The circuit will often tend to oscillate at some tens or hundreds of kilohertz, depending upon what op amp circuit you use. I ignored the oscillation because it did not seem to do any harm,

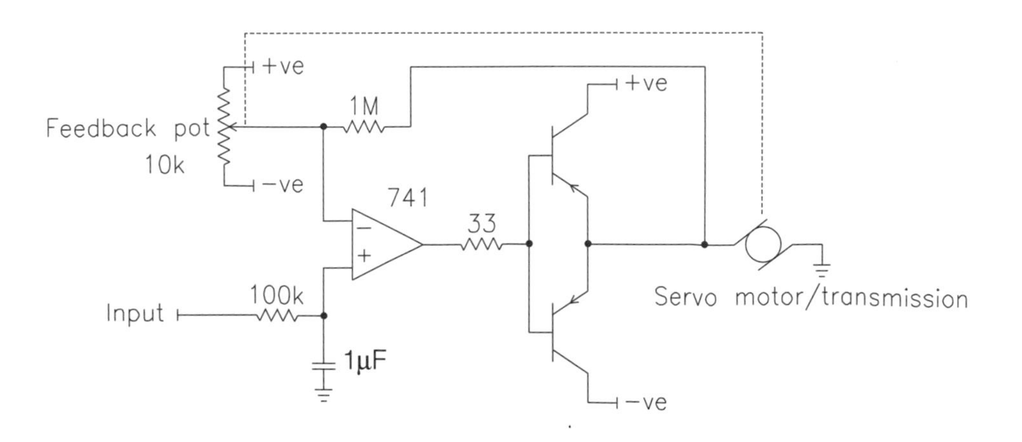

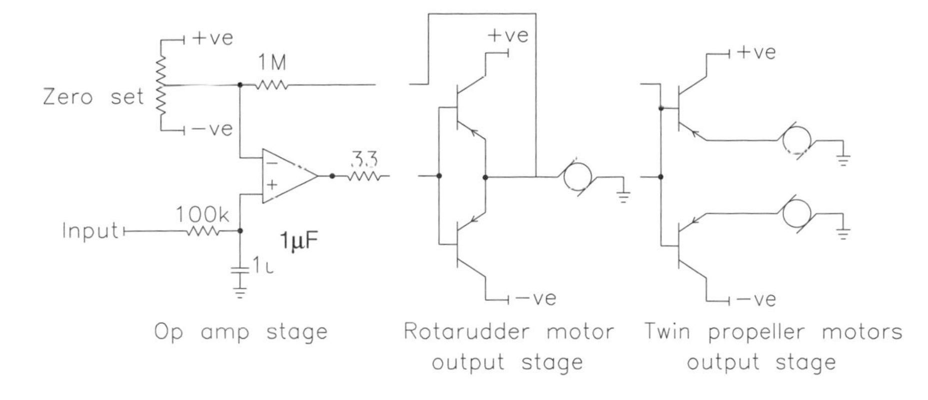

and I wanted to keep the circuit as simple as possible. The servo may also tend to "hunt" a little—jig to and fro by a few degrees—depending upon the mechanical linkage (the mechanical linkage includes all the gearwheels between the input and output). Again, this doesn't really matter.

As a rather simpler alternative, you could try using a "rotarudder," which is described in my earlier book, *Vacuum Bazookas*. The rotarudder is simply a rotating cylinder that dips into the water where the rudder would normally be, underneath the stern of the boat, after the propeller. In this case, the drive circuit is a fairly straightforward amplifier, which simply applies to the rotarudder motor a voltage that is proportional to the input voltage. Another alternative is to find a model boat with two propellers and motors spaced widely apart, so that the amplifier can switch between them.

THE SCIENCE AND THE MATH

The oscillator of the field-follower boat creates an asymmetrical pulse waveform voltage: 30 percent on and 70 percent off. Although a DC voltage could in principle be used for the electric field lines, this simple approach is unsuitable, for a rather subtle reason. The water in which the boat sails contains salts (alum and hypochlorite in swimming pools, calcium carbonate and sulfate in freshwater ponds, or sodium chloride in seawater). When dipped into this slightly salty water, the electrodes will provide the necessary conductive connection. However, the electrodes and the salt ions will also act, to some extent, as an electrochemical power source, a battery. Small differences in the sensor electrodes and any contamination on them will lead to differences in the voltage between them (several millivolts is typical); they act as a micropower battery. However, the electric field created by the oscillator electrodes is only a few volts. With the boat, say, 10 m from the electrodes and just a few degrees off course, the probes will have only a few millivolts potential on them, and the electric battery effects will cause interference and prevent the boat from steering correctly. The use of the asymmetrical AC eliminates this problem: the electrodes can be connected via a capacitor that prevents any DC drift due to electrochemical effects from reaching the sensitive amplifier. The capacitor does not affect the waveform on the output of the amplifier, however, and the averaging circuit produces a steady DC voltage from the AC input voltage, which then functions exactly as direct DC would have done.

The amplifier circuit is designed to boost even small input signals very strongly. The averaging circuit then takes this asymmetrical square wave and turns it into a DC shift from zero, a shift that is proportional to the input, unless the input is relatively large (tens of millivolts). The output of the averaging circuit in the absence of any input is designed to be zero, so that the rudder on the boat will give straight ahead in that case. The averaging-circuit output is positive if the left-hand electrode is closer to the "positive" output of the oscillator, since there is then a waveform that is 70 percent on and 30 percent off at the amplifier input. If, however, the right-hand electrode is closer to the positive output of the oscillator, then the waveform seen is 30 percent on and 70 percent off, and the averaging-circuit output is negative.

DC offsets on the order of 10–30 mV arise from the differences in electrochemical potential between the metal electrodes, differences that arise from surface conditions and ions local to those electrodes. Surprisingly these differences arise even with fairly clean and apparently identical electrodes. If these offsets were constant, they could simply be subtracted off, but they vary in an apparently unpredictable way. Particulate matter on or near the electrodes can give rise to some of these potential differences, and this effect has been used for practical purposes.*

*Basil Brook patented a device that shows how these offsets can be used. I spent a few happy days assisting Brook and his partner Mike Blanchard setting up a prototype control system using the electrode offset voltages to control the flow of trivalent salts used to enhance the settling out of particulates in a water treatment plant.

An AC field, which we use here, is a better alternative. By amplifying the electrode signals that this AC field gives with an AC coupled amplifier, we can make the circuit very sensitive and also make it give out a more or less full voltage swing (0 to 6 V) waveform. The waveform, can be averaged using the resistor-capacitor filter to give a voltage that swings reliably from about 2 V to about 4 V, which can then swing the rudder.

The electric potential (voltage) field, U, due to a single charge, Q, in free space is given by

$$U = kQ/R,$$

where R is the distance from the charge and k is a constant. The electric potential lines, lines of constant U, are like the contour lines on a map, whereas electric field lines, which our boat will follow, are gradients. In fact, the electric field, $\mathbf{E}$, is equal to $-\nabla U$. The dipole voltage field, V, created by two electrodes of opposite charge, Q, in free space can be obtained by adding up the potential due to each electrode:

$$V = V_0(1/R_1 - 1/R_2),$$

where $V_0 = kQ$ and R_1 and R_2 are the distances to each electrode. These distances are given by Pythagoras' formula,

$$R_1 = \sqrt{[(X - X_1)^2 + (Y - Y_1)^2 + (Z - Z_1)^2]},$$

where X, Y, and Z and X_1, Y_1, and Z_1 are the Cartesian coordinates of the position in the field and of electrode 1, respectively.

The voltage distribution in a homogeneous conducting medium will be the same as the distribution in free space. In our case, we have, rather than a very large (mathematically infinite) volume of conducting medium, something more like half of a large volume of conducting medium, with the other half being nonconducting air. However, symmetry arguments tell us that the voltage distribution in our homogeneous conducting medium will be the same as the distribution in free space.

With a shallow pond, we have in effect a two-dimensional sheet. In our case, the field pattern will be affected by the fact that there is no deeper water to "short-circuit" the water near the surface: we expect to see a voltage distribution than is more even over its middle portion. In fact, the distribution is just like that near a line charge extending in the z-direction, which we can derive from Maxwell's equations.

James Clerk Maxwell's master equations for electromagnetism, derived in the late nineteenth century, still stand today:

$$\text{curl } \boldsymbol{H} = \boldsymbol{J} + \partial \boldsymbol{D}/\partial t,$$

$$\text{curl } \boldsymbol{E} = -\partial \boldsymbol{B}/\partial t,$$

$$\text{div } \boldsymbol{D} = \rho,$$

and

$$\text{div } \boldsymbol{B} = 0,$$

where $\boldsymbol{H}$ is magnetic field strength, $\boldsymbol{J}$ is current density, $\boldsymbol{D}$ is electric displacement, t is time, $\boldsymbol{E}$ is electric field strength (E-field), $\boldsymbol{B}$ is magnetic flux density (B-field), and ρ is electric charge density. In free space, $\boldsymbol{E} = \boldsymbol{D}/\varepsilon_0$, where ε_0 is the electric permittivity of free space, so div $\boldsymbol{E} = \rho/\varepsilon_0$. Since $\boldsymbol{E} = -\nabla V$, we have

$$\text{div } \nabla V = \nabla^2 V = \rho/\varepsilon_0,$$

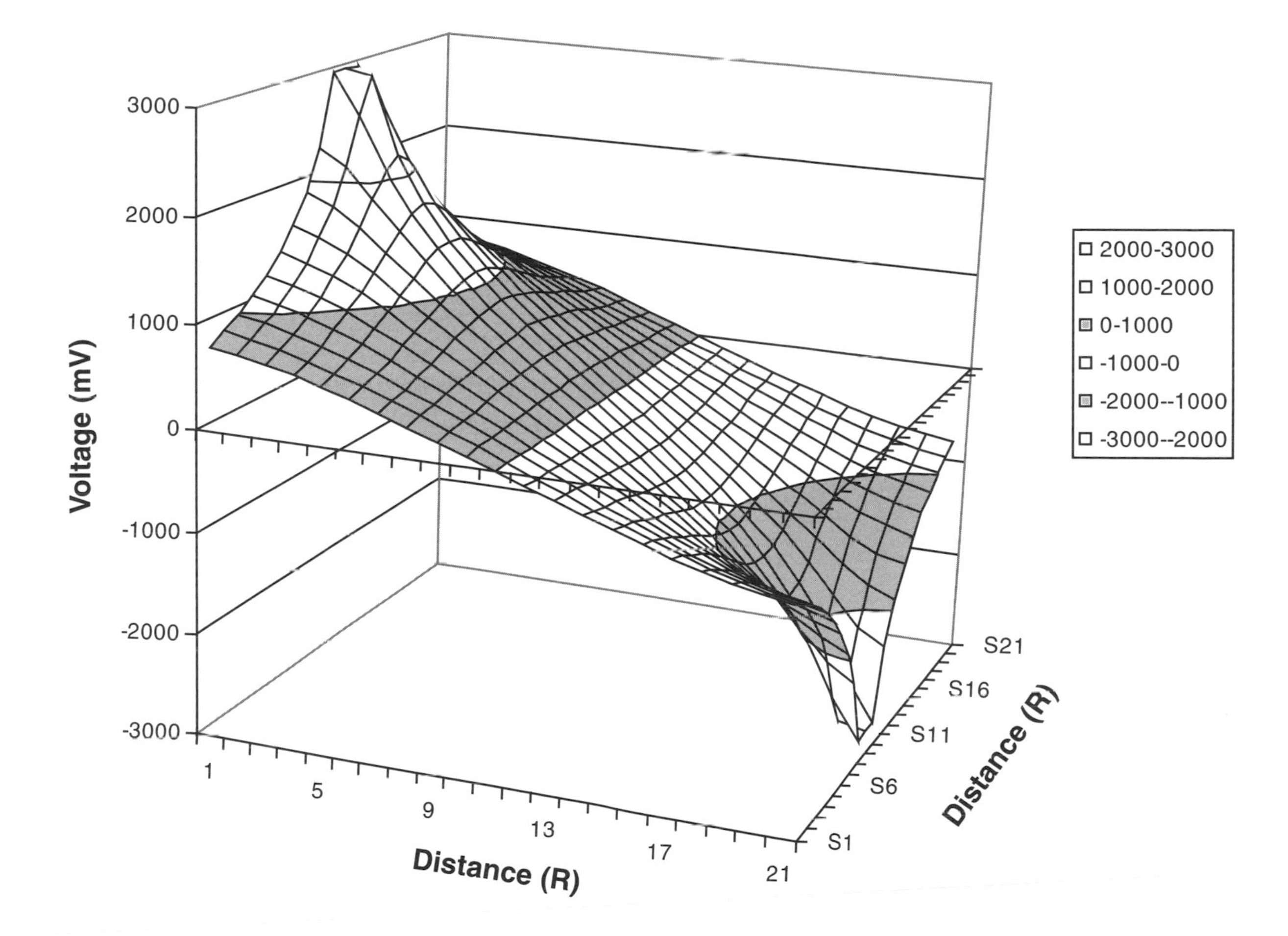

which is Poisson's equation for the electric potential in a space containing charge.

$$\nabla^2 V = \partial^2 V/\partial x^2 + \partial^2 V/\partial y^2 + \partial^2 V/\partial y^2 = \rho/\varepsilon_0,$$

which is

$$\nabla^2 V = 1/r\ \partial/\partial r(r\partial V/\partial r) + 1/r^2\ \partial^2 V/\partial \theta^2 + \partial^2 V/\partial z^2 = \rho/\varepsilon_0$$

in cylindrical coordinates. By symmetry, the z variation and the θ variation of V must be zero, so we can remove the partial differentials and use the ordinary differential equation

$$1/r\ d/dr(rdV/dr) = \rho/\varepsilon_0.$$

A solution to this equation is $V = \ln r$, which describes the equation for a single line charge. This equation says that the electric potential, the voltage, will rise, albeit slowly, as you go farther from the line of charge. On the face of it, this is very odd behavior: you don't expect voltages to rise as you move away from a charged object. However, this voltage distribution is a curious beast, since it is the potential around an infinitely long infinite charge! You can think of the potential rising with distance because as you go to larger distances you are "seeing" a longer line, with more charge in it, although that charge is more distant.

With two equal and opposite charges, the voltage becomes a more tractable mathematical object. It is simply the sum of the two potentials:

$$V = K(\ln R_1 - \ln R_2) = K \ln(R_1/R_2),$$

where R_1 and R_2 are the distances from each of the charges, located at $+d$ and $-d$. This sort of voltage distribution is seen over a cross section of the twin-feeder type of cable used in some radio installations. In it, lines of equipotential are circles of radius $R_c(V)$ offset from the electrodes by distance $D(V)$:

$$D(V) = 2dM^2/(1 - M^2)$$

and

$$R_c(V) = 2Md/(1 - M^2),$$

where $M = \exp(V/K)$.

This is the kind of electric potential that the hammerhead shark knows well. This shark has a curious head on which the eyes are spaced at a distance that is much larger than the width of its body. The streamlined eye stalks also contain the shark's electric-field sensors, the ampullae of Lorenzini. They can, according to Mark Denny, pick up electric signals as small as 7–10 μV, which allows them to swim along just above the seafloor and detect the electric field of the heartbeats of a fish such as a plaice, even if that fish is hiding 10–15 cm deep in the mud.*

*The hammerhead is a kind of inversion of the magnetic mine. These military mines detect magnetic field rather than electric field but otherwise amount to a kind of hammerhead shark lying in the mud on the seafloor. Magnetic mines are large bombs that pick up the tiny magnetic field of a passing ship and then explode with devastating force. The terrible problems they created for the unprepared navies of the free world when they were first unleashed by the Nazis is described by Winston Churchill in Volume 1 of his *The Second World War.* Only by the ingenuity and brave actions of men such as Lieutenant Commanders John Ouvry and Roger Lewis, who dismantled live mines, and unsung heroes such as Philip Hunter of Callender's Cable & Construction Co., which made a thousand miles of special floating cable for exploding the mines at a distance, was this deadly hazard reduced.

And Finally . . . Practical Uses?

Could this system be of any use? Could lost yachts be guided into a safe haven by the electrical equivalent of a lighthouse and an automatic pilot? Could this in fact be the yachter's new instrument landing system, analogous to the system used to steer aircraft into airports in fog?

REFERENCES

Black, Robert M. *The History of Electric Wires and Cables.* London: Peter Peregrinus, 1983. Chapter 12 describes the curious role of wires and cables in detecting underwater mines.

Brook, Basil William, and Michael John Blanchard. "Determining Charge in Water using Matched Electrodes." U.K. Patent no. GB2356257, May 16, 2001.

Churchill, Winston. *The Second World War,* vol. 1, chap. 28. London: Cassell & Co., 1948.

Denny, Mark W. *Air and Water.* Princeton, N.J.: Princeton University Press, 1993.

Downie, Neil A. *Vacuum Bazookas, Electric Rainbow Jelly, and 27 Other Saturday Science Projects.* Princeton: N.J.: Princeton University Press, 2001.

Transports of Delight

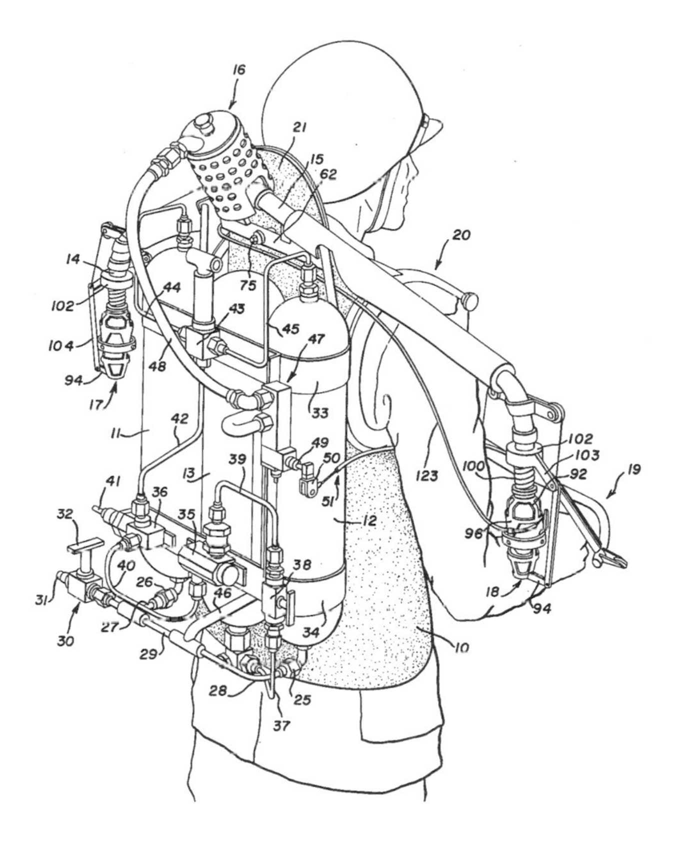

Along the Queen's great highway
I drive my merry load
At twenty miles per hour
In the middle of the road . . .
The . . .
Big six-wheeler
Scarlet-painted
London Transport
Diesel-engined
Ninety-seven horse-power
Omnibus.

—Michael Flanders and Donald Swann,
"A Transport of Delight"

Even the briefest study of the history of the past two centuries will reveal humankind's preoccupation with transportation. As soon as it was conceivable to imagine building an engine into a moving vehicle, pioneers like Nicolas Cugnot and Richard Trevithick came along and tried it. Cugnot's military steam tractor (1769) could pull a cannon at 3 mph for only 15 minutes. Despite this unpromising start, many engineers began to work in the field, and forty years later the first practical steam railroads and steam carriages began operating, starting with the designs of Trevithick. Hundreds of types of transportation systems have stepped for a moment into the limelight, most of them doomed to step back into obscurity a moment later. In this part, we look at some possibilities that don't seem to have gotten even as far as the momentary limelight.

11 Electric Worms

> Do Androids dream of electric sheep?
> Do Android zoologists dream of electric worms?
>
> —With apologies to Philip K. Dick, *Do Androids Dream of Electric Sheep?*

When railroads were first proposed, many people laughed at the smooth steel rails and smooth steel wheels, predicting that the wheels would slip and that the locomotive would be incapable of moving off with heavy wagons. Even engineers—who should have known better, or should have experimented if they were unsure—were bamboozled by this prediction, and they fitted some early railroads with completely unnecessary rack-and-pinion systems. In fact, a 50-ton locomotive can easily haul even a huge 500-ton train: the loco can pull with a force of at least 5 tons before the wheels will slip, and this force is enough to accelerate the train up to a respectable 22 m/s (50 mph) in 3 or 4 minutes on a flat railroad track.

The electric worm relies on friction, which often obeys, with some accuracy, an astonishingly simple mathematical law:

$$F_f = \mu F_n,$$

where F_f is the frictional force, F_n the normal force (the force exerted at right angles to the direction of motion), and μ the coefficient of friction for the two surfaces. The value of μ varies with the roughness and composition of the surfaces and is difficult to predict but often easy to measure, allowing empirical prediction of the behavior of a particular device. The coefficient of friction for steel on steel, as used in railroads, is between 0.1 to 0.2. Remarkably, surfaces as hard as diamond or as soft as candle wax, as rough as sandpaper or as smooth as glass, all follow this simple law, at least roughly. The law of friction seems to take into account all the details of atomic interactions and the huge variations in the apparent contact area between sliding surfaces.

The explanation for the simplicity of the friction equation has hitherto involved calculations on tiny *asperities,* minute parts that project from the surface, rubbing against one another. The applicability of this classical theory of friction—first propounded by Charles Coulomb and Guillaume Amontons, among others—has recently been challenged, at least for some situations, such as the movement of the tectonic plates of the earth's crust. Michael Marder and Eric Gerde at the University of Texas have a theory involving the formation and healing of nanocracks in the rubbing surfaces. These scientists are looking principally at the geological applications of this phenomenon, in the sliding of the tectonic plates and thus in earthquakes and volcano formation. They claim that their theory correctly describes tectonic friction, where the heat generated by the friction is anomalously low. Their theory also accounts for friction in systems in which little wear takes place.

Within the apparent simplicity of the friction equation, there are a few wrinkles. There are actually two types of friction: static friction, the maximum frictional force just before movement occurs, and dynamic friction, the frictional force after movement has begun. The two types of friction have their own μ values, μ_s and μ_m, respectively.

The value of μ is affected by the presence of a liquid between the sliding surfaces. When a liquid is present, μ values are typically much lower. In fact, if a liquid film completely coats the surface, it is more appropriate to think in terms of viscous slipping in a liquid rather than in terms of friction between surfaces. Furthermore, when a liquid is present, μ_m will decrease drastically with increasing speed as an increasingly thick layer of liquid becomes trapped between the surfaces. (This phenomenon, called *hydroplaning* or *aquaplaning,* can occur when auto tires run along wet roads, and it is exceedingly dangerous.) Friction coefficients will change with time, as the surfaces wear each other, and with temperature. At high speeds, μ_m values often decrease because the effective normal force cannot be so well defined, owing to dynamic inertia effects.

What You Need

- ❑ 2 motors
- ❑ 1 weight ("dummy motor," equal to the weight of the motor)
- ❑ Batteries and a holder
- ❑ 4 button switches
- ❑ Piano wire (tough springy wire, about 0.5 mm in diameter, AWG 30)
- ❑ 3 sliding pads (I used the plastic tops from milk containers, 40 mm in diameter.)
- ❑ Wood
- ❑ Glue
- ❑ Connecting wires

What You Do

The worm comprises a set of identical pad-motor units that slide on the ground and are linked by arched wires that push the units apart and by twisted-string actuators that pull the units together. The arched wires that link the worm

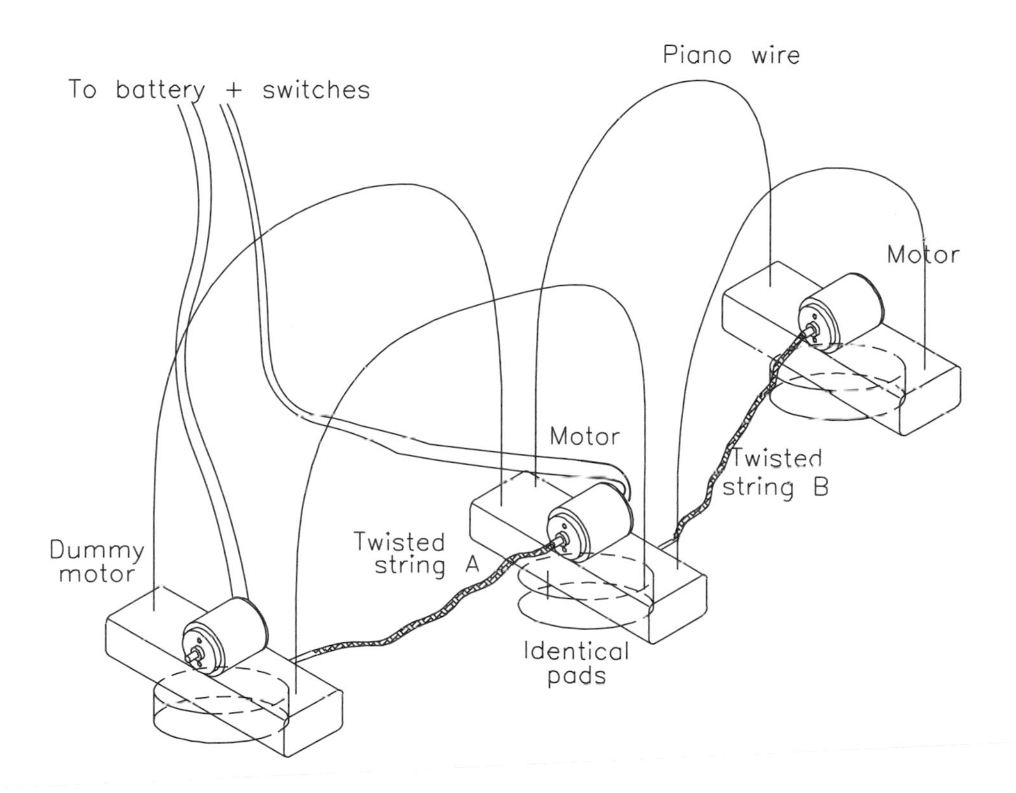

segments act as accurately balanced springs. (It is difficult to make helical springs that are as accurately balanced, and the rather weak springs that would be needed here are difficult to buy.) In a two-segment worm, each of the first two sliding pads bears a button switch–activated motor that twists the string pairs to pull the units together. The worm must have equal weight on each sliding pad, so place the weight ("dummy motor") on the third pad. Be careful not to unwind the string beyond zero twist; if you do, the controls will reverse and cause confusion.

You can make the two-segment worm inch forward by means of the steps diagrammed below.

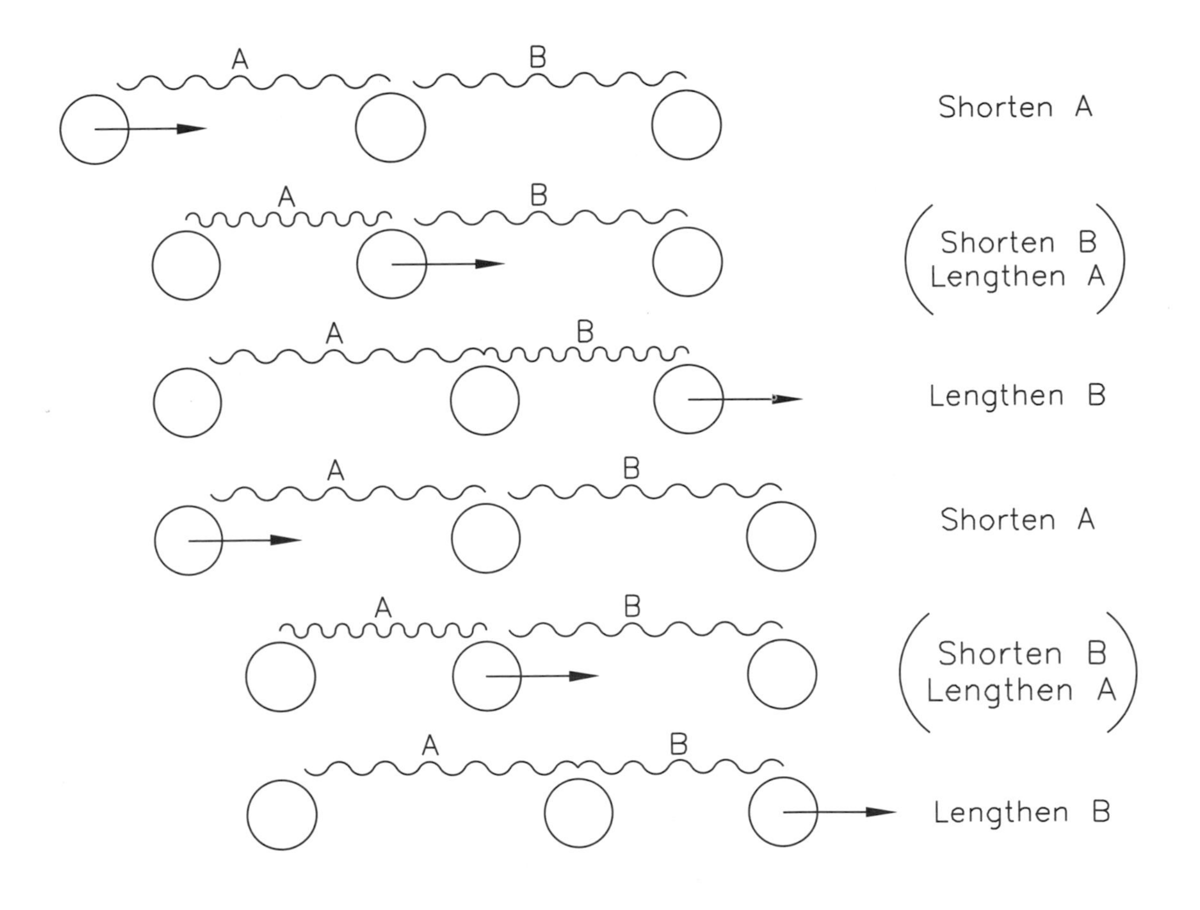

If you are having trouble phasing the movements, you may find it helpful to take a look at a drawing showing how a real worm moves, for example, in Barrington's *Invertebrate Structure and Function,* listed in the references.

Try the sequence with your fingers on the table before actually pressing the buttons. With deft button pressing you can make steps at about one per second. By changing the phase of the motors, you can make the worm travel backwards.

How It Works

When one of the worm's three identical pads is moving and the other two are stationary, the stationary pads provide a frictional force given by

$$F_f = 2\mu_s F_n,$$

and the moving pad exerts a frictional force given by

$$F_f = \mu_m F_n,$$

which is typically less than 33 percent of the force of the other two combined, because $\mu_s > \mu_m$. When only one pad is moved at a time, the worm can inch forward by means of simple friction. Because only one pad is moved, there will be little or no backsliding by the two stationary pads. The worm doesn't need to lift its pads, change their frictional properties, or employ backwards-pointing fibers (if it did, its movement would be restricted: it could not go backwards). It simply slides. And it will slide on virtually any surface.

What happens when the worm encounters a slope? Will backsliding start to occur when part of the worm is on the slope while the rest is still lying on the flat?

THE SCIENCE AND THE MATH

The electric worm uses twisted-string actuators, which are described in my earlier book, *Vacuum Bazookas*. The length, S, of the strings after N twists is given by

$$S = \sqrt{[L^2 - (N\pi d)^2]},$$

where L is length of the strings before twisting, and d is the string's approximate diameter. The equation says that when the motor rotates—that is, when N increases—the twisted strings shorten, at an accelerating rate. The equation also indicates that thicker strings will shorten quicker than thinner strings.

What are the possible segment-movement sequences for multiworms? With a three-segment, four-pad worm, only one pad should be moved at a time to avoid slipping. With four or more segments, there are five or more pads, and two pads can be moved at once, since there will still be three stationary pads to pull against. With six or more segments (seven or more pads), three pads could be moved at once, although there is now some risk of backsliding, since there will be only four stationary pads to anchor three moving pads.

Now let's consider what happens when the worm encounters a slope of angle θ. With all pads on the slope, the worm functions as it does on the flat, except that the normal force is reduced to $F_n \cos \theta$. If the slope is too steep, however, the normal force will be reduced so much that the worm will slide down to the bottom of the slope. What happens when the first pad is on a slope and the others are on the flat? Because of the reduced normal force on the first pad, it will tend to backslide when the others move. Backsliding will happen for certain when

$$F_n(1 + \cos \theta) < F_n,$$

that is, when the slope is vertical. This is not a limitation, since the steepest slope the worm can attempt is given by

$$\mu_s(F_n \cos \theta) > F_n \sin \theta,$$

which will normally give a θ value of 30 degrees or less.

And Finally . . . Advanced Electric Worms

If you could find suitably floppy helical wire springs, you could try making a creature with a more wormlike appearance. Of course, you could use the radio system from a two-channel radio-controlled toy to control the two segments and achieve a free-sliding creature without using umbilical wires.

Another possibility is to use two motor units alongside each other in the first segment to allow steering. You could add more segments, maybe a half dozen would look good, but you would need to get good at button pressing. Perhaps you could come up with an "autoworm," a creature with a programmable microprocessor? A simple PIC processor silicon chip (Allegro Microsystems, Inc.) could no doubt be pressed into service. Using two digital outputs for each motor would allow for the use of the off state as well as the contracted and the expanded states. You would still have to be careful to keep the autoworm from winding past its fully unwound state and winding up again in the other direction.

Maybe there are other actuators that would lend themselves to electric worm propulsion: pneumatic actuators, which are famous for their speed and robustness, are most readily available in linear-motion format, and they might be a good choice. But they are expensive and need a complex power source, and their sudden acceleration might lead to unwanted slipping, which might upset the balance of friction on which the electric worm relies.

REFERENCES

Barrington, E.J.W. *Invertebrate Structure and Function,* chap. 6. London: Thomas Nelson, 1967.

Downie, Neil A. *Vacuum Bazookas, Electric Rainbow Jelly, and 27 Other Saturday Science Projects,* chap. 7. Princeton, N.J.: Princeton University Press, 2001.

Gerde, Eric, and Michael Marder. "Friction and Fracture." *Nature* 413 (September 20, 2001): 285–288.

12 Vacuum Railroad

> I have no hesitation in taking upon myself the full and entire responsibility for recommending the adoption of the atmospheric system [vacuum railroad] on the South Devon Railway.
>
> —Isambard Kingdom Brunel, quoted in *Isambard Kingdom Brunel: Engineering Knight Errant*

Today's railroads are almost perfectly standardized: 99 percent use wheels on rails set 4 feet 8 inches apart with an engine-driven car that hauls freewheeling cars. It has not always been so. In the early and mid–nineteenth century, there were other systems. There were wide-gauge steam railroads such as the London-Bristol Great Western Railway, built by the great engineer Isambard Kingdom Brunel, which had rails that were set an extraordinary 7 feet apart. There were tiny narrow-gauge railroads that carried mineral ores inside mines and carried passengers through mountainous areas such as Wales. There were even monorails. Starting in the 1860s, backwoodsmen in Ireland enjoyed monorail transport on the extraordinary Listowel to Ballybunion railroad.* Here we try out a vacuum railroad that might still work well today in a subway system such as that in New York or London.

*The Listowel to Ballybunion monorail did not have futuristic, streamlined bullet cars speeding across a skyscraper cityscape: it was a steam monorail with a maximum speed of only 20 mph. Its carriages and engines came in pairs that hung down on either side of its single rail 3 feet up in the air. This pioneer monorail carried livestock as well as goods and people. Cows had to travel two by two à la Noah's Ark, one on each side of the rail, because otherwise the carriage would have been unbalanced!

The vacuum-railroad principle was used for quite a few full-size railroads before being relegated to a role in conveyor tubes for mail and the like. In the most straightforward systems, a tubular train was simply sucked along through a tunnel or tube. This technique was adopted in an early demonstration railway in 1840s New York. In a more subtle variation seen several times early in the Victorian era in Britain, the carriage did not itself go down the vacuum tube. Instead, the tube, which was about 12 inches in diameter, had a piston inside and a slot along the top via which a metal fin connected the piston with the train. The train ran conventionally on wheels on a track outside the tube. To avoid massive leakage of air into the slit, the slit was closed by a continuous flap valve that was held down by the vacuum itself. The valve sealed the slit except when a train came by. This was the technology behind Isambard Kingdom Brunel's system.

At one time, vacuum-railroad carriages were the fastest vehicles on Earth: one of Brunel's drivers was once accidentally pushed up the line near Bristol at 85 mph, the carriages normally attached having been uncoupled by mistake. In their heyday, these systems were called atmospheric railroads, which is a more correct term since they are driven by the pressure of the atmosphere. Although a number were built and operated, some for many years, Brunel's showpiece South Devon Railway was doomed by materials problems: the leather seal along the top of Brunel's vacuum tube froze in winter, caused the iron plates to rust by retaining moisture, and was also eaten by rats. After a few years of erratic operation, the railway was closed.

What You Need

- ❑ 5 plastic buckets, 10-l size or larger
- ❑ 5 clear plastic lids to fit the above (e.g., pieces of "unbreakable glazing")
- ❑ 15–20 m of plastic 70- or 100-mm-diameter drainpipe
- ❑ Drainpipe T-piece to match drainpipe
- ❑ Cap for drainpipe (e.g., a flat piece of light plastic or a metal lid from a jar, large enough to cover the end of pipe from the T-piece)
- ❑ Vacuum cleaner
- ❑ Vacuum cleaner hose
- ❑ Short (100–150 mm long) pieces of pipe one size smaller than the drainpipe, for the train cars
- ❑ Cloth wrapping (e.g., velvet) to make a seal around the cars

What You Do

The construction of the system is basically straightforward. Pay attention to achieving a reasonably good vacuum seal between the buckets and the drainpipe.

I did this by marking a circle on the bucket where the pipe would fit, heating the area inside the circle almost to the melting point using a small gas burner, and then forcing a spare piece of the pipe through the bucket wall. The result, if you do it right, is a hole whose edges stretch to form a conical skirt that blends smoothly with the pipe so that the whole system can be made to push fit. If any joints leak significantly, a little duct tape should help—fortunately, the tape does not even have to stick particularly well. As the vacuum is increased, the tape will tend to pull down and seal even better onto the walls of the pipe. Connect the vacuum cleaner to the T-piece in the system. The easiest way to do this is to use the vacuum cleaner's regular hose and seal the joint using well-compressed sponge rubber packaging material and duct tape to finish it off.

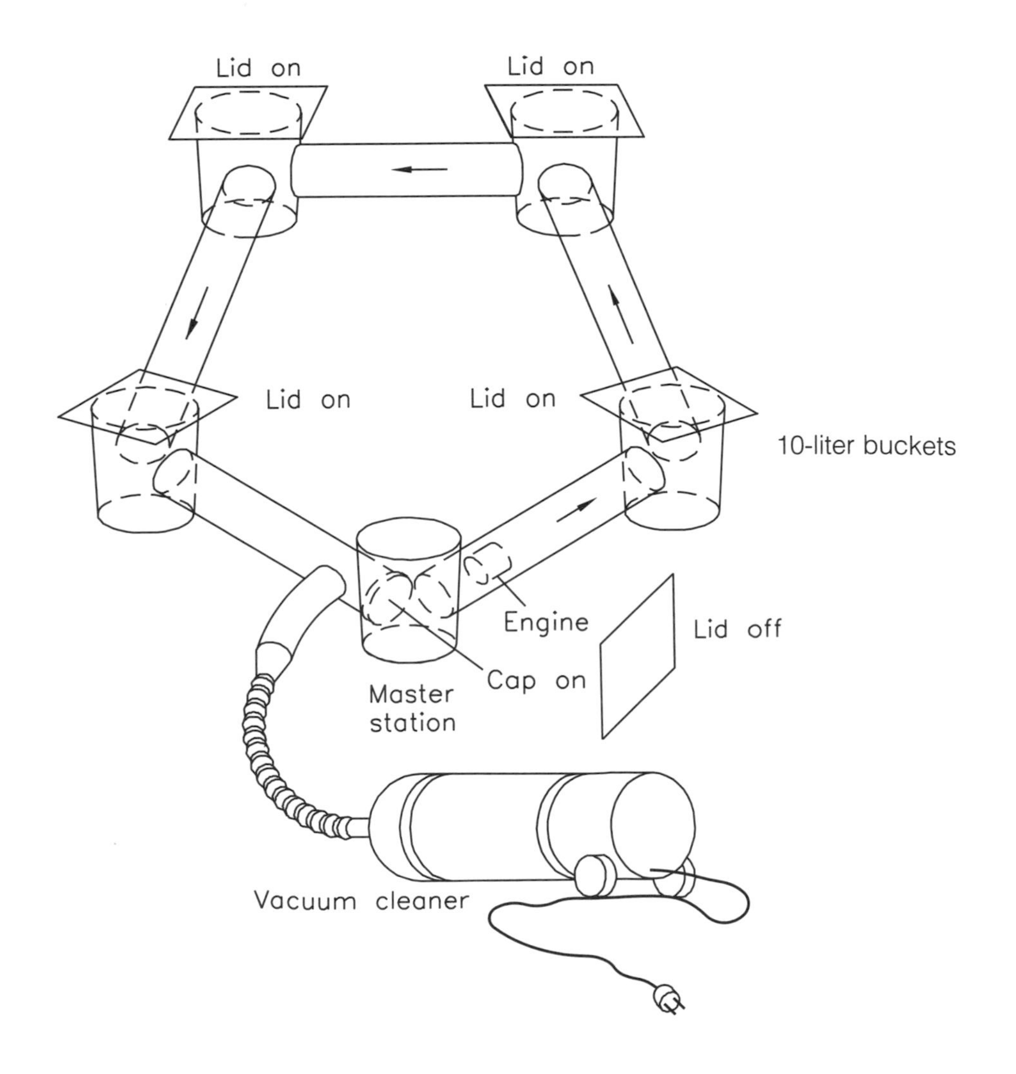

The train can consist of an "engine," which seals well against the pipe walls, and two or three "carriages" trailing along behind or pushed along in front. These can be as simple as lengths of tubing that fit loosely in the drainpipe, perhaps with cutouts to accommodate toy "passengers." You can achieve a simple seal by wrapping the engine with a band of cloth—heavy velvet drape material is good—until you get a good fit.

Our trains are completely enclosed and thus do not need Brunel's slot-sealing mechanism. However, you do have to seal up the station the train is heading toward, otherwise vacuum will be lost. So the basic protocol for operation is as follows:

1. Seal all the bucket stations except the originating one.
2. Place the cap over the end of the T-piece pipe in the master station (where the vacuum is applied).
3. Push the front of the engine into the tube.

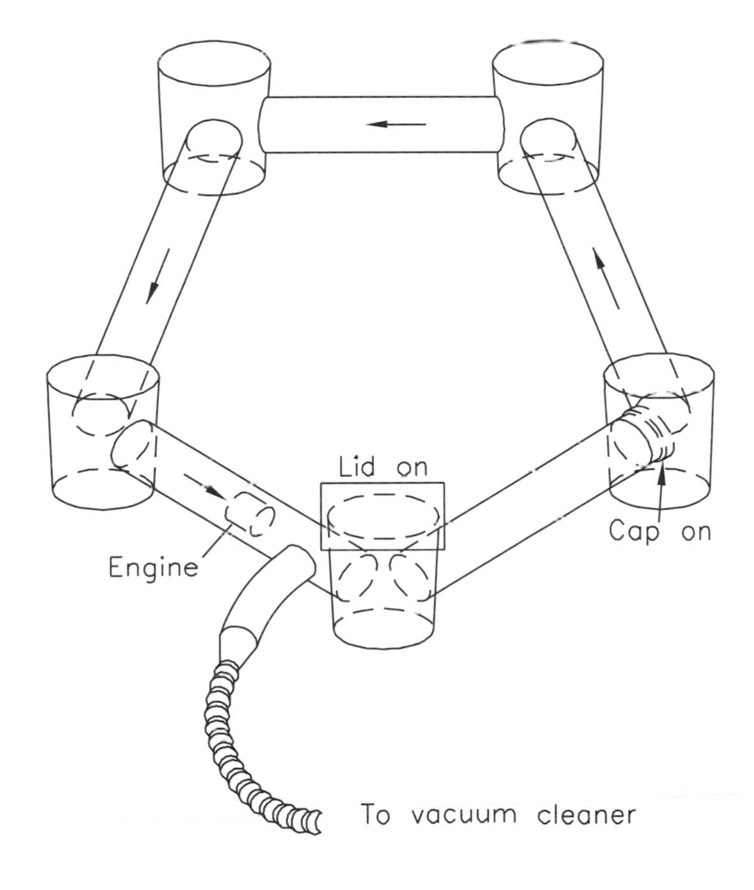

4. Apply a vacuum.
5. Wait for arrival at the next bucket station, remove lid.
6. Return to step 3 for the next stage of the journey.

There is a subtlety to getting the train to arrive back at the master station. The T-piece allows vacuum to be applied to the system to propel the train into the master station, completing the circuit. Without the T-piece, the train could be drawn only toward and then into the master station and could not be propelled out of it. You might expect the presence of the T-piece to be a problem. Wouldn't the train be sucked into the T-piece, where the vacuum connection is? This does not usually happen when the train is running at high speed because it has enough inertia to run past the T-piece and on down the short piece of pipe leading to the master station.

How It Works

When there is a train in the pipe and the vacuum cleaner is switched on, air behind the train will be at atmospheric pressure, about 100,000 Pa, and the air in front of the train will be at about 80,000 Pa. The differential pressure, which is the effective driving pressure, is thus on the order of 20,000 Pa. The net force exerted on the train is proportional to the area of the pipe multiplied by the differential pressure available from the vacuum cleaner. The buckets provide a "vacuum reservoir," increasing the available acceleration of the train. Without the buckets, the train would compress the air ahead of it as it moved, increasing the pressure and reducing the differential pressure that drives the train.

THE SCIENCE AND THE MATH

The vacuum railroad is capable of very high speeds. The acceleration, A, of the train is given by

$$A = \Delta P(\pi/4M)D^2,$$

where D is the tube diameter, M the mass of the train, and ΔP the differential pressure created by the vacuum cleaner. The maximum velocity of the train, V_{max}, is given by

$$V_{max} = \sqrt{(2LA)},$$

where L is the distance between bucket stations. Substituting for A gives the speed of the train as it arrives at the station:

$$V_{max} = D\sqrt{[2L\Delta P(\pi/4M)]},$$

which we might dub the "vacuum railroad equation." Increasing the distance between stations, the differential pressure, or the pipe diameter will increase the speed of the train. The effect of diameter, being a direct proportionality, is greater than the effect of pressure or tube length. As you might expect, increasing the mass of the train will slow it down.

The absence of a heavy locomotive engine allows the train to accelerate rapidly. With a 250-g train in a pipe 3 m by 65 mm and in the absence of friction or other slowing effects, the vacuum railroad train ought to be able to accelerate at 25*g*. Of course, the acceleration will depend upon the vacuum produced by your vacuum cleaner; 150–200 mbar is typical. Furthermore, the train ought to reach a final speed of 38 m/s, or 88 mph! (I have measured more than 60 mph, so this calculation is not so far from reality.)

One problem with early tube railroads was that they heated up excessively because of friction. In long-distance mail systems, delivered trains would get up to such high temperatures that the unfortunate clerks who pulled them out at their destination had to handle them like hot potatoes. You might think you could avoid the heating problem, and the inefficiencies it represents, by using a looser seal, but then you have to consider the power losses associated with leakage around the seal. How much power do we actually lose to friction and leakage?

Let's first calculate the power dissipated by seal friction. The seal will press with a certain force against the walls of the tube, and as the train slides along, will produce a frictional force, even if the train itself weighs very little and slides freely. With a train moving at a constant speed, V, and with a frictional force, F_f, on the wall, then the power, $P(f)$, dissipated will be

$$P(f) = F_f V.$$

Now let's figure out how much power, $P(l)$, we lose to leakage around the seal. With differential pressure ΔP across the seal, the leakage rate, Q (m^3/s), past the seal is given approximately by

$$Q \sim kA\sqrt{(\Delta P)},$$

where k is a constant and A is the area of the effective orifice formed by the seal. Let us assume that A is given by

$$A = A_0/F_n,$$

where F_n is the normal force applied and A_0 is a constant, so the leakage rate, Q, past the seal depends on F_n:

$$F_f = \mu F_n,$$

where μ is the coefficient of friction. The energy in a compressed gas is given by

$$U \ln(P_i/P_f),$$

where U is the volume, P_i the initial pressure, and P_f the final pressure. Using this in the leakage rate equation, we arrive at

$$P(l) = Q \ln(P_i/P_f) = \left[kA_0\sqrt{(\Delta P)}/F_n\right] \ln[(P_0 + \Delta P)/P_0],$$

where P_0 is atmospheric pressure.

We must also consider the simple sliding friction, F_f, of the carriage (which could be reduced by using wheels). The power, $P(g)$, needed to overcome sliding friction is proportional to the carriage weight:

$$F_f = \mu Mg.$$

Now we can put it all together:

$$P(f) = F_f V = \mu F_n V,$$

$$P(l) = \left[kA_0\sqrt{(\Delta P)}/F_n\right] \ln[(P_0 + \Delta P)/P_0],$$

and

$$P(g) = \mu MgV.$$

The total power dissipated is equal to the sum of $P(f)$, $P(l)$, and $P(g)$.

Can this total power be minimized? In engineering, there are often times when a compromise must be reached to optimize system performance. The optimization can be done mathematically, if the system is fully characterizable and if, ideally, its behavior can be expressed mathematically, by using a surface in a hyperdimensional space. You express the system performance, S, as a function of parameters $x_1, x_2, \ldots, x_n$; draw an n-dimensional surface of S in an $(n + 1)$-dimensional diagram with S and the x_i as axes; and then find the biggest peak of S in this hyperspace.

In one dimension, this process is straightforward. To optimize a parameter such as power, all you need to do is find the derivative with respect to the most significant controlling parameter, such as seal tightness (or maybe F_n is the correct choice here), and then look for a zero in this derivative. Finally determine whether it is a maximum or minimum, as desired, and—Bingo!—you have your optimum compromise.

The compromise in our vacuum railroad is a fundamental one in engineering situations in which gas-tight seals are needed: tightening the seal increases friction, which wastes power; loosening the seal increases leakage, which also wastes power. Much engineering ingenuity is devoted to this quandary: great precision of manufacture is part of the answer, but there are economic limits to how far precision be carried. Low-friction materials for the seals and lubrication or both also can help. Pressure-actuated seals, such as the those designed around an annulus with a C-shaped cross section can also be helpful: the very air pressure that the annulus is sealing out is used to inflate it, pressing it harder against the pipe walls when the pressure is high and leakage through gaps would be great, and pressing it gently when pressure is low.

In the case of vacuum delivery systems, much of the power lost will end up heating either the tubes or the vehicle. If you assume that half goes to the vehicle and half to the tube, then the vehicle heating will be the problem, because the tube is hundreds of times more massive than the vehicle. This explains why the mail train vehicles exited the tubes hotter than roast potatoes!

And Finally . . . Controlling Your Vacuum Railroad

You could try using magnetic detectors or metal detectors to locate your trains in the opaque tubes. You could also measure their speed, perhaps using an electronic stopwatch that is stopped and started by means of magnetic detectors (magnet plus Hall detector, or magnet plus simple coil). You could measure pressure in the tubes, using a (very tall) manometer or an electronic pressure gauge. You could also pull along a train outside the tube, most easily by means of two powerful magnets of opposite polarity, one inside and one outside the tube.

As long as you stick with plastic drainpipe, the train could be radio controlled, because radio waves will penetrate the pipe. In their heyday, vacuum railroads were controlled by the driver, who applied the brakes. A radio-controlled brake—perhaps an expanding shoe to fit in the tube like an auto drum brake—is a possibility. Another option, with radio control available, would be to include a large-bore valve that would allow atmospheric pressure air into the tunnel ahead, releasing the vacuum and allowing you to stop the train more easily. I found that a 22-mm valve would work in a 70-mm tube system, stopping a train that had some degree of friction between it and the tube walls.

REFERENCES

Dempsey, G. Drysdale. *A Rudimentary Treatise on the Locomotive Engine in All Its Phases.* London: John Weale, 1857. Reprint; Bath, U.K.: Kingsmead Reprints, 1970. Reveals just how standardized railroads had become even as early 1850.

Newham, A. T. *The Listowel and Ballybunion Railway.* Oxford, U.K.: Oakwood Press, 1989. An amusing tale about the first monorail in Ireland.

Vaughan, Adrian. *Isambard Kingdom Brunel: Engineering Knight Errant.* London: John Murray, 1991.

13 Naggobot, or Reverse Ice Vehicle

These rough notes and our dead bodies must tell the tale.

—Robert F. Scott, *Scott's Last Expedition*

Robert Scott's scientific expedition to the South Pole in 1912 was a disaster. Scott, and all those who went with him to the pole, perished on their return journey, having reached their goal just a day or two after Roald Amundsen, the leader of the first successful expedition. Scott's diaries tell us how well prepared the explorers were and reveal the tremendous heroism of the polar party, but, mysteriously, preparation and heroism were not enough, and no one survived. We have the scientific observations made by Scott—under the most arduous conditions imaginable—and other Antarctic explorers, but these have not hitherto yielded the solution to the enigma. In a recent book, atmospheric chemist Susan Solomon has reviewed all the weather data, as well as other scientific data, and she offers a compelling new explanation for what doomed Scott and four of his men. It was not the explorer's incompetence, as several popular accounts have suggested. It was the cold, highly abnormal cold. The polar party died during the coldest March on record, a month when temperatures plunged as low as –60 °C. The cold winds dealt Scott's team a triple-whammy disaster: head winds that thwarted their progress, low temperatures and wind chill factors that sapped their stamina, and, perhaps worst of all, friction. At temperatures lower than about –30 °C, friction and pressure no longer melt snow or ice into a slippery film of water beneath a sledge's runners. Under these conditions, pulling a sledge over ice is little different from pulling a sledge over concrete. Scott's team found that they could barely pull their sledges. Scott knew that he might get into trouble with abnormally high friction, and he describes how the men tried stitching sealskin over the runners of the sledges to see whether friction could be reduced to more normal levels.

The formation of a film of lubricating water on ice is what allows us to enjoy winter sports. Normally an ice vehicle works by means of a metal runner that slides on top of flat ice or packed snow. Devices from Santa Claus's sleigh to skis and ice skates work in exactly this way. However, things don't have to work this way. Here, we have reversed the toboggan and produced the naggobot.

What You Need

- ❑ U-shaped plastic rain-gutter pipe
- ❑ Electric motor (the powerful miniature 1.5- to 3-V variety often seen in small toy cars)
- ❑ Small propeller or fan blades
- ❑ NiCd batteries and a battery holder
- ❑ Polystyrene foam
- ❑ Ice cubes

What You Do

Cut rectangular holes in the bottom of the polystyrene foam such that the ice cubes will fit tightly in the holes. I used the small rectangular ice cubes made in an ice cube tray rather than the finger-shaped pieces that you get from an ice machine. Trim the fan or propeller blades short so that they will miss the sides of the U-channel. Mount the motor and the battery on the polystyrene and connect them together.

Once you've got the naggobot constructed, set it in the channel without the ice cubes, just to satisfy yourself that the motor and fan assembly is completely incapable of propulsion. Now insert the ice cubes. You need to get the ice cubes installed swiftly, so I recommend just four cubes, certainly no more than six. They melt quickly, so use the largest cubes you can, at least 25 mm, although really huge cubes may make the vehicle too heavy for the fan to shift it. Once you've got the cubes installed, try the naggobot again. With luck, the vehicle will accelerate smoothly; with a long enough channel, it would ultimately reach quite a high speed, at least with the channel horizontal.

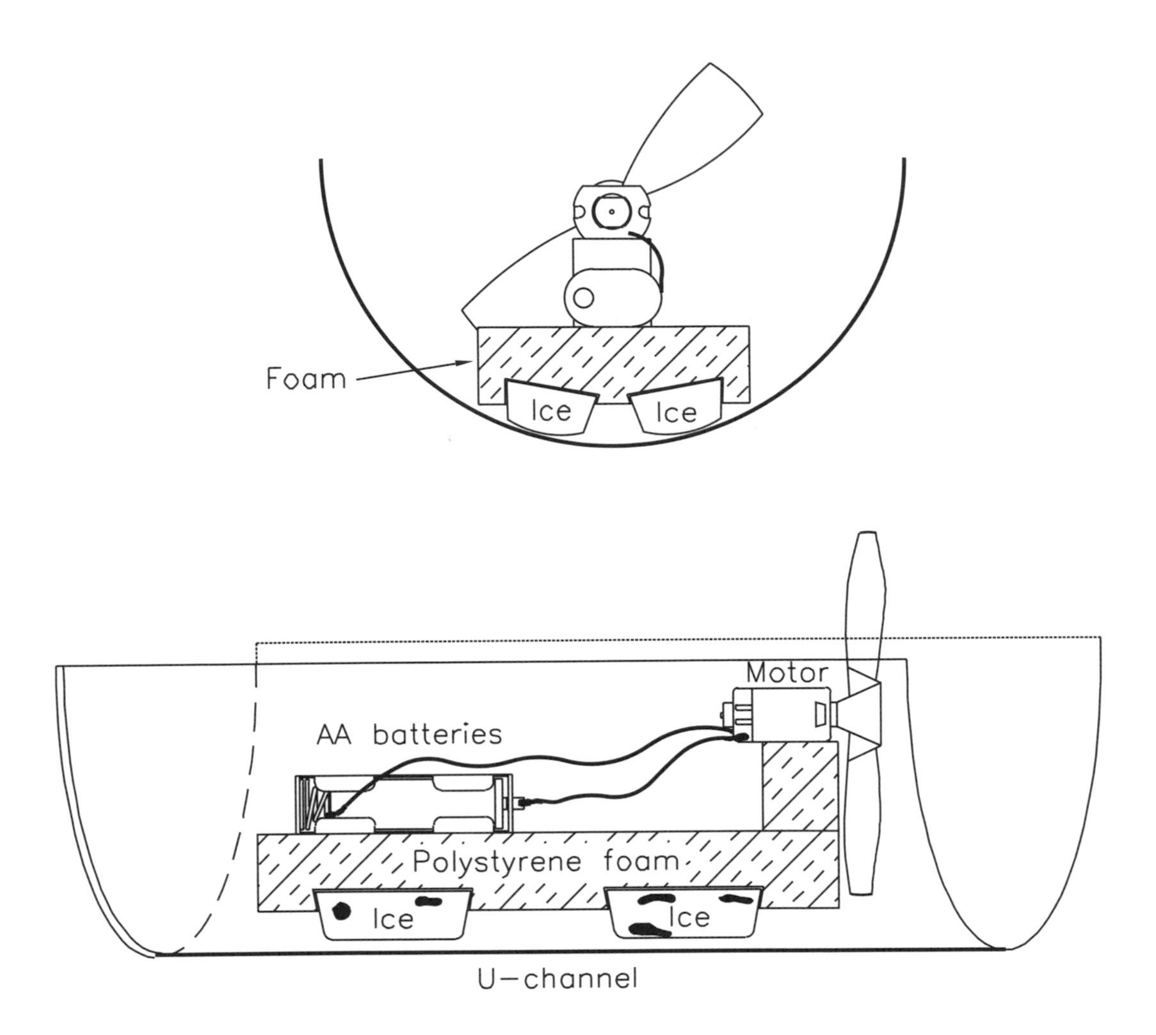

How It Works

The main reason that ice slides so well is that it continuously provides its own lubricant by obligingly melting at the interface. The lubricant film is extruded only slowly from the interface and remains thick enough to largely prevent contact between surfaces, hugely reducing the coefficient of friction. The coefficient of friction of a rigid material like steel or plastic on ice is very small, no more than 0.01; normal friction coefficients are ten to thirty times as large.

The process by which ice melts when subjected to pressure is known as *regelation.* The process occurs because ice has a larger volume than the water that freezes to make it: exerting a high pressure must melt the ice. Normal substances experience the opposite effect: the solid has a smaller volume than the liquid, so pressure tends to freeze them. In the naggobot, the regelation effect, which requires high pressure, probably cannot occur all over the ice cube; instead it will occur at high points, at least to start with. However, because regelation occurs only when the ice is only slightly below its freezing point—no lower than –3 °C or so—the principal mechanism involved in making ice friction so low is not regelation but the production of a thin water film via frictional heating.

D.C.B. Evans and his colleagues in Bristol, England, carried out careful measurements on ice from –15 °C to –1 °C and showed that the heat produced by the friction of sliding produces a film of meltwater just a fraction of a micron thick. Their theory of frictional heating actually explains sliding on ice almost perfectly. They showed that their theory agreed closely with measurements and correctly predicted the linear increase of friction with decreasing temperature.

THE SCIENCE AND THE MATH

Some mystery still remains about exactly why devices slide so easily on ice. As of the 1970s, there was controversy about the significance of the regelation effect, and this undoubtedly stimulated a team of scientists in Bristol, England, to study sliding friction on ice. I think they have at least 90 percent of the answer to the mystery of friction. Let's take a look at the elements of their theory.

The viscosity, η, of water at 0 °C is 1.8 mPa s, and the drag force, F_d, due to viscosity is given by

$$F_d = \eta A (dv/dx),$$

where A is the contact area and dv/dx is the gradient of velocity between the surfaces. Let us assume that the water film is 0.3 µm thick, as indicated by the Bristol team. (This leaves open the question of what happens when there is slight roughness on the sliding surface of the extruded plastic U-channel, roughness that would penetrate the film, obviating its beneficial effect.) For 1 m/s, for example, dv/dx is approximately equal to 3×10^2 s^{-1}.

So the expected "friction" force (actually viscous drag) with a water film lubrication for a 1-mm^2 contact area is given by

$$F_d = (1.8 \times 10^{-3})(3 \times 10^{-6})(10^6) = 0.0060 \text{ N}.$$

The Bristol team used a F_n of about 10 N, so μ is approximately 0.0060/10, or 0.0006. This sort of value for the coefficient of friction is small, as you might expect for ice, but in fact it is too small—by a factor of fifty. The rest of the friction must come about from the high spots, the parts of the surface that project above the 0.3-µm-thick water layer and that are responsible for almost all the friction, and the heat. The Bristol team showed that in fact the rest of the coefficient of friction can be calculated by considering heat conduction: the ice conducts heat away from the interface. Heat conduction increases friction because it thins the water layer, which increases drag and high-spot friction. The team's calculation depends upon the observation that the very top of the ice is melted and thus must be near its 0 °C melting point. Since they knew the bulk temperature of the ice in each experiment, they could calculate the required heat flow and, hence, how much friction was needed to generate it. We can approximate their complex calculation by means of the following equations. The rate at which heat is conducted away, dQ/dt, is given by

$$dQ/dt = kA\,dT/dx,$$

where k is the thermal conductivity of ice, A is the contact area, and dT/dx is the temperature gradient in the ice. (We assume here that the heat conducted away by the metal is small.) Because dQ/dt must be equal to the work done per second, we arrive at

$$dQ/dt = F_d(\text{effective})v,$$

where v is the speed. So

$$\mu = F_d(\text{effective})/F_n = dQ/dt/(vF_n),$$

and

$$\mu = [kA\,dT/dx]/(vF_n).$$

This equation has some of the correct properties for a "skating equation." It shows that the frictional force increases with decreasing temperature and is linear with respect to dT/dx. It even shows why the coefficient of friction decreases with increasing speed—although the measured decrease is actually not inversely proportional but rather goes as $\sqrt{(1/v)}$.

However, this preliminary skating equation does not allow for the fact that dT/dx will decrease with increasing speed because heat can diffuse away only at a certain speed and because the distance to which diffusion can occur varies as $\sqrt{t}$. In other words, the dx we should use is proportional to the time available, t:

$$dx = \sqrt{(Dt)},$$

where D is the diffusion coefficient for heat in ice. So

$$dT/dx = \Delta T/dx = \Delta T/\left[\sqrt{(Dt)}\right]$$

and

$$\mu = \left[kA\Delta T/\sqrt{(Dt)}\right]/(vF_n).$$

Because t equals the contact-area length divided by the velocity of sliding, L, we arrive at

$$\mu = \left[kL^2\Delta T/(\sqrt{(DL/v)}\right]/(vF_n).$$

This gives us our final skating equation:

$$\mu = (kL^{1.5}\Delta T)/\left[\sqrt{(Dv)}F_n\right].$$

This equation has the same dependence upon the temperature differential as our preliminary equation, and it also has the correct speed dependence. However, it is clearly incomplete because the length of the contact area, L, and the normal force, F_n, are related to each other. But we can try the equation out nevertheless. If we assume that the ice is 5 °C cooler than the surroundings, then the gradient dT/dx must be around (5×10^{-3}) with a contact area of 1 mm^2. Using these numbers, we arrive at a μ value of around 0.01, which is much closer to observed values.

We neglect here the friction due to heat conduction by the metal and the plastic. This seems reasonable, since the material of the skate makes little difference to the friction seen. There are a few exceptions; aluminum, for example, does not slide well on ice. I once made a toboggan with aluminum runners and was disappointed to discover that it performed poorly. Aluminum has extraordinarily high thermal conductivity, five or ten times as high as that of steel, so it conducts away the heat needed to create the film of lubricating water. The Bristol experiments showed that copper, which conducts heat even better than aluminum, has double the coefficient of friction of steel or plastic, and we might expect aluminum to be the same.

It is also obvious that the temperature is critical, perhaps much more critical for our naggobot than for a conventional vehicle. My observation that the plastic U-channel worked fine at room temperature (although the ice did melt quickly) but not so well on a cold day seems to fit nicely with the theory above.

It is also fairly obvious that the finish of the sliding surface is important. Coarse sandpaper won't work as well as a smooth polished surface. With ordinary friction between dry surfaces, the finish does not have a big impact: a sandpaper surface might have a coefficient of friction that is twice that of a smooth surface. Try pulling wood blocks over a polished wood table and then over sandpaper if you don't believe me—the difference is not as great as you might think. However, with the naggobot, we have the lubricating film of water with a limited thickness, so we want to minimize any surface roughness. Some Olympic Games contestants (speed skaters, bobsledders) use diamond paste and other polishing materials to finish the hardened steel runners to a mirror finish before competitions.

The regelation effect may not have as much influence on the naggobot as it does on an ice skate, which has a very small contact area and can exert high pressure on the ice; the naggobot is more akin to a broad ski or snowboard, with a large contact area and low pressure. However, as I noted above, regelation is probably not the key effect in ice sliding.

Ice is an extraordinary material with many curious properties. Consider, for example, the reversal in adhesion seen in metal that is frozen to ice at very low temperatures. Air Products & Chemicals Inc. has a patent based on this reversal, which is the basis of a technology used in the frozen-food industry to produce more finely molded and interestingly shaped popsicles and the like. And Victor Petrenko of Dartmouth College Ice Research Lab has filed a patent on controlling ice adhesion strength using electrodes in the skin of the surface of sliding objects, such as skates or aircraft wings, where ice adhesion must be reduced. Interestingly, Petrenko has recently claimed that ice friction can be increased greatly by the application of modest electric fields to electrodes on the surface of snowboards: you can, in other words, fit brakes to your snowboard. An article about Petrenko by Ian Sample in *New Scientist* refers to this. Could the effect be applied to our naggobot somehow?

And Finally . . . Directional Stability and Naggobot Racing

There is another fundamental difference between a naggobot and a toboggan: the latter will tend to go in a straight line, but the former will not. The forward progress of the conventional ice vehicle carves a minute U-shaped channel, allowing directional stability. Imagine for a moment trying to ski or ice skate on a round board: the inability to control direction would add a whole new dimension of difficulty. Yet this is the situation we have with the naggobot. To gain directional stability, we use a U-shaped channel and a long vehicle that fits only in one orientation in the channel.

There might be other ways to add directional control. Could a naggobot be steered with twin fans, like a hovercraft? And could a directional surface, a kind of plowed field surface, provide good control? The small U-shaped channels of corrugated plastic roofing material would clearly work, and the material could be used as a track for competition racing of different naggobots. But would much smaller corrugations work too? And would narrow ridge corrugations (you can buy drainage channels with these) improve the performance of the U-channel naggobot by giving it better directional stability?

REFERENCES

Evans, D.C.B., J. F. Nye, and K. J. Cheeseman. "The Kinetic Friction of Ice." *Proceedings of the Royal Society A* 347, no. 1651 (1976): 493–512.

Petrenko, Victor. "Systems and Methods for Modifying Ice Adhesion Strength." U.S. Patent no. 6,027,075 (2000), February 22, 2000.

Sample, Ian. "Skis Get Electronic Brakes." *New Scientist,* February 6, 2002.

Scott, Robert F. *Scott's Last Expedition.* 2 vols. London: Macmillan, 1913.

Solomon, Susan. *The Coldest March: Scott's Fatal Antarctic Expedition.* New Haven, Conn.: Yale University Press, 2001.

14 Boadicea's Autochariot

"Up my Britons, on my chariot, on my chargers, trample
them under us."
So the Queen Boadicea, standing loftily charioted,
Brandishing in her hand a dart and rolling glances
lioness-like,
Yell'd and shriek'd between her daughters in her fierce
volubility.
Till her people all around the royal chariot agitated.

—Alfred Lord Tennyson, "Boadicea"

Was Alfred Lord Tennyson the first to imagine the queen of the ancient tribes of the British Isles leading chariot charges against the invading legions of ancient Rome? In his "Boadicea," Tennyson doesn't mention the scythelike knives on the wheels of the queen's chariot, knives that other traditions accord to chariots at the time, but he is pretty specific about the other unpleasant punishments that Boadicea's army would inflict on the Romans, given half a chance.

During World War II, a war machine, code-named the Panjandrum, that rivaled Boadicea's scythe-chariots was devised.* The idea was that the Panjandrum, a kind of explosion-driven Ferris wheel, would be set rotating and then released in shallow water to roll up onto enemy beaches. It was powered by gunpowder rockets in the manner of a Catherine wheel and could be steered by means of cables attached to its axles.

*The word *panjandrum,* meaning a self-important person, comes from a nonsense rhyme composed by Samuel Foote to test actor Charles Macklin's vaunted powers of memory: "and the Grand Panjandrum himself, with the little round button at top, and they all fell to playing the game of catch as catch can, till the gunpowder ran out at the heels of their boots" (according to an 1854 article in the *Quarterly Review*).

In *The Secret War,* Gerald Pawle describes how the Panjandrum rolled along successfully, though erratically, but proved impossible to steer properly. The device was tested by a small naval department manned by a set of oddball engineers and scientists including novelist and engineer Nevil Shute. Why its inventor thought it such a good idea as an engine of war we can only guess. I think we can safely assume that the absence of the Panjandrum from the military arsenals of the world today means that its effectiveness is doubtful. However, using a small tip jet to set in motion a large and heavy device is a rarely used but nevertheless effective concept. Helicopters driven by tip jets have from time to time been proposed, and some have been built and operated, although without any lasting success. Here we investigate a vehicle that is half chariot, half Panjandrum.

What You Need

- ❑ Bearings on which to mount a vertical shaft
- ❑ 4 wheels
- ❑ Wood for chassis and pylon
- ❑ Wire for axles
- ❑ 3-V electric motor
- ❑ Battery and holder
- ❑ Balsa for whirling arm
- ❑ 2 pulleys
- ❑ Rubber band
- ❑ Propeller

What You Do

The pylon and its bearings must support the vertical shaft with the whirling arm so that it rotates freely but not too loosely. The position of the battery must be adjusted until it exactly counterbalances the motor. To do this, turn the autochariot on its side and use the whirling arm as a balance, gluing the battery holder in place when balance has been achieved.

Once everything is constructed, set the gadget going. Determine how fast the whirling arm will rotate when the autochariot is held off the ground and the arm can rotate freely. Then test the driving-on-the-flat rotation speed of the arm.

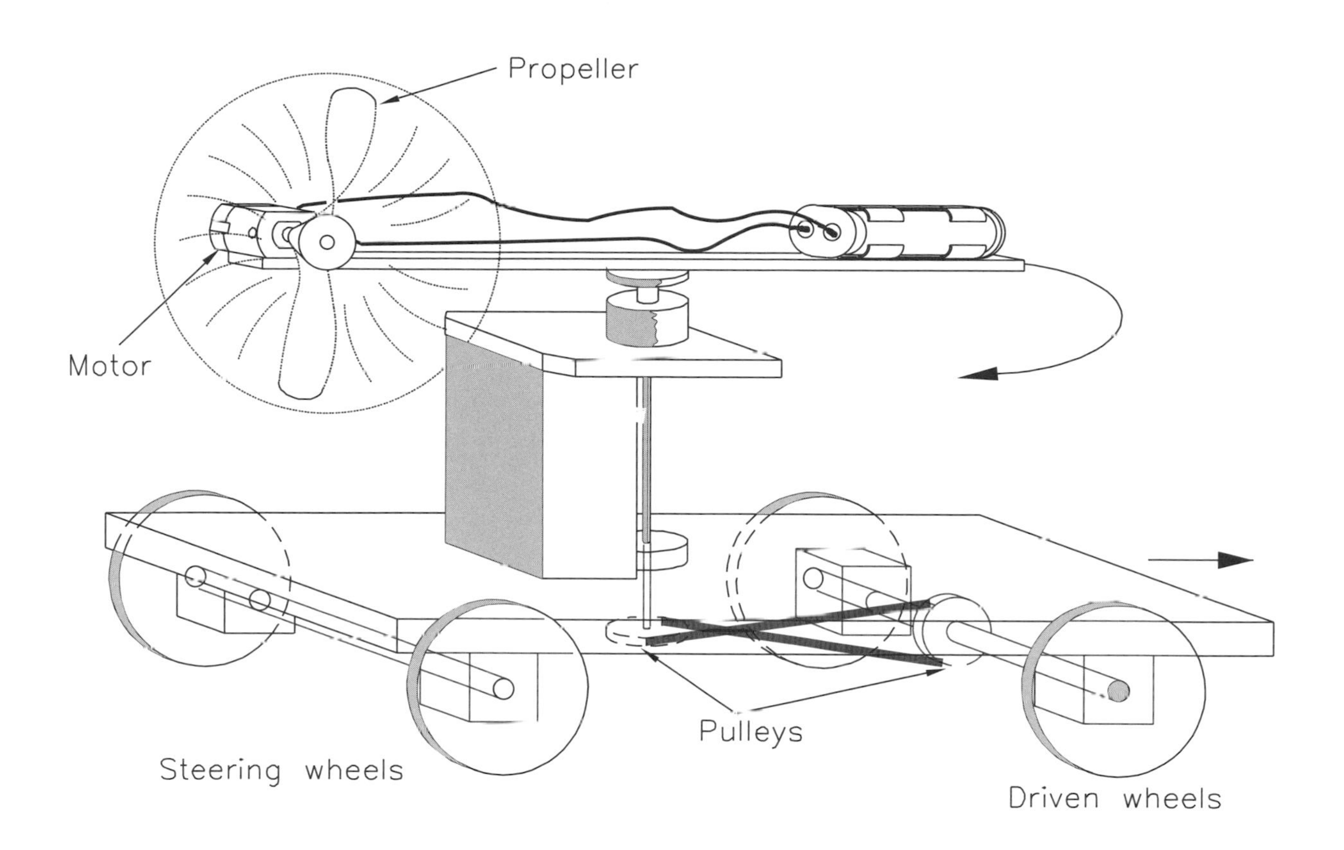

How stable is the autochariot when going over bumps? How much does it slow down on encountering a slight slope? How steep a slope can it manage? If you have the time, you could build a similar vehicle chassis and equip it with the same motor powering a simple backward facing propeller by way of comparison. Try this new vehicle against the autochariot on slopes and you will probably see a dramatic difference.

Stopping the vehicle is a bit of a problem because the rapidly rotating arm and its even more rapidly rotating propeller will inflict bruises on the incautious. If the armies of ancient Rome were impressed with the rotating scythe blades fitted to Boadicea's horse-drawn chariots, they would have had real problems with the autochariot. Probably the best way to deal with the problem of stopping is to use radio control. However, failing that, approach a runaway autochariot by picking it up from underneath, slowing the whirling arm by braking the pulley wheel on the bottom with your fingers, and then disconnecting the battery once the whirling arm has halted.

How It Works

The propeller-motor combination we use here is not capable of providing a thrust force of more than a few thousandths of a newton. With typical plastic wheels running on typical wire axles through plain "bearings," actually just holes in metal plates, the friction force restraining the autochariot is much higher than the thrust force. If we replaced the wheels with more efficient pneumatic tires and the bearings with high-quality ball bearings, we could probably make a vehicle that would move along on flat and smooth surfaces. However, a rough surface or the slightest slope would defeat it. Our autochariot can move, even on rough surfaces or on a slope, because we translate that small thrust force into a large twisting force, a large torque, by using the long whirling arm. This large torque, which is transmitted to the wheels via the pulleys, provides a much larger propulsive force, much larger than the friction force opposing it. And the autochariot works in spite of the very obvious power losses in the system. The pulleys and belts and the bearing of the whirling arm are all inefficient, stealing some of the power coming from the motor and propeller, but the system is still capable of exerting a considerable force on the wheels and moving the autochariot, even up a slope. The price we pay for this extra force is speed: the autochariot is considerably slower than a simple airscrew-driven vehicle using the same power.

THE SCIENCE AND THE MATH

The length, R_1, of the whirling arm is the key parameter in the design of the autochariot. The traction force, F, is given by

$$F = (fR_1R_2)/(R_3R_4),$$

where f is the thrust from the propeller, and R_2, R_3, and R_4 are the radii of the first pulley, second pulley, and drive wheels, respectively. In exchange for getting a greater traction force, you have to go slower; the velocity of the vehicle is given by

$$v = (uR_3R_4)/(R_1R_2),$$

where u is the speed of the motor on the arm. This equation is the inverse of the traction force equation. With typical values, you will get, as I did, something around a fifteenfold magnification in tractive force, but there are of course considerable frictional losses that reduce this magnification.

With R_1 equal to R_4 and with a vertical plane of orientation, you could dispense with the whole of the vehicle except the wheel, and mount the propellers on the wheel: such a device would be a kind of panjandrum, although the wartime Panjandrum used gunpowder rockets rather than propellers for propulsion.

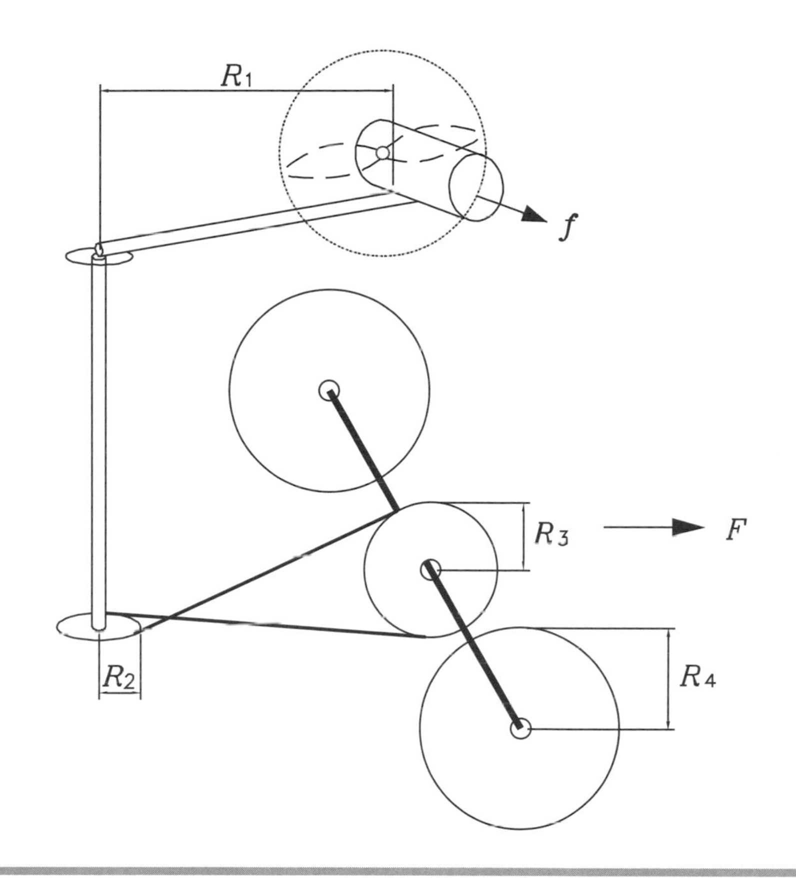

And Finally . . . Could the Ancient Romans Have Built Automobiles?

You could enclose the rotor arm to reduce some of its potential for bruising your fingers. Could you improve the efficiency of the vehicle by mounting vanes in an enclosure around the rotor arm? Design practice in the water turbine industry indicates that this ought to be possible—in essence, you improve the efficiency by providing something for the airstream to push against. Would the whole system work better with two propellers?

The vertical axis of the whirling arm is not an essential feature. Horizontal-axis operation should be equally possible, although the device would be bulkier and less stable (with its higher center of gravity), and on a full-scale vehicle there might be gyroscopic problems. A horizontal-axis whirling arm would need to be more carefully balanced, or starting would be difficult.

Could the ancient Romans have invented the automobile in Rome, two thousand years before Gottlieb Daimler and Carl Friedrich Benz? Most people would say definitely not, for the following reasons: cylinders and pistons could not have been accurately bored, so engines could not have been constructed; there was no electricity for creating a spark; and there was no gasoline.

But Romans of old did have wheels, they did make gearwheel transmissions (see "Vitruvius' Odometer" in *Scientific American*), they could certainly make horse-drawn and ox-drawn vehicles, and, finally, they did have some liquid fuels, principally vegetable oils. (Vegetable oils actually burn reasonably well: diesel engines can be run on sunflower oil, although the exhaust smells like that from a restaurant cooking french fries.)

The history of the locomotion has been dominated by the piston-and-cylinder model, and I would be the first to admit how hugely fruitful this paradigm has been, first with steam, then with internal combustion. But the concept we have here—of an engine that produces a weak force that is able to produce a powerful traction force via the long length of the whirling arm on which it is mounted—is one the Romans could have used. The whirling-arm technique has the advantage of not needing high-speed bearings or large numbers of gearwheels to give a high reduction-transmission ratio. Maybe a simple steam jet mounted in such a way could be made to attain real power, unlike the aeolipile, the toy engine of ancient Alexandrian scientist Hero. The secret is probably to ensure that the steam jet rotates at a speed comparable to the jet efflux—at least several hundred miles per hour from a high-pressure orifice. A rotating steam-tight joint would be necessary, however, and a pressure-resistant boiler. Riveted copper is used in some model steam engines and would have been attainable by the Roman smiths. But even without a rotating joint and a high-pressure boiler, perhaps Roman technology could have provided tip-jet power—via a small pulse jet or a ramjet at the end of the whirling arm.

A pulse jet powered the V1 cruise missiles of Hitler's army, and similar jet engines are still occasionally used today in high-speed model airplanes. They function at speeds ranging from 50 mph to 500 mph. The pulse jet is a tube, with a spray of kerosene or a similar fuel, and a simple butterfly check valve at the air inlet: the valve is the only part that needs to be made of reasonably high-quality steel—the rest could have been crudely made by a blacksmith anywhere along the legendary Appian Way. Maybe the Romans could have made such a pulse-jet engine and run it on vegetable oil. But the valve would have been a problem.

An even simpler power source might be possible, even with vegetable oil fuel: the ramjet. It functions only at speeds above a few hundred miles per hour—ideally near sonic or even supersonic speeds. But if the whirling arm were forced into rapid rotation (a small legion of Roman soldiers would have been handy here), then maybe a ramjet could have been operated in ancient Rome.

Could a road vehicle made from Roman materials—wood, brass, copper, and some iron—and fueled by sunflower oil somehow have been made to function in a ramjet, using the whirling-arm technique? I think so. But if the Romans had done it, would history have been much different? It's certainly possible. The ancient Britons are said to have been experts at handling their primitive chariots with those scary rotating knives mounted on the wheels. Perhaps, commanded by the fearless Queen Boadicea, the ancient Britons could have gone to battle in their whirling-arm chariots and held off the Roman legions.

REFERENCES

Pawle, Gerald. *The Secret War, 1939–1945.* Foreword by Nevil Shute. London: Harrap, 1956.

Sleewyk, Andre Wegener. "Vitruvius' Odometer." *Scientific American,* October 1981, 188–200.

15 Tubal Travelator

> Each bottle could be placed on one of fifteen racks, each rack, though you couldn't see it, was a conveyor.
>
> . . .
>
> The faint hum and rattle of machinery still stirred the crimson air in the Embryo Store. Shifts might come and go, one lupus-coloured face gave place to another; majestically and for ever the conveyors crept forward with their load of future men and women. . . .
>
> Slowly, majestically, with a faint humming of machinery, the Conveyors moved forward, thirty-three centimeters an hour.
>
> —Aldous Huxley, *Brave New World*

Conveyors have never gotten very good press. Aldous Huxley imagined them in a factory for manufacturing humans—the Central London Human Hatchery of his dystopic future world. His science fiction book is not the only one of its genre to paint conveyors into a future dismal metropolis. However, conveyors are a logical solution (as Dr. Spock of Star Trek might have it) for large-scale short-range transport. And personally, I quite like them, particularly the moving sidewalk rubber belt–type travelators you sometimes find at airports. Some of them bounce slightly underfoot, giving you a kind of "walking on the Moon" feeling, particularly if you stomp along them putting your feet down heavily.

Transportation systems can be placed on a continuum from single-person systems that can go anywhere (like a single-seat helicopter) to systems that move hundreds of thousands of people between a limited number of fixed points (like an escalator). An automobile clearly lies closer the helicopter, and the railroad lies closer to the escalator. Because the demand for transportation grows each year, the helicopter end of the spectrum is unlikely to be able to expand much. So future generations will need more systems nearer the escalator-travelator end of the continuum.

Shifting goods, or people, short distances in large quantities has, over the years, spawned a spectacular assortment of inventions: conveyor belts, conveyor chains, chutes, rows of rollers, arrays of roller balls, chains with dangling hooks, air-blast suspended pneumatic conveyors (for sand and similar materials), chain-hauled trolleys, and so on. Within the conveyor belt category there are still more variations: top-loaded and bottom-loaded conveyors, upturned-edge belts, belts with handrails.

Here we look at a conveyor that will smoothly propel prodigious quantities of material through a simple tube. The operation of the device does require a slight slope, but the slope can be small indeed, as little as 1 degree. And things don't necessarily have to *down* the tube; you can make certain shapes and designs of object to go up the tube in an almost magical way.

What You Need

- ❑ Tube (e.g., plastic or cardboard tube 30–100 mm in diameter and 0.5–2 m long)
- ❑ Motor, ca. 60–300 rpm (e.g., cordless drill or electric screwdriver)
- ❑ Round flange to fit the diameter of the tube and motor shaft (e.g., hole saw)
- ❑ Wooden blocks (e.g., from a child's building block set)
- ❑ Stopwatch

What You Do

Connect the motor to the tube using the hole saw. (The hole saw is a particularly useful adaptor device because it allows you to use tubular blades of various sizes to fit tubes with different diameters.) Make sure that the unpowered end of the tube is immobilized—perhaps by placing heavy books either side of it—so that it can rotate freely but cannot wander around the room. Now raise the powered end to the height you wish to try and put a block in the tube at the raised end.

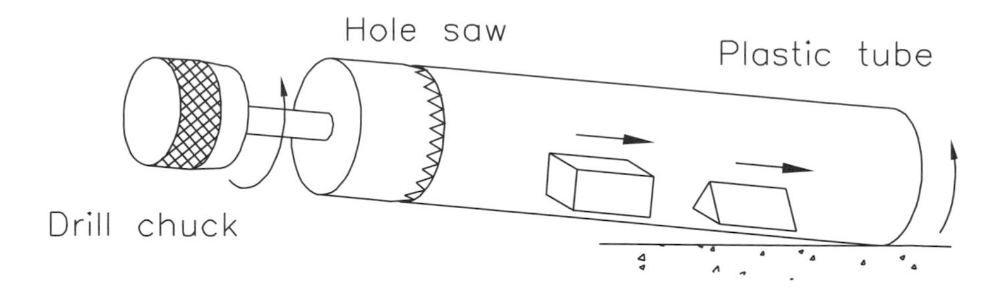

You will find that at about a certain critical angle, θ, the block will slide down the tube even in the absence of rotation, and that at angles near θ, the block will progress in fits and starts rather unpredictably. At smaller angles, say, less than 10 degrees, the block will be deflected (by maybe 20 degrees or so) off to the side that it is driven to by the rotation and will slowly progress down the tube, oscillating from side to side as it goes. After a time ranging from a few seconds to a minute or so, the block will emerge from the lower end of the tube.

The rate of conveying may vary, depending on whether the block slides or rolls or tips as it goes down the tube, and the rate may also depend upon whether the block slides irregularly from side to side or progresses smoothly down the tube. The degree of axial wobble in the tube, in fact any motion other than straightforward axial rotation, may also slow progress or, if nothing else, waste power. I chose flattened-profile blocks to get fairly smooth sliding.

Most objects will travel only down the conveyor tube. The law of gravity is not going to be repealed any time soon, so it doesn't seem likely that freely rolling or sliding objects can go upward. Nevertheless, rolling devices of the correct design will in fact roll up our conveyor tube. That's right, up! You may need to try it to believe it. Find some wheels or narrow rollers 3–5 cm in diameter and place them at the bottom end of the rotating tube. Orient them so that they begin to roll along around a circular path at right angles to the tube axis to start with. You should find that at least some of them will begin to roll and then begin to climb the tube, eventually exiting at the top.

With some assistance from the kids at the Saturday science club I run, I found that the fastest movers were wheels with tires on which the outer tread had a slightly larger diameter than the inner tread. When the wheels were placed so that they immediately began to roll around a circumference of the tube, they did not progress up the tube at first. But after a few seconds they quickly picked up sideways speed and accelerated up the conveyor to the exit with a rapid sideways movement a few seconds later. Different wheels will behave a little differently: a quick way to investigate a range of shapes is to try a molding them out of modeling clay. Perhaps you can persuade small toy cars to run up the tube by twisting their steering wheels in the correct direction.

How It Works

It is easy to understand how the tubal travelator works if you imagine yourself in the following rather unlikely scene. You are sitting on an extremely small block in an extremely large tubal travelator in a pea-soup fog, which means that you can see only the block you are sitting on and a small portion of the plastic wall of the tube near you. You are sliding down a gentle slope on an apparently flat plastic surface at a steady speed. Now the fog begins to disperse, and you see that you are actually sliding down a plastic slope that becomes shallower in front of you and steeper behind, and you figure that you must be in a large horizontal rotating tube and that you are simply sliding down around its inner periphery. The fog disperses more, and you can see clearly that you are in a rotating tube and, furthermore, that the light from one end of the rotating tube is becoming dimmer while the light from the other end is becoming brighter. You finally figure out that you are not in a horizontal tube but actually in a tube that slopes slightly toward the brighter light. In sliding downward along the line of the steepest slope, you slide toward the lower end of the tube.

THE SCIENCE AND THE MATH

Geometry of Sliding

Once the tubal travelator starts to rotate, the block inside will tend to be displaced by an angle θ to the vertical. The frictional force on the block, F_f, is given by

$$F_f = Mg \sin \theta = \mu Mg \cos \theta,$$

where μ is the coefficient of fiction, M the mass of the block, and g the acceleration due to gravity. Tan $\theta = \mu$ and $\theta = \tan^{-1} \mu$, so if, for example, $\mu = 0.3$, then $\theta = 17$ degrees. The same critical angle, θ, is involved in calculating the maximum slope, the slope beyond which blocks will slide freely down the tube in the absence of rotation.

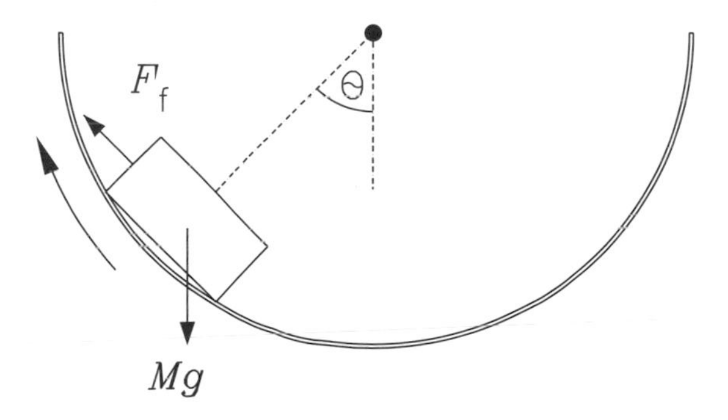

Roughly speaking, the block slides mostly tangentially with just a small component down the slope. In fact, the block takes a helical path down the tube. Imagine that the tubing is a tightly rolled rectangle of paper with a length greater than the block's helical path, length S, and a width equal to the tube length, L. The block's helical path traces a diagonal line across this paper, at an angle α to the tube. Angle α will in fact be the same angle that the tube makes with the horizontal, θ; in this way the block essentially slides down the steepest slope available to it. The geometry of this arrangement yields the following equation for angle α:

$$\alpha = \tan^{-1}(H/L),$$

where H is the height difference between the ends of the tube. And S is given by

$$S = L/(\sin \alpha).$$

Maximum Speed of Conveying

If the tube rotation speed, f ($= \omega/2\pi$, where ω is the angular velocity), is too high, then the load inside the tube will tend to rotate around on the inside of the tube without moving. At high speeds, the device behaves a bit like a Wall of Death ride at a fair. This ride consists of a rotating vertical cylinder with a floor that drops out once the ride gets going. Sometimes the cylinder even tips sideways from vertical. But the riders are held firmly in place on the wall: the wall gives them the centripetal force needed to keep them going round in a circle. In the high-speed travelator, the block riding down the tube will begin to loop the loop in a vertical plane at one place in the tube rather than slide down the tube. This will happen when the centripetal force is greater than the force of gravity—that is, when

$$M\omega^2 R > Mg,$$

where R is the radius of curvature. Thus, the highest rotation speed at which the conveyor can operate is given by

$$f > (1/2\pi)\sqrt{(g/R)}.$$

Using this formula, we can obtain the fastest possible conveying speed for a given angle to the

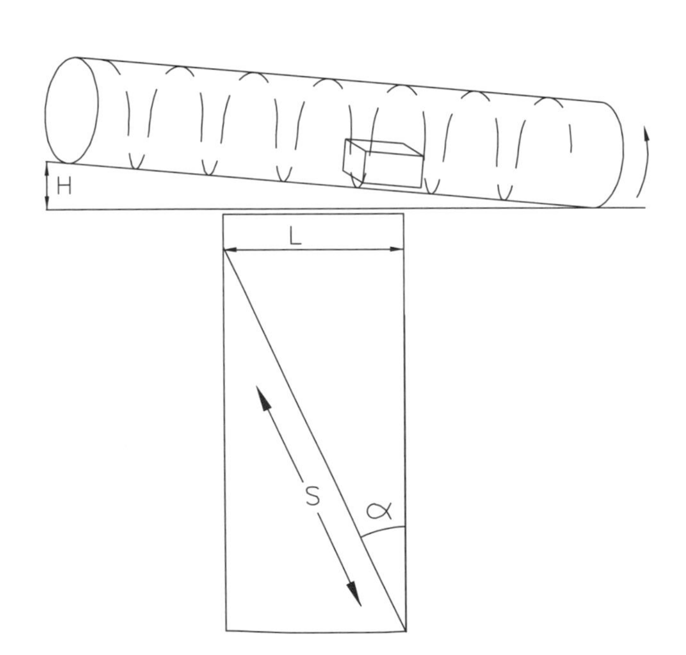

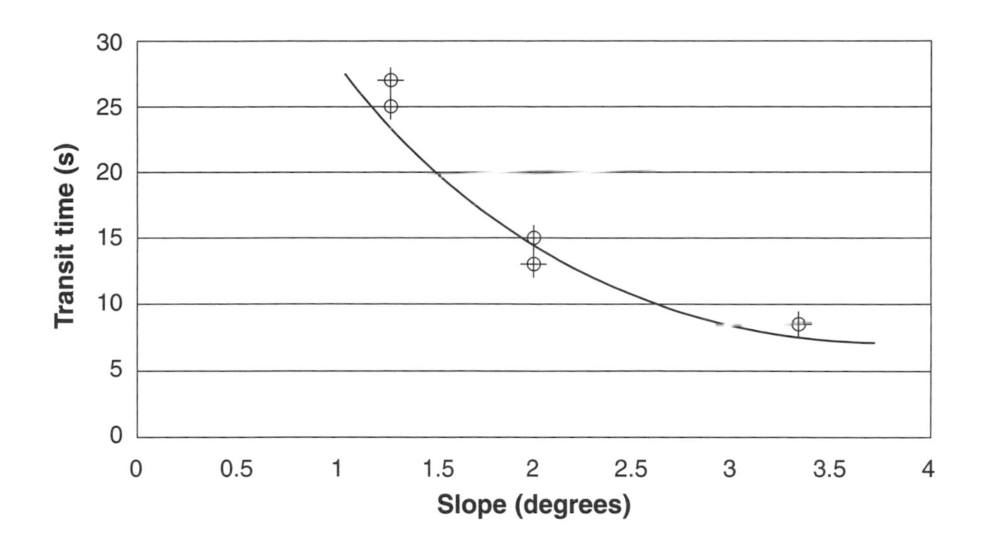

horizontal, α. The speed along a path of length S is given by

$$V = 2\pi f R,$$

and the maximum speed along the cylinder axis is

$$V_{max} = \sqrt{(g/R)} \sin \alpha.$$

So, for the fastest conveying speed, we should use the largest tube radius and the steepest slope α, and we should turn the motor revs up until the block is just about to loop the loop. How does this compare to what we actually measure on a real tubal conveyor? The points on the graph illustrate how the transit time varies with slope, versus theory. (The theoretical values have been divided by two to fit the data.)

Power

The power consumed by a conveyor system is of interest to any potential user. With rolling motion of any kind, the power consumed should be small (otherwise the systems would be uneconomical to run), and this is indeed the case with ideal conveyor chains and the like. However, with the sliding motion seen in the tubal conveyor, power consumption is inevitably going to be higher. The energy used, E, is probably mostly due to the friction of the transported block sliding over the tube in a helical pattern, with a smaller amount, MgH, due to the fact that the block must be raised in height by H to allow conveying. Following its helical path, the block takes a route that is ($1/\tan \alpha$) times the tube length, L. Furthermore, the block exerts a frictional force equal to $\mu Mg \cos \theta$, where $\theta = \tan^{-1} \mu$. Hence the energy consumed for a block of mass M is

$$E = \mu Mg \cos(\tan^{-1} \mu)(L/\tan \alpha) + MgH.$$

If we wanted to compare our tubal conveyor with other ways of conveying goods, we might look for a "figure of merit" (FoM), a number that would allow quantitative comparisons. Power stations, for example, look at "thermal efficiency," that is, electrical energy output divided by thermal energy input. A suitable FoM for the conveyor might be E/MLg, so

$$\text{FoM} = E/MLg = \mu \cos(\tan^{-1} \mu)(1/\tan \alpha) + \tan \alpha.$$

Rolling Uphill

There is no magical source of energy that allows the wheels to roll uphill. In their own reference frame, the wheels are actually rolling downward, but in our reference frame, they are rolling upward. The sliding block takes a path that is helical on the tube. However, the wheels roll along a path that it is a helix of increasing angle on the tube: their path, downward in their reference frame, is upward in our reference frame. They are actually trying to roll on a path that would be an arc of a circle if the tube were unrolled; hence the illusion that the wheels are accelerating upward.

We can calculate the geometry of this phenomenon if we think about the rolling of a thick wheel or roller whose rim is slightly conical. Obviously, the wheel or roller will roll not in a straight line but rather around a circle, since the outer-radius side will travel farther than the inner-radius side in the same number of rotations:

$$(R_w + W/2)/(R_w - W/2) = R_{out}/R_{in},$$

where R_w is the radius of the wheel path, R_{in} and R_{out} are the inner and outer radii of the wheel, and W is its width. So the radius of the wheel path on a flat surface is given by

$$R_w = W/2[(R_{out} + R_{in})/(R_{out} - R_{in})].$$

Hence, when R_{out} is nearly equal to R_{in}, as here, the path of the wheel is a large-diameter circle, many times the wheel width. Now imagine the wheel rolling along the x-axis of a graph: the displacement, Y, on the y-axis will be the distance up the tube. Taking the equation of a circle to be

$$(Y - R_w)^2 + X^2 = R_w^2,$$

where X is the distance traveled on the x-axis, we have, for small Y,

$$Y \sim X^2/2R_w.$$

This is the equation for a parabola. If the tube is rotating at a constant speed, V, on its periphery, then the wheel's progress along the x-axis will be a constant times T^2, where T is time, and the equation describes an accelerating sideways motion:

$$Y \sim (V^2T^2)/2R_w.$$

The wheels sometimes fall over once they are going substantially in an axial direction. In fact some wheels don't accelerate quite like this and instead, after an initial acceleration, progress smoothly up the tube at a constant axial speed. They seem to be skidding slightly so as to alter direction, the effect being to maintain a roughly constant angle to the axis. What determines which possible rolling and skidding modes are important, and which wheels will roll uphill in this way, is an interesting subject for further analysis.

Penultimately . . .

What happens if you try to convey, as is often a requirement in industry, a powdered material such as sand using your tubular conveyor? Will a set of loose granules behave the same way the blocks do? Certainly if the sand was contained in a loosely stuffed cloth bag, you might reasonably expect similar behavior. So the average motion of the sand might be similar to that of the block? But what happens to fast grains of sand? Do they accelerate ahead, leaving their slower cousins behind and rushing down to the exit. Certainly if the sand comprises tiny, nearly spherical grains, as is sometimes the case with beach sand, then the fast grains might well be expected to form a vanguard arriving well in advance of the average. Once they escape from the pack and can roll more or less freely down the tube, they will rush ahead. Those in the pack, even the rounded grains, will progress sedately like the block, as they cannot roll freely over their companions. But what about regular angular-grained sand? Will it travel in a tidy heap down the tube and arrive in a lump, or will spread out, in a process that physicists call *dispersion,* and dribble out over a period of minutes?

This phenomenon is actually studied in devices like lime kilns (see "Residence Time Distribution" by Ang et al., *Shreve's Chemical Process Industries* by George Austin, or "The Transient Response of Granular Flow" by Richard Spurling et al.). In lime kilns, limestone and clay raw materials are placed in a massive, slowly rotating tube. The tube, often as much as 200 m long and 4 m in diameter, is heated strongly, which slowly converts the materials into cement clinker, which is then ground to the powder we know as cement. Although much of the research is devoted to the heating of the rotating tube and to the mixing of the material inside, some of the experiments do study material transport issues. Rotary kilns, correctly adjusted, turn material over continuously, with sand or lime on the top of the heap being overlaid by material avalanching down on top of it, and the material underneath being carried off and up the side of the tube, so that there is overall a circulation of the sand or lime.

And Finally . . . Magic Carpets

Could a giant version of the tubal travelator be made? Imagine a steadily rotating smooth-walled tube with a diameter of 2.5 m. Bring along a rug, and step into the end of the tube, sit down on your rug, and relax against the wall as the rug slides effortlessly along. Could this really work? How fast would that tube have to go to keep up a pace of, say, 4 or 5 mph? Could intrepid "tube surfers" zoom down (or up) the tube faster than the tube's peripheral speed, like a surfer riding sideways along the crest of a breaking wave? If toy cars can be made to go up the tube, why not skateboards? Maybe Disney can try it out in its next theme park, although there might be some safety issues.

REFERENCES

Ang, H. M., M. O. Tadé, and M. W. Sze. "Residence Time Distribution for a Cold Model Rotary Kiln." *Australasian Institute of Mining and Metallurgy (AusIMM) Proceedings* 303 (1): 1–6.

Austin, George T. *Shreve's Chemical Process Industries.* 5th ed. New York: McGraw-Hill, 1984. See Chapter 10 on limestone and cement kilns.

Spurling, R. J., J. F. Davidson, and D. M. Scott. "The Transient Response of Granular Flow in an Inclined Rotating Cylinder." *Transactions of the Institution of Chemical Engineers A* 79 (January 2001).

Centripetal Force and Centrifugal Projectiles

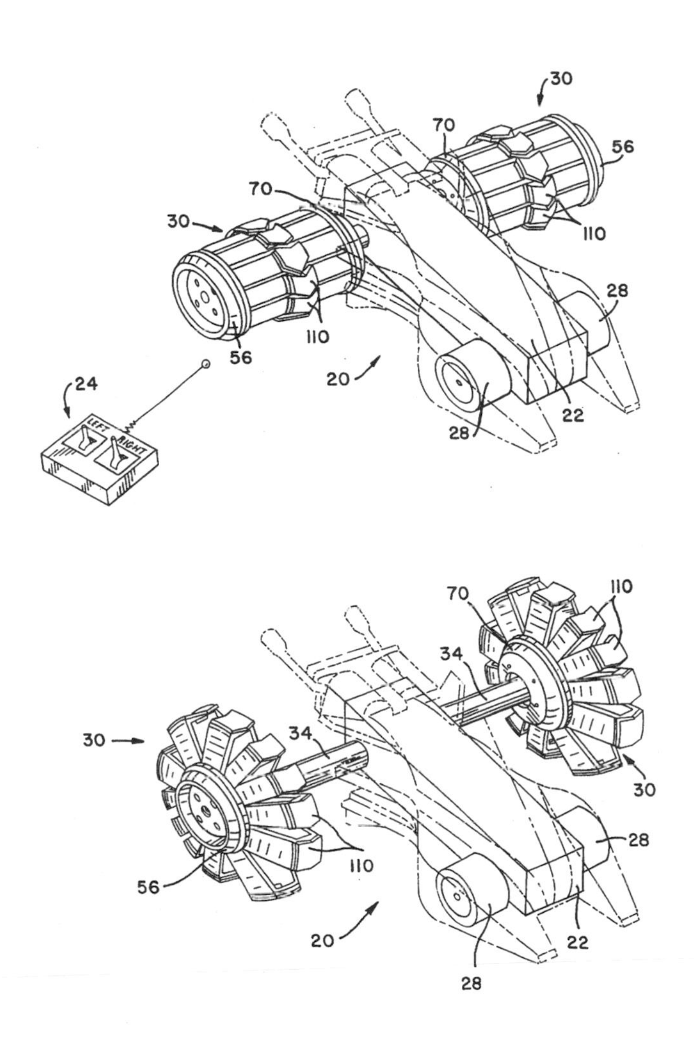

"The Inflection of a Direct Motion into a Curve by a Supervening Attractive Principle"

—Robert Hooke, title of a lecture to the Royal Society of London, May 23, 1666

Centripetal force arises from the acceleration toward the center of rotation that is needed to cause a body to travel in a circle; centrifugal force is the apparent force that a rotating body exerts on the radial arms that attach it to the center of rotation. Why does centripetal force exist? If you hide yourself in a closed shed with a bucket of water and look at the surface of the water, it will be flat—unless the shed, with you and your bucket in it, is rotating, in which case, centripetal force is provided by the water's forming a parabolic surface. What if the shed were made of a metal that screened electric and magnetic fields; how would the bucket know that it was rotating? The only force penetrating the shed is gravity. In fact, it is rotation relative to the distant stars and galaxies of the universe that counts, rather than rotation relative to nearby things like the shed, the earth, or even the galaxy. If you are rotating relative to the distant stars, then the water surface is curved. If you are not, then it is flat. What if the rest of the universe didn't exist? Then you obviously could not be rotating relative to it, so the need for centripetal force would disappear, and the water surface would be flat, right?

Believe it or not, that is what experts in cosmology and gravitation are pretty sure is the truth. In addition to Newton's gravitational force—which is a static force between objects and is related to the mass of the objects and the distance between them—there is another gravitational force that appears only when masses are accelerated and that is related to the mass of the objects, the distance between them, and the rate of change of their velocities relative to each other.

Centripetal force is a bit like magnetic and electric forces. Magnetism can be considered to be a force that arises from the velocity of electric charges: it is simply an extra electric force that arises between charges when they are moving. In fact, physicist Richard Feynman has shown how you can derive magnetism from electrostatics simply by noting the minute shortening of objects that occurs when they travel at a fraction of the speed of light—the shortening follows from the Einstein's special theory of relativity (*The Feynman Lectures on Physics*, 2:13.6–13.12).

In a similar way, the need for centripetal force and inertia appears when the equations of gravity are applied to moving masses. The force between massive objects that are accelerated relative to each other is almost vanishingly weak. However, there have been some attempts to measure it directly by tracking orbiting reflecting satellites in space, as Bernard Schutz reports in an article on the work of Ignazio Ciufolini of the University of Rome ("New Twist on Gravitational Spin"). Ciufolini believes that he has actually measured these forces using satellites.

REFERENCES

Feynman, Richard P. *The Feynman Lectures on Physics*. Reading, Mass.: Addison-Wesley, 1964.

Inwood, Stephen. *The Man Who Knew Too Much: The Strange and Inventive Life of Robert Hooke*. London: Macmillan, 2002. A biography of the extraordinary Robert Hooke.

Schutz, Bernard. "New Twist on Gravitational Spin." *Physics World*, November 1997, 23–24.

16 Centripetal Chaos

> We have also here an acting cause to account for that balance so often observed in nature,—a deficiency in one set of organs always being compensated by an increased development of some others—powerful wings accompanying weak feet, or great velocity making up for the absence of defensive weapons; for it has been shown that all varieties in which an unbalanced deficiency occurred could not long continue their existence. The action of this principle is exactly like that of the centrifugal governor of the steam engine, which checks and corrects any irregularities almost before they become evident; and in like manner no unbalanced deficiency in the animal kingdom can ever reach any conspicuous magnitude, because it would make itself felt at the very first step, by rendering existence difficult and extinction almost sure soon to follow.
>
> —Alfred Russel Wallace, "On the Tendency of Varieties to Depart Indefinitely from the Original Type"

The centrifugal governor, invented by the Scottish engineer James Watt, regulates the amount of fuel supplied to a steam engine and thus regulates its speed. In an 1858 article in the *Journal of the Proceedings of the Linnaean Society,* Alfred

Russel Wallace, a colleague of Charles Darwin, expresses a touching faith in the centrifugal governor, but the truth is that such gadgets simply don't work all that well without considerable effort on the part of the designing engineer. Sometimes the centrifugal governor either will not work at all or, most often, will "hunt" (go up and down rhythmically), causing the engine to speed up and slow down.

In engines, and in nature, what looks like balance is often actually wide oscillations with huge unbalanced deficiencies in both directions: only an average over a long period of time or a large area shows balance. Put a den of foxes in a field of rabbits and see what happens: in the field (and in even a rough computer model), you will get not a balance but rather oscillations, often somewhat chaotic, in the populations of both animals. Like populations of foxes and rabbits, the simple centrifugal device in this project behaves in a chaotic and rather interesting way.

What You Need

- ❑ Water
- ❑ Soda bottle (e.g., 0.5-l or 1-l size)
- ❑ Bolt, large (e.g., 8 mm), with nut and washers
- ❑ Oil (e.g., clear corn oil, olive oil, or mineral oil)

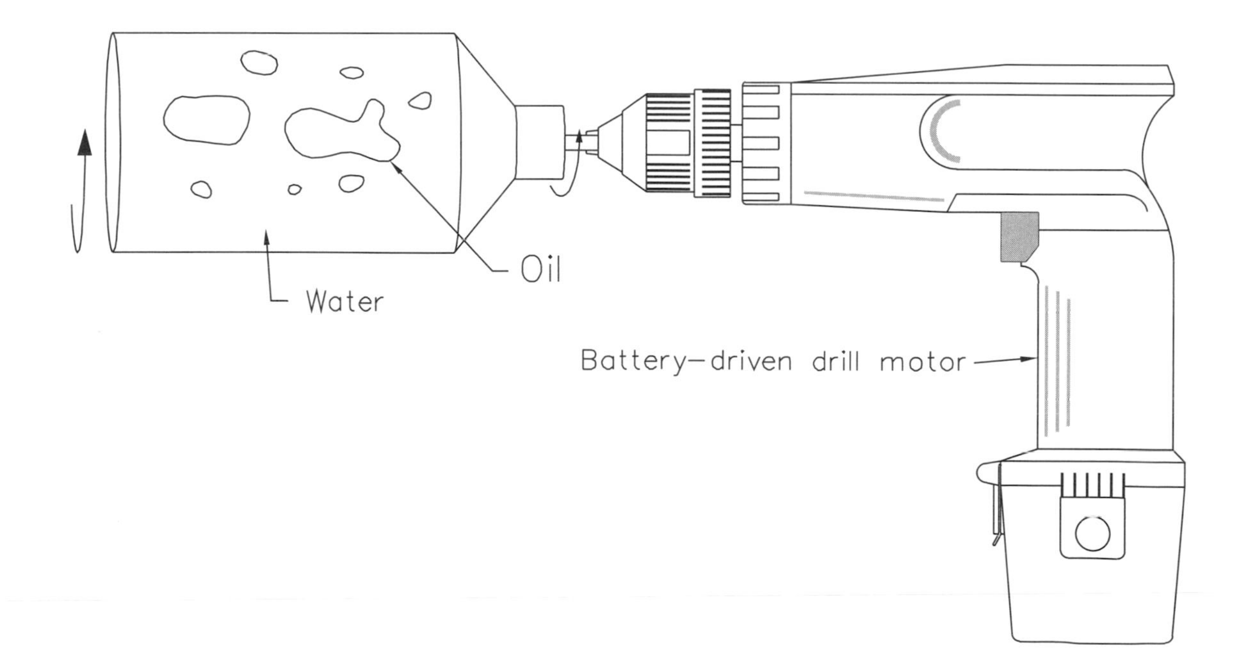

- ❑ Coloring material for the oil, optional (Choose one that dissolves readily in the oil but is immiscible in water.)
- ❑ Electric motor with variable-speed drive transmission giving 0–1,000 rpm (e.g., a battery-driven electric drill)

What You Do

If your oil is colorless, then color it with the coloring material. Then fill the soda bottle almost to the top with water, and add just a teaspoonful or so of the colored oil. Mount the bottle on the motor, using the bolt. Once you've got the device constructed, turn on the motor and rotate the bottle at a constant speed (100–200 rpm, at least until you have gathered some experience with the device).

The oil does not form a mayonnaise-like emulsion with the water but instead separates into a series of sausage-shaped globules that continuously coalesce and divide, with a few small globules orbiting around the center. You must rotate the bottle sufficiently fast that the force of gravity is overpowered by the centrifugal force. Otherwise, the oil will accumulate at the top of the bottle. Note how at lower speeds the oil does not align exactly with the axis of rotation; in fact, it lies slightly off to one side of the axis. Does the side of the axis it accumulates on depend upon the direction of rotation?

How It Works

The device is similar to devices used for various purposes in industry. For example, oil is separated from water commercially in hydroclones, the liquid equivalent of the cyclone filters used on factory roofs to separate dust from air. (Air cyclones are also used in the new generation of bagless vacuum cleaners pioneered by the Dyson company.) The relatively low speed and horizontal orientation of the centripetal chaos device, however, give it additional features, since gravitational effects are allowed to contribute. These effects explain the accumulation of oil near but not exactly along the axis and the curious random wriggling motions of the oil globules: they are being pushed up by gravity because of their buoyancy, but pushed down by the circular flow of liquid in the rotating bottle.

If you leave the chaos machine going long enough, it will settle into a steady state, at least for certain speeds and geometries and relative quantities of the two liquids. However, there are transient effects that lead to secondary circulation currents in the rotating bottle, currents that typically tend to sweep the oil out of the center of the bottle. These currents arise because of so-called end effects: the inertia of the liquid in the middle of the bottle and the central liquid's remoteness from the ends cause it to rotate more slowly at first than the liquid near the ends. The liquid at the end is thus pumped, as if in a centrifugal pump; the pressure at the periphery of the ends is higher than that near the middle. In this way, a pressure gradient is set up that allows water to circulate, starting from along the axis in the middle toward the ends and returning at the periphery.

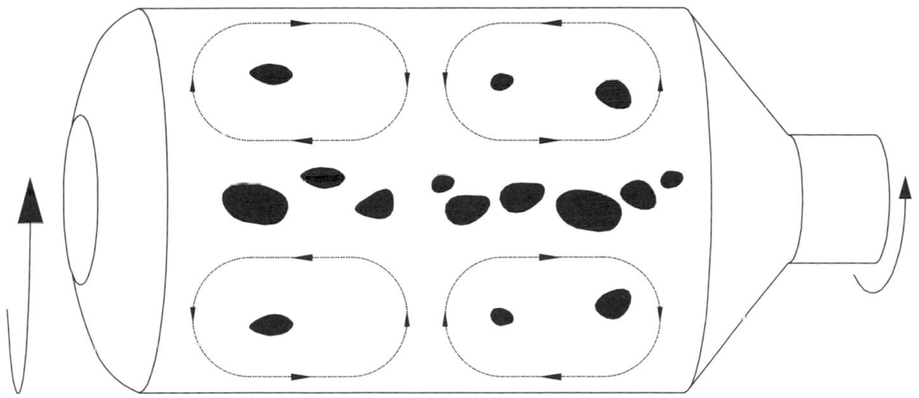

The secondary circulation often causes the oil to be swept out of the center when the bottle is accelerated. However, varying the bottle speed will allow you to control this behavior. If the bottle is slowing down, the secondary circulation is reversed. In addition, at certain speeds the oil will tend to reach a steady state in which globs of oil remain in the middle along the axis, because the secondary circulation eventually disappears if a constant rotation speed is maintained for long enough.

THE SCIENCE AND THE MATH

The buoyancy force, F_g, on a volume, V, of a material of density ρ_1 in a liquid of density ρ_2 is given by

$$F_g = (\rho_1 - \rho_2)Vg,$$

where g is the acceleration due to gravity. For a rotating body of water, the effective g is given by $R\omega^2$, where ω is the angular velocity in radians per second, and R is the radial distance of the material from the axis of rotation. So the centripetal force, F_c, is given by

$$F_c = (\rho_1 - \rho_2)VR\omega^2.$$

From this centripetal force we must subtract the downward buoyancy force, F_g. However, the centripetal acceleration will be big enough to oppose gravity only at high speed—that is, when

$$R\omega^2 > g.$$

Thus the angular velocity, ω, must be greater than $\sqrt{(g/R)}$.

You might at first think that this is all there is to it: we have a gravity gradient within the bottle, with a zero-gravity area just above the middle. Hence the globules of oil will tend to accumulate just above the center. The gentle gradient of the field near the center means that the globules are not highly constrained, and they don't immediately form a cylindrical constant surface. Instead they are propelled by small vortices and other random processes and will float around and form and re-form different shapes in the zone just above the axis.

Our formula for ω would seem to indicate that the bottle must rotate at more than 250 rpm to cause the liquid to migrate halfway to the middle, 15 mm off axis. This is not what you find in practice. I found that 160 rpm was enough to produce nicely centralized sausages of oil and, furthermore, that they were then just a few millimeters off axis. What is going on here? As I've already hinted, the truth is more complex and is bound up with the drag forces on the oil globules and the viscosity of the water. The rotation of the bottle, as it speeds up, sets the whole mass of water circulating, and the oil globules tend to position themselves alongside the axis on the side where the water is moving downward—that is, where viscous drag forces will push the globules down with a force equal to the buoyancy force.

And Finally . . . A Centripetal Chaos Lamp?

You could try oils of different densities. If the oil and water have similar densities, then rotation at an even lower speed will be possible. If you can get hold of two mutually immiscible oils, you can make a three-liquid chaos device. How would that behave? At low speeds, the heavier oil should migrate to the middle, leaving the lighter oil orbiting and dodging around; whereas at higher speeds, both oils should fight for the same on-axis territory.

Mounting a powerful source of light so that it shines into the rotating chamber—a lamp could be mounted at the end of the bottle opposite the rotating motor—gives you an interesting variation on the lava lamp, which is currently undergoing a fashion resurgence. Could a centripetal chaos lamp follow the lava lamp to commercial nirvana as a desirable objet d'art, I wonder? Potential manufacturers may contact the author.

REFERENCES

Austin, George T. *Shreve's Chemical Process Industries,* 56 et seq. 1945. Reprint; New York: McGraw-Hill, 1984. A discussion of hydroclones, which are used in the manufacture of starch and in separating oil from water in the petroleum industry.

Dunnet, Donald. "A Display Device using Liquid Bubbles in Another Liquid." U.K. Patent no. GB703,924, February 10, 1954. Possibly the original patent for the lava lamp, although there are lot of similar patents in this area.

Wallace, Alfred Russel. "On the Tendency of Varieties to Depart Indefinitely from the Original Type." *Journal of the Proceedings of the Linnaean Society* 3 (July 1, 1858). 53–62.

17 The Rotapult

And there went out a champion out of the camp of the Philistines, named Goliath, of Gath, whose height was six cubits and a span. And he had an helmet of brass upon his head, and he was armed with a coat of mail; and the weight of the coat was five thousand shekels of brass. And he had greaves of brass upon his legs, and a target of brass between his shoulders. And the staff of his spear was like a weaver's beam; and his spear's head weighed six hundred shekels of iron. . . .

And it came to pass, when the Philistine arose, and came and drew nigh to meet David, that David hasted, and ran toward the army to meet the Philistine. And David put his hand in his bag, and took thence a stone, and slang it, and smote the Philistine in his forehead, that the stone sunk into his forehead; and he fell upon his face to the earth.

—I Samuel 17:4–7, 48–49, King James Version

Ever since David and Goliath, people have wondered whether more couldn't be done with the sling. Perhaps army gunners wondered why they should bother with all that smelly gunpowder? Why not just whirl the shells around on a special military carousel and let go of them at the right moment? Well, that might be all right up to a point, but such a weapon would be severely limited. To attain the capabilities of modern guns, you would require twenty-first-century materials of truly heroic strength.

There is another problem: How do you aim the weapon? The weapon must grip the projectile powerfully enough to keep it from flying off, but the projectile has to be released with millisecond or microsecond precision to get accurate aiming. John Wyndham, in his doom-laden sci-fi novel, *The Day of the Triffids,* envisaged weapons that launched whirling disk-shaped blades. The weapons were needed for fighting off the Triffids, deadly walking plants, but Wyndham did not specify how the disks were to be aimed accurately.

The materials problem might well be fundamentally intractable, but the aiming problem could be avoided by eliminating the need to unhook the projectile. Arranging for the swinging arm to hit a free projectile—in a kind of military motorized golf—would be a good solution. If an elastic or nearly elastic collision could be achieved, the projectile could also be made to travel much faster than

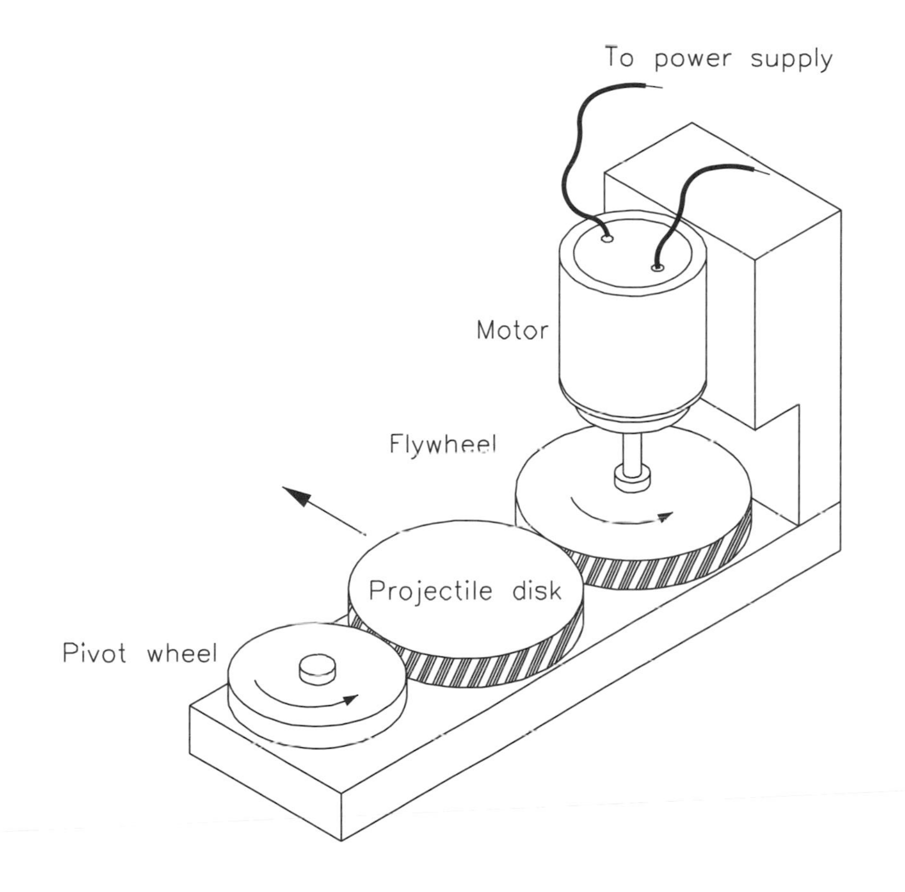

the arm while still conserving energy, although achieving the necessary elastic collision becomes much more difficult at high speed.

Storing energy in a rotating flywheel is relatively efficient compared with other ways of storing mechanical energy, such as springs. The simple spring-operated launchers used in clay-pigeon shooting are limited in exactly the same way that the trebuchets and catapults of the ancient Roman army were: the amount of energy that can be stored in and extracted from a spring is very limited relative to the weight of the spring. Full-size transport equipment such as railroad cars and even buses in Switzerland have used energy stored in a flywheel, but no one has ever made a clockwork vehicle big enough to ride around in.

With a continuously whirling arm, arranging for the projectile to fly off the arm at the right moment is difficult, and in this project we get around that problem by means of a cunning launch arrangement. Instead of the whirling arm, we use a wheel (which eliminates some air drag problems), and the projectile is not hit but gripped between the flywheel and a pivot wheel. Anyone who has had a piece of wood snatched by a high-speed woodworking machine such as a planer or circular saw will appreciate the principle. A large part of the wheel's rotational inertial energy is given to the projectile in the form of linear kinetic energy.

What You Need

- ❑ Electric motor
- ❑ Battery or other power supply
- ❑ Flywheel
- ❑ Foam plastic or plastic container tops for disk projectiles
- ❑ Baseboard
- ❑ Pivot wheel and pin

What You Do

Assemble the parts on the smooth baseboard. Use a round projectile to start with, made from circles of foam plastic or plastic container tops. The projectile should be made of a compressible material that is also lightweight; a semirigid foam plastic is ideal. Cut it just a little smaller than the gap between the flywheel and the pivot wheel. If the projectile is round, then it will be stabilized by the rotation given to it on launch. Turn on the motor and feed the projectile gently into the gap: it will be snatched away by the wheel and projected forward, almost directly along a line perpendicular to the line between the flywheel and pivot wheel.

If you use a rigid plastic disk such as a milk-carton top as a projectile, you may find it useful to put a rubber band around it so that the wheel can get a better grip on it. You could put the rubber band on the motor wheel—but be aware of the large centripetal force needed to stop the band from flying off. I calculate that even with a tiny electric motor running at 6,000 rpm and with a wheel diameter of 38 mm, the force required is equivalent to an acceleration of 600 g!

How It Works

The rotational kinetic energy built up in the flywheel over a second or two is partially transferred to the projectile in a few milliseconds. The stored energy is transferred partly in the form of rotational kinetic energy but mainly as linear kinetic energy.

THE SCIENCE AND THE MATH

The amount of energy, E, stored in a flywheel of mass m_f rotating at rotation speed ω with most of its mass in the rim at radius r is given by

$$E = \tfrac{1}{2}I\omega^2,$$

where I is the moment of inertia ($\sim m_f r^2$), and $\omega = 2\pi f$. (This is strictly true only if all the mass of the flywheel is concentrated in the rim.) So with a flywheel of mass of 3 g rotating at 6,000 rpm (f – 100 Hz), we have 0.3 J of energy stored, which is comparable to the amount of energy stored by a child's catapult. To equal the performance of medieval archers, the stored energy would need to equal the draw energy of a longbow: perhaps 300 N of force over a distance of 1 m, or approximately 300 J. By sticking to a 6,000-rpm motor and increasing the wheel mass to a more sturdy 100 g, perhaps we could store 300 J in a wheel with a radius of just 100 mm. Maybe this is what John Wyndham had in mind for the weapons that launched whirling blades to resist the onslaught of those fiendish, homicidal plants, the Triffids.

Knowing the flywheel and projectile sizes, we can calculate roughly how fast the projectile leaves the device, and we can also estimate how much energy is transferred from flywheel to projectile. The periphery of the flywheel is moving at speed ωr and has mass m_f, and we can think of the impact of the flywheel as equivalent to the impact of a mass, m_f, moving linearly with linear momentum $m_f\omega r$, on the edge of the projectile. The impact on the periphery of the projectile give it a combination of forward speed, v, and rotation. For calculation purposes, think of the projectile as being composed of two masses, each $m_p/2$. Finally, assume that the launch is an elastic collision in which energy is conserved. After launch, the flywheel will be moving at a new lower speed, $\omega' r$.

Now we write the equations for conservation of energy and momentum:

$$\tfrac{1}{2}m_f(\omega r)^2 = \tfrac{1}{2}m_f(\omega' r)^2 + \tfrac{1}{2}(m_p/2)v^2$$

and

$$m_f \omega r = m_f \omega' r + m_p v/2.$$

By substituting the expression for ω' from the second equation and solving the first equation for v, we arrive at

$$v = 2\omega r/[1 + (m_p/2m_f)]$$

and

$$\omega' = \omega\{1 - m_p/m_f[1 + (m_p/2m_f)]\}.$$

If you graph this last equation, you will find that it is nearly 1 for small values of m_p, that it is 0 at $m_p = 2m_f$, and that it becomes negative for heavier projectiles.

The special case in which the projectile has twice the mass of the flywheel results in what is known in billiards as a stun shot.* The stun shot is a collision in which the incoming mass is stopped, or stunned, and the outgoing mass proceeds at essentially the same speed as the incoming mass. In this case, 100 percent of the flywheel energy is transferred to the projectile, if there are no inelastic effects.

The other interesting special case occurs when the diameter of the flywheel is much bigger than that of the projectile. In this case, the projectile can fly out at twice the speed of the flywheel periphery:

$$v = 2\omega r.$$

However, this equation describes only the initial motion of the projectile. After the projectile experiences the initial impulse, the other half of its mass will be pulled backwards. If we assume that immediately after the launch, the center of mass of the projectile instantaneously moves with velocity $v/2$, and the periphery rotates with speed $\omega_p r_p/2$, where ω_p and r_p are the angular velocity and the radius of the projectile and where

$$\omega_p r_p = v/2$$

and

$$\omega_p = v/(2r_p),$$

then the projectile can be considered to be rotating and translating in such a way as to leave the backwards-going part of its periphery instantaneously stationary, rather like the periphery of the wheels on a wheeled vehicle.

In the case of the very massive flywheel and small projectile, the velocity of the center of mass, v_{cm}, and the angular velocity of the projectile, ω_p, are given by

$$v_{cm} = \omega r$$

and

$$\omega_p = \omega r/(2r_p).$$

Although these equations give you a good feel for the operation of the launcher, they are not accurate. You could calculate the behavior of the rotapult more accurately by deriving conservation equations for energy, angular momentum, and linear momentum without the simplifying assumptions we made.

The rotapult launcher pushes the projectile out along a defined path, which makes the device relatively accurate, but it has the disadvantage of relying on the efficient transfer of momentum to the projectile in a brief moment rather than on the gradual building of momentum. This is probably the Achilles' heel of the rotapult. If we launched a

*In the laboratory, colliding bodies—for example, air-track suspended devices—if they are of equal mass, undergo stun-shot collisions as a matter of course, and linear momentum and kinetic energy are conserved as we have calculated. In contrast, stun shots don't normally happen on the pool table. Pool balls usually possess both rotational and linear kinetic energy, and when they collide, the incoming cue ball will normally continue rolling, and the outgoing ball doesn't get all the cue ball's energy: that is, the rotational energy of the cue ball cannot be transferred in a straightforward collision. In a stun shot, the incoming cue ball stops. A player using a stun shot simulates laboratory air-track conditions by skidding the cue ball to its target ball, so that there is no rotational energy in the incoming ball.

1-kg projectile to go 1,000 m, the flywheel or the projectile or both would likely be damaged by severe shear stress. The damage is a problem in itself, but it also prevents us from assuming an elastic collision, and thus more launch energy will be lost. However, a more obvious idea, the whirling arm, avoids this problem but creates a number of other pitfalls.

What's wrong with a simple whirling arm? Imagine yourself a modern-day David lobbing a fairly slow cannon shell (say, 1 kg in mass) at Goliath, who is 1,000 m away from you and driving a tank. If you fire the shell at an initial velocity v_i m/s and at an angle of 45 degrees to vertical (to give maximum range), the vertical and horizontal components its velocity will be equal, at $v_i/\sqrt{2}$. By equating the gravitational potential energy of the shell with its kinetic energy, we can determine the height, H, to which the shell will rise from the ground as a result of the vertical component of its velocity:

$$[m_p(v_i/\sqrt{2})^2]/2 = m_p gH,$$

so

$$H = v_i^2/4g.$$

But if the path of the shell describes a parabola with an initial slope of 1, the height reached is one-quarter of the range, s, so

$$s = 4H = v_i^2/g.$$

Because $v_i = \sqrt{(sg)}$, you need a v_i approximately equal to $\sqrt{(1{,}000 \times 10)}$, or 100 m/s, to hit your Philistine opponent. If you had a "sling" with a diameter of 2 m mounted on the top of your Jeep, you would need to whirl the sling around at about 900 rpm. No problem, you say, and you pick up the phone to order your new whirly-armed Jeep. Hold it! At 900 rpm, the centripetal force needed to stop your cannon shell (or any other part of the whirling arm) from launching prematurely might present a slight problem. Centripetal force is given by

$$F = mv^2/r,$$

so if the mass of the whirling arm plus the 1-kg shell is, say, 3 kg, then

$$F = (3 \times 100 \times 100)/1 = 30{,}000 \text{ N, or 3 tons.}$$

This is not impossible. An ordinary material would be sufficient, just barely. But if the arm were made from carbon fiber with a 1-cm^2 cross section and a tensile strength of 1,000 MPa, for example, the arm could take 100,000 N (10 tons) tension. But your custom-auto-parts dealer might have a problem with this. And if you wanted an longer range, say, 10 km, then your poor whirling arm would need to be up to resisting, with a good safety factor, 300,000 N (30 tons)! Try materials books such as Willam Smith's *Principles of Materials Science and Engineering*, and you might find materials strong enough to be safe at these heroic speeds.

There are other problems with the centrifugal launcher, of course. With a gun, you have a parallel-sided barrel, and you can be pretty sure that the projectile you launch will head off in the direction in which you point the gun, at least initially. With a centrifugal launcher, you get no such assurance. Taking our 900-rpm Goliath-fighting example above, a 1-m error in the idealized target zone requires a pointing error of just 1 in 1,000 at 1,000 m and 1 in 10,000 at 10 km. Spinning at 900 rpm, the timing needs to be accurate to within $1/(2\pi 15)$ ms to get a 1 in 1,000 error—you have to launch at the right moment, plus or minus 11 μs. This is no problem for an electronically controlled launcher—you could measure the time of launch to within nanoseconds—but for a mechanical system, such accuracy presents a big problem. And with a 10-km-range launcher rotating at 9,000 rpm, you would need to be one hundred times better again at judging the moment of lift-off—just 0.11 μs!

Although lateral accuracy is a problem with the rotapult, the device could offer a very low range error. The range error—the error in the distance a projectile flies—of explosive ordnance is considerable, since it depends upon the uniformity of the explosive chemicals, the state of the barrel, the temperature, and so forth. A rotapult rotating at around 900 rpm, known to 0.1%, will have a range error of only 0.2%, just 2 m at 1,000 m range. Also, the errors in aiming in the plane of rotation should be small.

And Finally . . . Two-Flywheel Rotapults

You could try using two motors and two flywheels, which obviates the need for the pivot wheel and allows you to control the spin. I found that this worked fine, at least for some combinations of motor speeds. With the two motors at the same speed, the projectile spin could be minimized. With one motor rotating backwards, but slowly, the projectile would perhaps be spun to a higher speed at launch. Would such an arrangement allow you to produce predictable rotation speed and perhaps use aerodynamic rotating disks, something like high-speed Frisbees?

Using projectile accelerators in series might also be possible, particularly if the spin speed were low. A set of wheels in which each wheel pair had a greater speed than the preceding pair would pass the projectile on, each giving an increment of speed and achieving more than a single wheel could.

REFERENCES

Smith, William F. *Principles of Materials Science and Engineering.* New York: McGraw-Hill, 1986.

Wyndham, John. *The Day of the Triffids.* 1951. Reprint; London: Ballantine Books, 1986.

Exotic Amplification

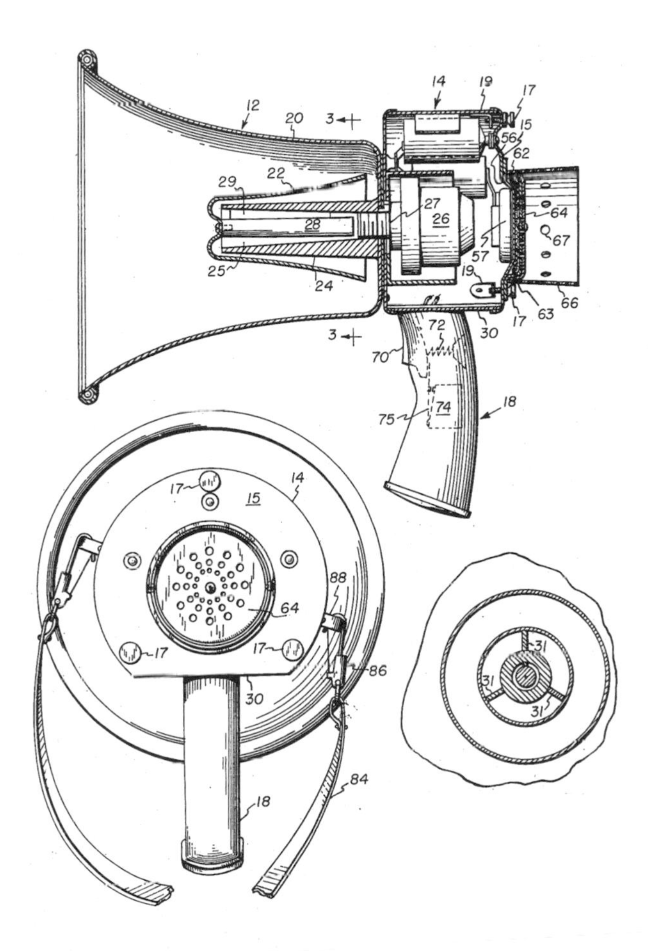

Who made this circuit up for you anyway? . . . You've got your negative feedback coupled in with your push-pull output; take that across your red-head pickup to your tweeter, and if you're modding more than eight you're going to get wow on your top—try to bring that down through your rumble filter to your woofer and what have you got? Flutter on your bottom!

—Michael Flanders and Donald Swann,
"Song of Reproduction"

Although the amplifier is arguably the most fundamental invention of instrumentation or process-control engineering—it allows a tiny "signal" input to be converted into a larger "actuator" output—the concept is a surprisingly recent one. Even the word *amplifier* seems to have referred in the past not to a device but rather to a person. The former meaning occurs in the language only in the twentieth century. (The same is true for the word *computer,* which, until the late 1940s, also referred to a person rather than a machine.)

Until the nineteenth century, science was done either by direct observation or by means of measuring instruments whose function was to transform rather than amplify. Observers first had to utilize the exquisite sensitivity of their senses to "amplify" effects transduced by instruments and then had to record the results with paper and pencil. It is only in the past century that we have had the ability to truly amplify effects, an ability that makes the observer's task easier and even allows automatic recording devices to take over some tasks.

Why was the amplifier so long in coming? One answer might be that we already have biological amplifiers, our senses, built into our nervous systems and, furthermore, that these biological amplifiers already possess the greatest possible degree of amplification. This is almost true. If the human ear were more sensitive, it could detect thermal noise, the random jiggling of air molecules. Under ideal circumstances, our eyes can detect just a couple of photons of light. Another answer might be that amplifiers introduce noise, which might nullify any advantage that they confer. Certainly many ordinary amplifying devices generate some noise, although most electronic-amplifier designers would claim that the added noise is small relative to the noise in the signal.

There was, then, no great clamor for amplifiers in Victorian times. Only the few scientists working with signals outside the range of human senses, signals such as infrared or ultraviolet light, might have wished for amplifiers. These scientists could have used them to increase the signals from crude transducers such as thermopiles or early photoelectric devices. Even here, however, the ingenious researcher could often avoid the need for an amplifier. Ultraviolet light, for example, could be converted to visible light with a fluorescent material, and the visible light could then be efficiently detected by the human eye or with a photographic emulsion.

The imperatives that drove amplifier development seem to have come from commerce rather than from science. The two great inventions of electronic communications, telephone and radio, were the driving force. Telephone systems needed signal boosters on the longest (and hence the most commercially valuable) cables, and without amplifiers, radio was functional only at the crudest level.

18 Transformer Transistors

> In spite of [the correctness of the cycle] the tests were a failure, since the first machine never ran independently by itself. I returned to my home, at that time Berlin, very depressed, and made drawings for a complete rebuilding of the engine, which lasted for five months.
>
> —Rudolf Diesel, letter to his wife

Even in technological endeavors, where quantitative measurement is preeminent, there often is no hard and fast measure of success or failure. However, for an engine there is an obvious figure of merit: the engine must output more mechanical energy than it takes in. Engineer Rudolf Diesel's first engine, which failed to run independently, was a flop on these grounds. Although Diesel's engine certainly converted some of the fuel energy into mechanical energy, it also absorbed mechanical energy from the drive belt used to turn it for starting purposes. But over a complete rotation, the engine put back less energy than it absorbed. There is a similarly obvious figure of merit for an amplifier: it must have a gain factor (output divided by input) of more than unity—that is, more signal must come out of the output than goes in to the input. My first attempts at a truly simple magnetic amplifier made me sympathize with Diesel.

Amplifiers based on magnetic principles, called *magamps,* were once used for activating powerful servo motors with small electric inputs. Magamps are a little known but still viable technology. They take advantage of the nonlinearity of magnetic materials. Magamps could have been made in the Victorian era; the Victorians had access to suitable magnetic alloys and copper wire, and, at least by the end of the nineteenth century, they had alternating current (AC) suitable for powering such amplifiers. However, it was not until much later that these devices blossomed.

From the 1940s to the early 1960s, much effort was devoted to magamps, and books on them were published by the dozen (see the references for a couple of these). Hundreds of companies supplied myriad different types of magamps, although production was relatively small. However, almost all this activity died out as superior semiconductor transistor technology became available in the 1960s. Magnetic principles are still occasionally used today, though. The Royal Philips Electronics company of the Netherlands, for one, still markets a line of magnetic amplifying devices. These are mainly used in high-efficiency power supplies rather than as amplifiers, however; they are useful for the conversion of AC to DC using audio or low-radio-frequency techniques.

Transformers and the Magamp Principle

Here's how a transformer works. A coil, the primary, wrapped around a core and connected to an alternating current produces a powerful alternating magnetic field inside the core material, typically a magnetic iron alloy. When another coil, the secondary, is placed over the core, the alternating field produced by the primary will induce an AC voltage in the secondary. The strength of the AC magnetic field—which is proportional to the input current and the magnetic permeability of the core—determines the output voltage in the secondary coil. The output voltage induced in the secondary will be similar to the input voltage if the secondary coil has the same number of turns as the primary coil. However, if the secondary coil has more or fewer turns than the primary, then the output voltage will be higher or lower, respectively, than the input. In this way, electric current can be transformed between different voltages, hence the name *transformer.*

The transformer principle can be used to create an amplifying assembly. The basic principle here is that a DC magnetic field (a bias field) through an easily saturated magnetic material can be used to control the AC in a circuit that uses the magnetic material as a transformer core. A DC magnetic field reduces the magnetic permeability of the iron core in the transformer, making the core less magnetic. The reduced magnetism of the core makes the transformer less efficient, and it will transmit less current at lower voltage to the output. So increasing the DC field decreases the AC output, which is an "inverting" amplifier action. The DC field can be produced by another winding on the magnetic core of the transformer.

There is a problem with the simplest scheme one might devise here: the AC from the transformer will be imposed on the DC source, the batteries, in our case. Therefore, we use two transformers with opposite connections, and these have equal and opposite AC voltages, which results in a canceling out of the AC in the DC lines. We can use the AC output of the amplifier to drive any load that will function with an AC drive, or we can drive a DC load by adding a rectifier. However, it is perhaps simpler and more elegant to choose a load, such as a filament lamp, that will intrinsically function just as well on AC drive as it does on DC drive.

What You Need

- ❑ At least 3 small mains transformers of similar type (e.g., two 2 × 6-V output transformers and 1 12-V output transformer)
- ❑ Wire
- ❑ Filament lamps or other "loads" for output (e.g., 12-V, 5-W auto lamp)
- ❑ Batteries and battery holders for input (e.g., 1.2-V NiCd batteries)

What You Do: A Transformer Transistor That Only Just Works

My first magamp, which only just worked, used four transformers: two 12-V transformers provided the motive current to two 6-V transformers via their 6-V secondary coils. The lamp was connected to the other 6-V secondaries. The 55-V alternating currents on each of the control primaries were wired to oppose each other, to prevent AC from appearing on the DC control power supply. I then wired a string of three PP3 NiCd batteries (8.4 V each) and a few AA 1.2-V NiCd cells to provide the control current to the control primaries. With mains power applied, the 12-V, 5-W filament lamp could be powered up to about 6.5 V, whereupon it glowed fairly yellow. With 40 mA of current at 30 V through the DC control coils, I could reduce the lamp voltage to about 3.5 V, whereupon it was dim and red.

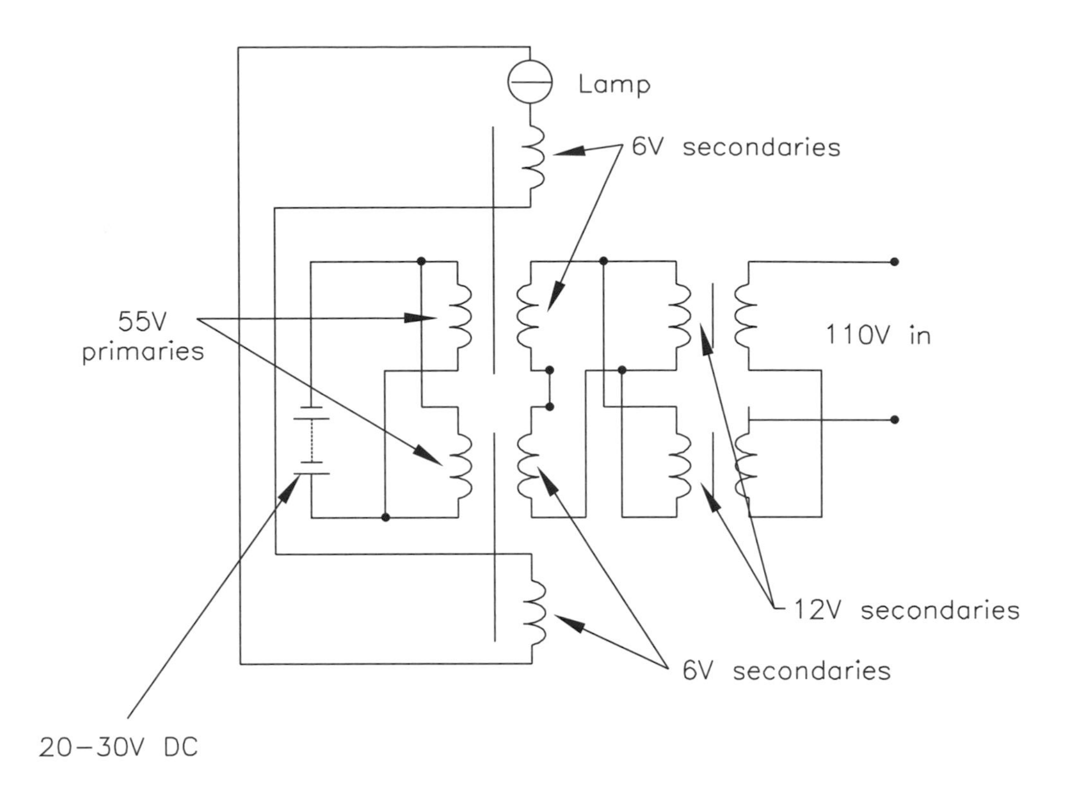

This achievement, I should explain, occurred after quite a bit of preliminary experimentation. I was pleased to have made the whole complicated rat's nest of wires work even slightly. I then calculated the gain factor of the magamp. Did I have a magamp, or was it just a mag? Because 6.5 V at 300 mA equals 1,950 mW, and 3.5 V at 200 mA equals 700 mW, the modulation of the output power was approximately 1,250 mW. But because the input power at 30 V and 40 mA equals 1,200 mW, the gain factor of my simple magamp assembly was, within the limits of experimental error, just unity. I tried other lamps—2 W instead of 5 W—and I tried even smaller lamps. But the gain factor remained stubbornly at 1 or lower.

What You Do: A Transformer Transistor That Really Works

A design variation yielded an improved transformer that really worked. I switched to a slightly different configuration, in which the lamp was powered by the AC, but via a couple of transformer primaries, which under AC conditions act as large inductors, blocking the path of the AC and keeping the lamp pretty much

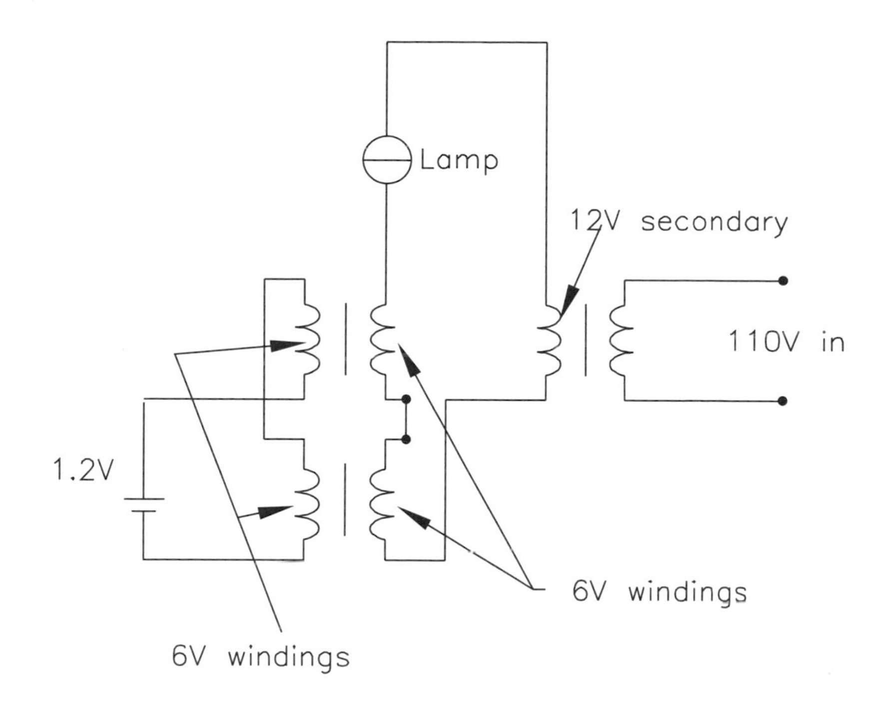

switched off. However, when DC is applied, the inductors become more or less like two pieces of plain wire, and they transmit much higher current. The result is an amplifier with a positive (noninverting) action and with increased AC out when increased DC is applied.

Note that a 6-V secondary from each control transformer is connected in series with the wire to the lamp. The DC control current is supplied through the other 6-V secondary on each control transformer, but is wired so that the AC voltages from each of these secondaries oppose each other (again, so that the battery does not see any AC voltage). With a control current of 400 mA at 1.2 V, I found that I could vary the output from a 2-W bulb from 5.5 V to 11.5 V. Because 11.5 V at 180 mA equals 2,070 mW, and 5.5 V at 100 mA equals 550 mW, the modulation of the output power was approximately 1,520 mW. But the input power was 480 mW (1.2 V at 400 mA), so the gain factor of this improved magamp assembly was approximately 3.

Possible Improvements: Square Your Loops

We are trying to get the control transformers in the working magamp to switch between being high-value inductors with no DC current applied to being very low value inductors with current applied. We would get a much bigger gain factor for the assembly if we could find a magnetic material that switched from one state to the other with smaller DC current applied. The problem is with the magnetic material of which transformers are typically made.

The magnetic material in standard transformers is designed to respond gradually to an applied magnetic field: small changes in input current don't lead to large changes in magnetization. This is also true of the output current. With special magnetic materials, such as certain "magnetic glasses" and ferrites, however, small changes in the applied DC field can lead to large changes in magnetization. If that large change "saturates" the magnetic material—that is, makes it unable to increase its magnetization—then changes in magnetic field don't give rise to changes in magnetization. Therefore, an applied AC field won't give rise to an AC voltage on the coils. The magnetized material, once saturated by the DC field, will no longer respond to an AC field. These special magnetic materials—called *square loop materials*—can be "switched," in effect, from being magnetic to being almost inert, by the application of small magnetic fields.

Possible Improvements: Positive-Feedback Magnetic Amplifier

You can increase the amplification factor of any amplifier by feeding the output back to the input. Perhaps we could effectively achieve this in our magamp by

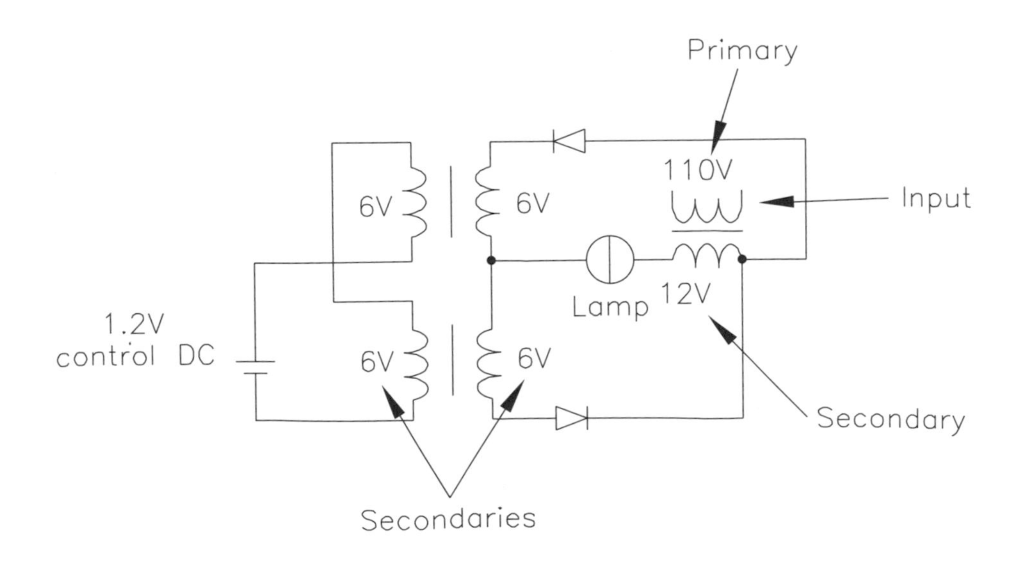

constructing our circuit with diodes. When a DC bias is applied to the two control coils of this new circuit, the output circuit also applies a DC bias, which further boosts the effective gain. But will this circuit really increase the gain factor of the magamp?

THE SCIENCE AND THE MATH

The Fundamentals of Magnetization

When two current-carrying wires are set close to each other, the current flowing through one of the wires will exert a force dF on the other, as described by Monsieur Ampère:

$$dF = [\mu_0/(4\pi R^3)]I_i I_j\, d\mathbf{S}_i \times d\mathbf{S}_j \times \mathbf{R},$$

where μ_0 is the permeability of free space, dS_i and dS_j are the lengths of the two wires, R is the distance between the wires, and I_i and I_j are the currents flowing through the wires. (The boldface letters denote vector quantities, and the × symbols denote vector cross product multiplication.)

You can split the equation into two simpler parts with the aid of mathematical substitution: think of the force on one of the wires as arising from the element of magnetic field, $d\mathbf{B}$, induced by the other wire, and you will get all the same numerical answers from the equations that result. The induced field is proportional to the field, $d\mathbf{H}$, produced by the current-carrying wire ($\mathbf{H}$ is technically called the magnetic field strength and has units of A/m):

$$d\mathbf{B} = \mu_0 d\mathbf{H} = [\mu_0/(4\pi R^3)]I_j d\mathbf{S}_j \times \mathbf{R}.$$

The element of force, dF, that a B field exerts on current I flowing in a wire element dS long is proportional to the value of $\mathbf{B}$:

$$dF = I(d\mathbf{S} \times \mathbf{B}).$$

The B field is what most people understand to be a magnetic field, although technically it is known as the *magnetic flux density.* You can picture the field as a set of three-dimensional lines running through space, the lines being curved and locally parallel to one another and having higher or lower density depending on the field strength.

Now picture a B field existing in a zone containing a magnetizable medium. We can think of such a medium as consisting of an array of tiny dipole magnets that can pivot around on weak hair springs. The dipoles will pivot until they face in the same direction as the B field, aligning perfectly in a strong B field, less perfectly in a weak B field. The alignment of the dipoles, each of which exert its own small magnetic effect, boosts the B field over and above the amount produced by $\mathbf{H}$ by an amount $\mathbf{M}$, which is called *magnetization.* The overall effect can be captured by the following equation:

$$\mathbf{B} = \mathbf{B}_{app} + \mathbf{B}_{dip},$$

where $\mathbf{B}_{app}$ is the applied magnetic field and $\mathbf{B}_{dip}$ is the field due to the dipoles. The following equations apply:

$$\mathbf{B}_{app} = \mu_0 \mathbf{H},$$

$$\mathbf{B}_{dip} = \mu_0 \mathbf{M},$$

and

$$\mathbf{B} = \mu_0(\mathbf{H} + \mathbf{M}) = \mu_r(\mathbf{H}),$$

where μ_r is the relative permeability of the medium.

If the springs of our imaginary tiny dipole magnets are weak, a large B field results from a small applied field, since the applied field lines up the dipoles, which then reinforce the field. In this case, we say that the relative permeability, μ_r, is high; a material of low permeability has stronger springs.

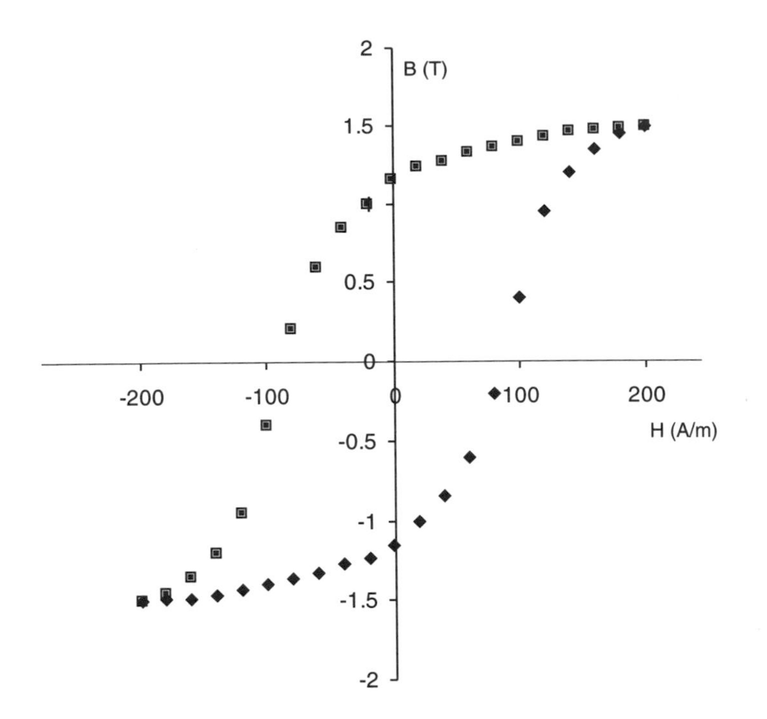

The dipoles of a "soft" magnetic material rotate freely on good bearings, and the dipoles of a "hard" magnetic material can be thought of as rotating on pivots that are somewhat rusty and that tend to stick unless a large force is applied. After the applied field, $\mu_0 H$, is removed from a hard magnetic material, the dipoles remain lined up, and the magnetic material is left as a "permanent" magnet. Saturation of a magnetic material occurs when all the dipoles have lined up almost perfectly at a certain value of applied field, $\mu_0 H$, and more applied field will be boosted only slightly by the rotation of the pivoted dipoles. You can visualize this phenomenon by graphing ***B*** versus ***H*** on a magnetization loop diagram for the material.

Square Loops

The basic problem with the transformers I tried was that the ***B/H*** magnetization loop diagram for the material, a soft iron, was not square enough. A way of viewing the advantage of square loops is to think of the inductance value as determined by the relative permeability, μ_r. This analysis nicely illustrates the idea of the magnetic amplifier: by changing the mean flux density, ***B***, we change μ_r and thus change the inductance, *L*, of the blocking inductors and control the transmitted power. (This analysis would be precise for small AC signals, but because our transformer experiences large AC signals, μ_r varies too much over a cycle to be regarded as a constant.)

Feedback Gain Boosting

The fact that the amplification factor of an amplifier increases when the output is fed back into the input can easily be understood from the relevant equations. Say that the voltage amplifier starts with basic gain, *G*, and that a fraction, α, of the output is fed back. Then the overall effective gain is given by

$$G_{\text{eff}} = G[1 + G\alpha + (G\alpha)^2 + (G\alpha)^3 + \ldots],$$

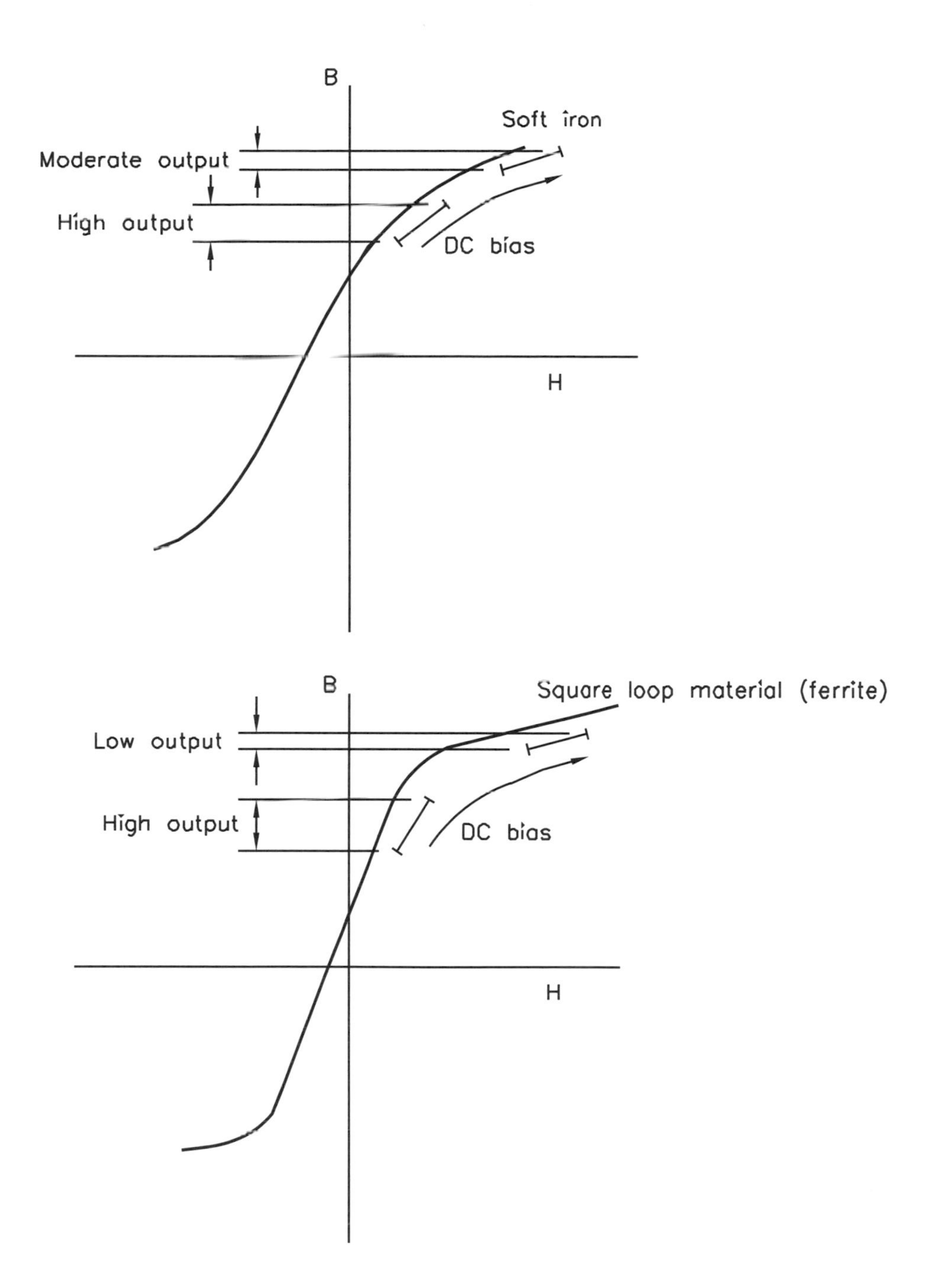
B
Soft iron
Moderate output
High output
DC bias
H
B
Square loop material (ferrite)
Low output
High output
DC bias
H

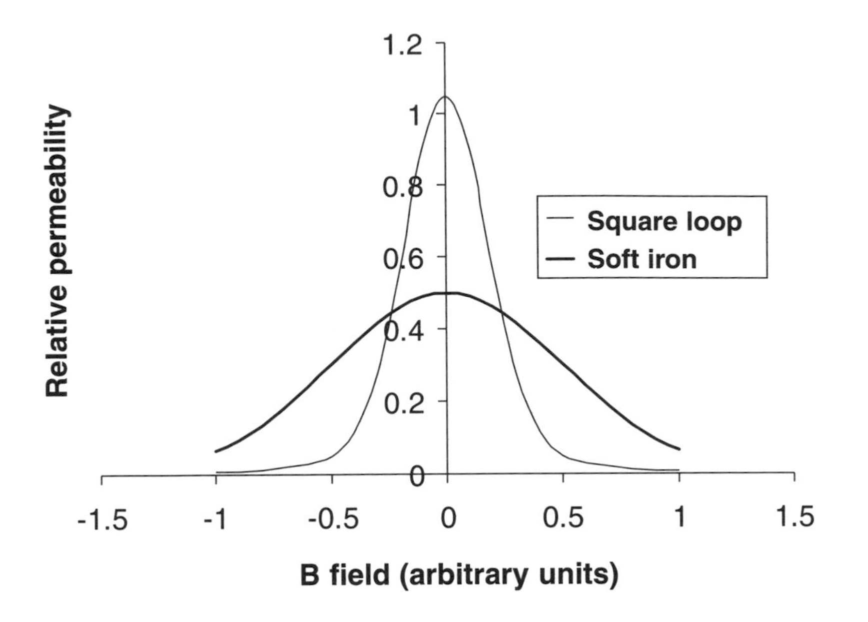

which one can derive simply by following an applied voltage around the loop. The summation of this geometric series is

$$G_{eff} = G/(1 - G\alpha).$$

G_{eff} diverges—becomes infinite—when $G\alpha = 1$. In this situation, the amplifier amplifies any arbitrarily small noise signal up to a large value, and, in fact, the amplifier will often become an oscillator.

The feedback concept is powerful one. By choosing the correct amount of feedback, you can choose the gain of the overall amplifier system—as large or as small as you like—independently of the intrinsic gain of the physical device. G_{eff} values can even be less than 1 if a negative α is chosen. (You might think that choosing a G_{eff} value less than 1 would be a bad idea, but negative feedback is actually a useful idea for amplifiers that have a large G value, because it improves other characteristics of the amplifier, such as its sensitivity to changes in component values with time or temperature.)*

*The use of negative feedback to improve an amplifier's frequency response and decrease its sensitivity to component characteristic changes was pioneered by Harold S. Black, an electronics engineer at Bell Labs who produced a number of important innovations in the 1920s. Black had been puzzling over the problem of how to reduce distortion in long telephone lines for several years. One day, he had a flash of inspiration on the way to work and immediately wrote the idea down on a piece of paper he had at hand—a page of the *New York Times* (see *IEEE Spectrum,* December 1977).

And Finally . . . Multistage Transformer Transistors

You could try some simple modifications to the basic transformer transistor. For example, you could cut slots in the iron laminations of the core. With slots cut, the magnetic flux will be forced to pass through a narrower piece of iron, which will saturate more easily. Saturation will result in an increase in the standing current that flows with zero DC control current. However, when the DC is applied to activate the device, the slots should allow smaller currents to control the device or should allow the same applied current to give a greater output. In this way, cutting the slots could give a higher gain factor. I tried this with the small mains transformers and showed how the lower inductance allowed more current to flow. However, the current that flowed when the transformer transistor was activated was only proportionally greater, so the overall gain factor was similar.

Can we feed the output from one transformer transistor to another, thus further boosting amplification? A multistage amplifier using transformer transistors must use diodes to convert the output AC into DC for controlling the next device in the cascade.* Using a full-wave bridge rectifier will allow you to make the most of the power from the first stage. The diodes have a forward voltage loss—0.6 V is lost when silicon diodes conduct, and with the bridge rectifier there are two diodes in series in the circuit—so the output voltage is reduced by 1.2 V. This voltage drop may actually help a little with the operation of the device overall, since the voltage drop reduces the amount of quiescent power. Instead of 3 V or so coming out of the first stage in the absence of an input signal, we have only 1.8 V, which means that less input power is wasted.

REFERENCES

Black, Harold S. "Inventing the Negative Feedback Amplifier." *IEEE Spectrum,* December 1977, 55–60.

Platt, Sidney. *Magnetic Amplifiers: Theory and Application.* Englewood Cliffs, N.J.: Prentice-Hall, 1958.

Storm, Herbert. *Magnetic Amplifiers.* New York: John Wiley, 1955.

Thomas, Donald E. *Diesel.* Tuscaloosa: University of Alabama Press, 1987.

*Using a diode might seem like cheating if we are trying to construct an amplifier that a Victorian could have built. A Victorian could not have included a silicon diode. However, it is possible to create a diode with the simple technology of Victorian times using selenium or cuprous oxide (Cu_2O, copper(I) oxide). Some of the equipment that I worked on in the 1970s occasionally still had selenium or Cu_2O rectifiers.

19 Electrolystor Amplifier

Anode
To understand, as in "anode that was right."
Cathode
A poem to a girl named Cathy.
Cathodic Reaction
The kiss or slap following cathode.

—From "The Corrosion Dictionary,"
Matter Realisations Web site

The electrolystor amplifier—an amplifier that uses only wires and salt—could have been the nineteenth-century transistor. Victorian inventors were well aware of the possibilities that a liquid conductor of electricity offered. They often devised inventions that used electrolytes. For example, the early telephone pioneers, such as Alexander Graham Bell and Thomas Edison, worked on electrolyte-based microphones before finding more tractable technology.* Here we use what might be thought of as a problematic side effect of AC conduction, electrolysis, to control that conduction.

*Inventors were also well aware of the potential of electrochemistry and were quick to develop applications such as electroplating. Electroplating developed rapidly once its superiority to other plating methods became clear. Other methods were often clumsy by comparison: the heating of mercury amalgam to drive off the mercury as vapor and release the alloyed metal as a permanent coating on a metal article was a particularly clumsy and hideously dangerous example of one of these earlier processes.

What You Need

- ❑ AC mains transformer (to take 110- or 240-V AC at 50 or 60 Hz to 9- or 12-V AC)
- ❑ 1-mm solid copper wire for the electrodes

- ❑ Batteries (2–3 V)
- ❑ Auto electric lamp (e.g., 12-V, 21-W type)
- ❑ Coffee cup (A clear glass cup will allow you to see what is going on.)
- ❑ Connecting wires
- ❑ Ammeter (An AC- and DC-measuring multimeter, 0–10 A, is ideal.)
- ❑ Resistance wire (A few ohms resistance is needed.)
- ❑ Sodium bicarbonate ($NaHCO_3$, baking soda)

What You Do

The controlling circuit is simply the battery (I used two NiCd cells, giving a total of 2.4 V) connected to the control electrode and the nearest output circuit electrode. The output circuit is the auto electric lamp connected to the AC transformer via the two output electrodes. Adjust the input (controlling) current using the resistance wire until you have about 0.1 A flowing. When you pulse the input current on and off, you should see a small but steady stream of tiny bubbles

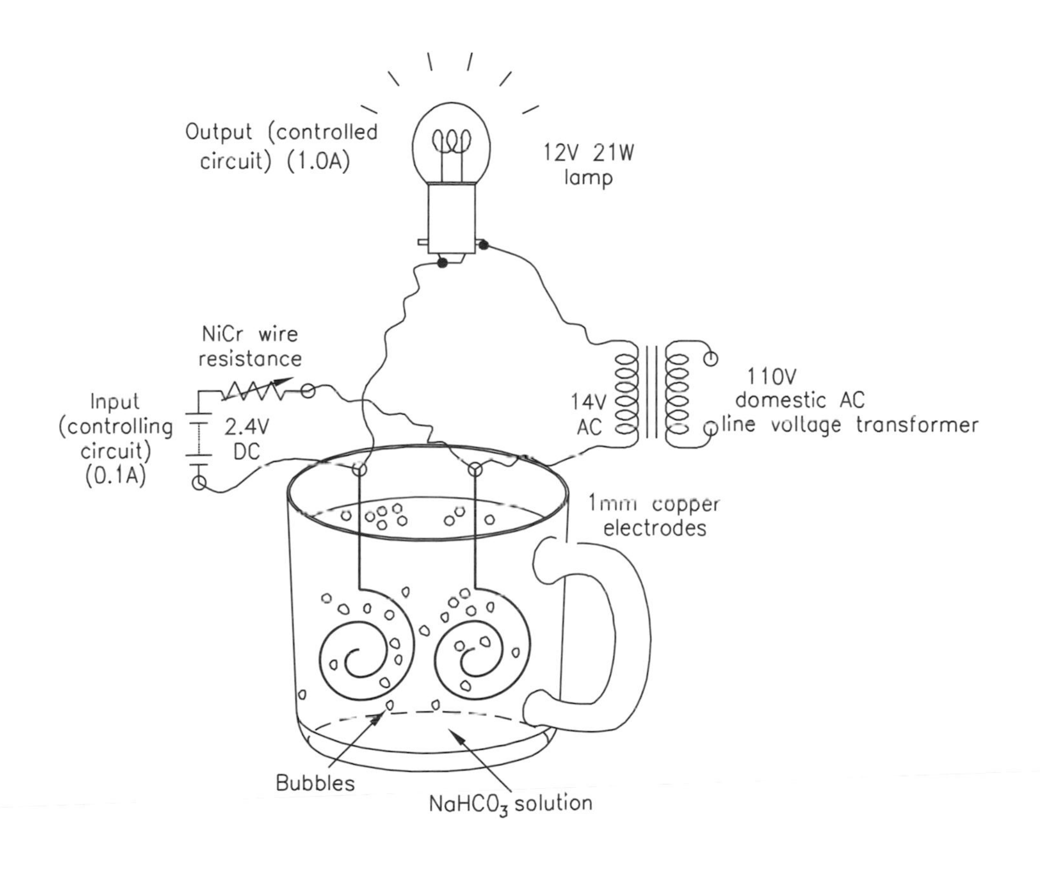

issuing from the electrodes. The bubbles of gas are insulators and will block an AC current flowing in the electrolyte. As in the case of the magamps of the previous chapter, we can thus use a DC input to control an AC output.

Now connect up the output circuit. You should find that the lamp lights up quite brightly, although not as brightly as when used on a full 12-V DC from a battery, since the current flowing will be only 1 A instead of the 1.8 A it would normally be. You should find that a current of 100 mA (0.1 A) or so flowing to the control electrode will reduce the lamp current from 1 A or so down to 0.5 A. In fact, because of the lamp's nonlinear response to current, you will find that you can pretty much turn the lamp from on to off, and to all stations in between, by programming the right amount of controlling current.

How It Works

The sodium bicarbonate solution is a conductor because it contains charged sodium (Na^+) and bicarbonate (HCO_3^-) ions, which can carry current. AC current flows fairly freely through the solution unless something blocks the path through the solution between the electrodes. An effective resistance to AC of a few ohms is typical for electrodes of a few square centimeters in area. The thin layer of tiny bubbles you see when you apply DC is hydrogen or oxygen gas. These bubbles arise from the electrolytic splitting of water:

$$H_2O \rightarrow O_2 + 2\,H_2$$

Gas is a perfect insulator for low-voltage electric current, and although the amount of gas is very small, the bubbles block most of the route for current flow through the bicarbonate solution. They can't block all the current flow, because they don't fit together perfectly—there are always gaps between them. However, with an effective resistance of 10 ohms or more with the bubbles partially blocking the way, the auto lamp will be dimmed substantially.

WARNING

Be careful with the domestic line voltage side of the circuit: it is best to use a transformer already wired up to a plug on the high-voltage side so that neither your fingers, nor anyone else's, will touch anything with more than 12 V.

THE SCIENCE AND THE MATH

The DC applied to the electrodes in the sodium bicarbonate solution produces mainly hydrogen and oxygen. The equations for the negative and positive electrodes are

$$4H^+ + 4e^- \rightarrow H_2$$

and

$$4OH^- - 4e^- \rightarrow O_2 + 2H_2O.$$

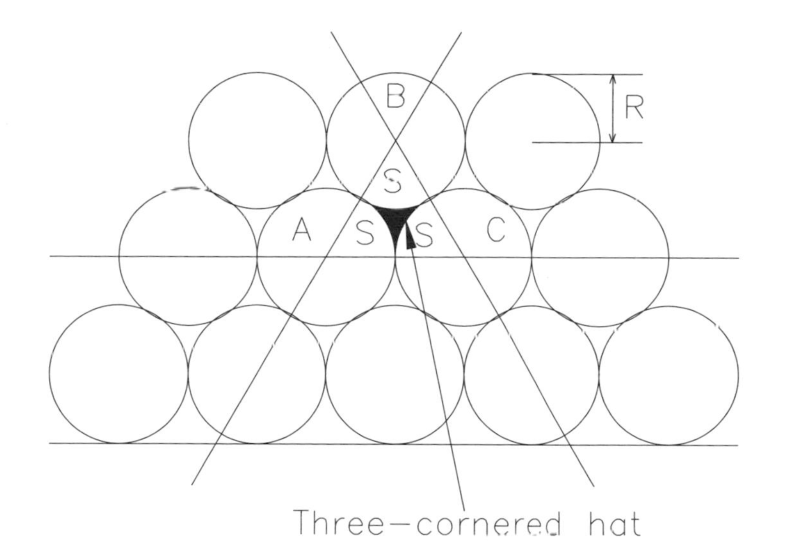

The sodium ion, Na^+, is not discharged in an aqueous solution. A little carbon dioxide may also be generated at the positive electrode:

$$HCO_3^- - e^- \rightarrow CO_2 + OH^-.$$

The bubbles evolve on the electrode surface in profusion in very small sizes at first, so that the surface looks as if it has a paper-thin coating of silvery fur. As these bubbles grow, or as small ones amalgamate with one another, bubbles of a millimeter or so in diameter form, shaped roughly like sections of a sphere; eventually a few bubbles reluctantly pull away and float up. As the bubbles continue to amalgamate, the surface of the electrode takes on a blistered appearance, and bubbles at last pull off the surface freely.

Think of the bubbles that form on the electrodes as hemispheres of radius R. Each bubble will insulate an area πR^2 of electrode, which removes a column, diameter R, of potentially conductive liquid. In fact, if all the bubbles crowd together to the maximum extent possible before they touch and then amalgamate, they will a form something like a hexagonal close-packed structure in two dimensions. When the bubbles are in this configuration, an area of the electrode is still functional. This contact area falls between the packed circles and is shaped like a three-cornered hat. We can determine the area of this shape in the following manner. The area, A_t, of triangle ABC is given by

$$A_t = bh/2 = [(2R)(2R \cos 30)]/2 = 2R^2\left(\sqrt{(3)/2}\right)$$
$$= 1.732R^2,$$

where b and h are the triangle's base and height, and R is radius of the bubbles. The area of the three sectors S, A_s, is

$$3(\pi R^2)(60/360) = \pi R^2/2 = 1.571R^2.$$

Therefore, the contact area, A_c, is

$$A_s - A_t = 0.161R^2.$$

The contact area makes up only 9.3 percent of the original surface area, so a significant drop in conductivity will occur as the surface fills up with bubbles; the drop may be as big as a factor of ten.

Looking at the output resistance versus the input current is perhaps the most obvious way of characterizing the device. However, this relationship doesn't take into account either the time required for the accumulation of bubbles or the fact that the bubbles eventually amalgamate and float to the surface of the solution. The bubbles don't appear immediately after DC is applied; they take time to build up. The loss rate of the bubbles depends not only on their rate of accumulation but also on the orientation of the electrode surface (I used approximately vertical surfaces).

We can use the following analysis to examine the changes in conductance. Suppose that the AC conductance, K, is given by

$$K \sim gA(1 - hx/A),$$

where x is the volume of gas accumulated on the surface (area A), and g and h are constants. In the absence of gas bubbles that inhibit current, the conductance (I/V) will be gA, so g is a constant having to do with the conductivity of the solution used. The constant h has the dimension of 1/length and is inversely proportional to the radius of the bubbles formed. If x becomes so large that it approaches A/h, then the conductivity will be modified significantly.

But x is given by summing the gas volume added by DC current I to the initial gas charge and then subtracting the gas bubble volume lost, L:

$$x = x_0 + \left(S\int I\,dt - \int L\,dt\right),$$

where x_0 is the starting value of x and S is a constant. The gas bubble loss rate, L, might be proportional to x itself, say, $L = \lambda x$, where λ is a constant; and we can consider a first case in which current I is constant:

$$x = (1 - \exp(-\lambda t))(SI/\lambda).$$

In this case, the gas builds up fast at first, then more slowly as the loss rate increases. The gas accumulates exponentially until it is leaving the electrode as

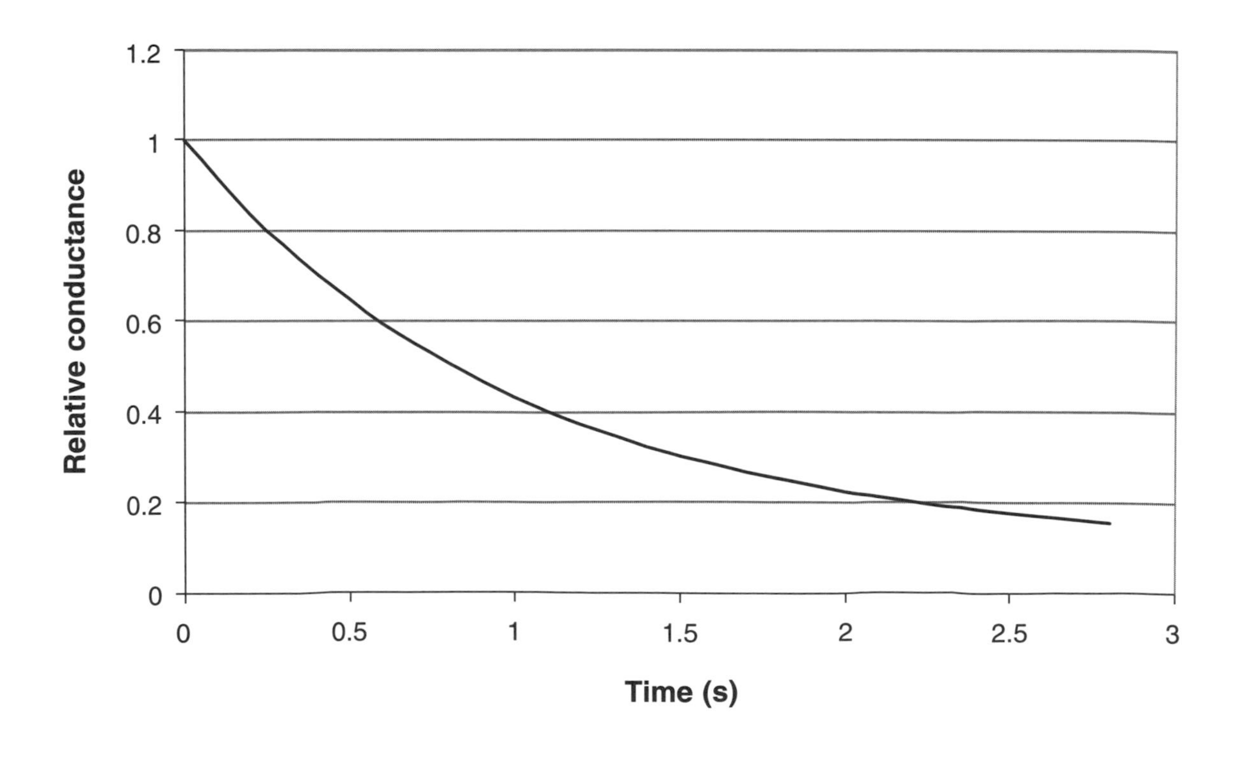

fast as it is being generated, and the volume of gas on the electrodes takes on an asymptotic value, given approximately by I/L. Thus we expect to see an exponential variation of the conductance with time,

$$K = gA[1 - (hSI/\lambda A)(1 - \exp(-\lambda t))],$$

and eventually an equilibrium conductance given by

$$K = gA[1 - (hSI/\lambda A)].$$

The AC conductance will tend asymptotically to the steady state value when the rate of bubble loss is exactly balanced by the rate of bubble creation by the DC. Increasing the DC will decrease the conductance, K, and decreasing the DC will increase K. But if the bubble loss rate, L, is increased, the conductance will decrease. Increasing the electrode area will increase the conductance, since the basic conductance factor is increased; and on top of that, the proportion of the electrode masked by bubbles will decrease.

Why Filament Lamps Are Nonlinear Resistors

Ordinary lamps, the kind with a coiled tungsten filament that glows white-hot inside a glass bulb, are particularly suitable for use with the electrolystor because they have a very nonlinear response to current. If you double the voltage to a filament lamp, you don't get twice as much current through it: its resistance will change, and not by a small amount. We can understand this nonlinearity by considering the fundamentals of black-body thermal radiation. A hot body emits heat, and if it gets very hot, it will also emit light. The total power radiated by a hot body, P, is given by an a relatively simple equation sometimes known as the Stefan-Boltzmann law:

$$P = e\sigma T^4,$$

where e is a constant, σ is the Stefan-Boltzmann constant, and T is the absolute temperature. The Stefan-Boltzmann law holds only for bodies that are "gray," meaning those that, owing to their molecular or atomic make-up, emit all colors of light equally. The constant e, which you might call "grayness," is equal to 1 for a black body—that is, one that when cool absorbs all incident light—and is equal to 0 for a white body, which reflects all light.

The radiated power, P, must be equivalent at equilibrium to the power given to the filament, VI, so

$$VI = \sigma T^4$$

or

$$T = (VI/\sigma)^{0.25}.$$

The temperature of the filament will increase only very slowly with increasing current applied. (We're neglecting, for the moment, the change in resistance of the tungsten filament, which, although large, changes the situation quantitatively rather than qualitatively.)

The spectral intensity, S, of a black body at temperature T is given by

$$S = K_0/[(\exp(ch/(\lambda kT) - 1)\lambda^5],$$

where K_0 is a constant, k is the Boltzmann (kinetic theory) constant, λ is the wavelength, h is Planck's constant, and c is the speed of light. The amount of visible light we perceive is only that portion of radiation with wavelengths between 400 nm (blue) and 700 nm (red). This amount varies sharply with temperature, due to the spectral intensity law, and thus also varies sharply with current applied. The maximum spectral intensity occurs at wavelength λ_{max}, which varies with T:

$$\lambda_{max} = a/T,$$

where a is a constant. In fact the maximum is at 2,000 nm or so for a hot-filament electric lamp, and the visible light we see is only the tail of the

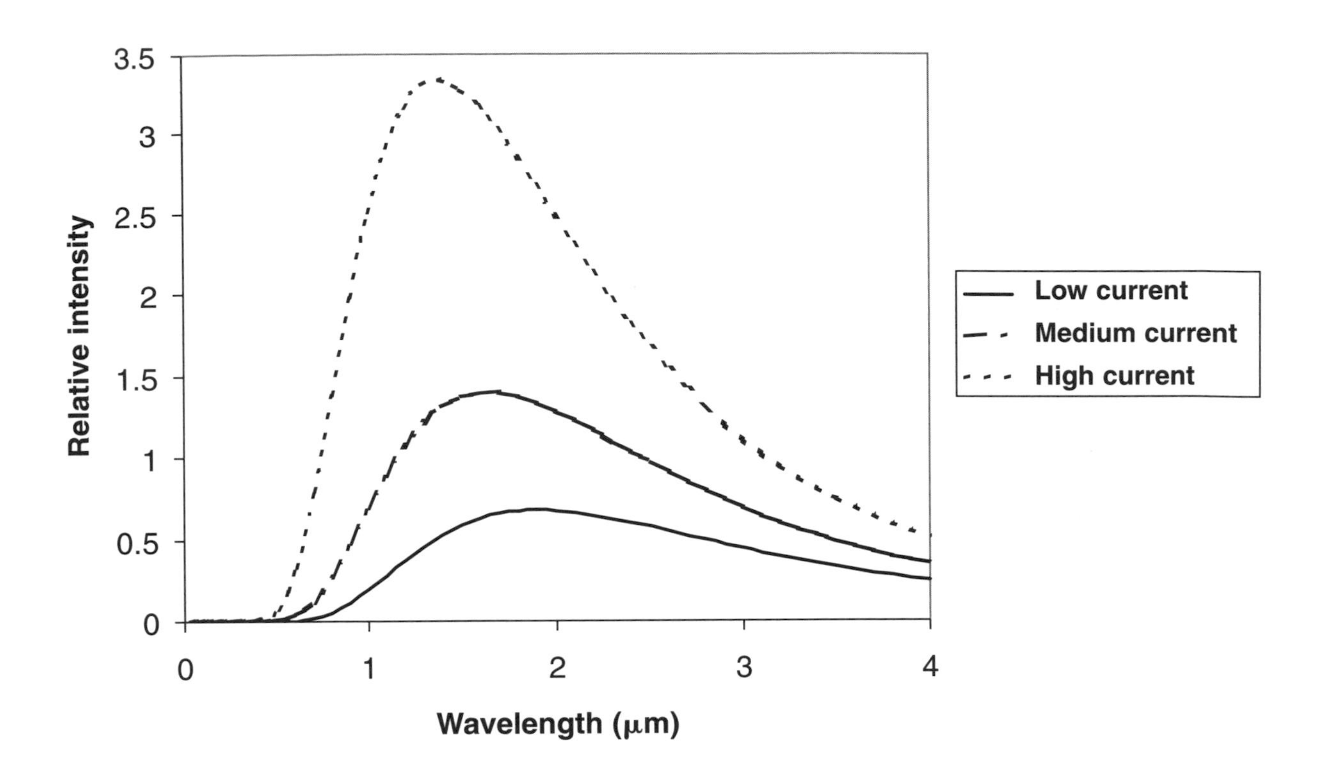

distribution, the tail that lies within the wavelength range visible to the human eye. As the current and the temperature are lowered, this tail slides off to the right of the peak, and the amount of visible light drops swiftly. As a result, lamps running at even half their normal current glow surprisingly dimly, which is why the electrolystor is seen at its best with an electric lamp as the output load.

Amplification (Gain) Factor of the Electrolystor

To measure the gain factor of the electrolystor you have made, you will need to measure what input power change, ΔP_{in}, is needed to produce a certain output power change, ΔP_{out}. Perhaps we can calculate the expected gain factor in the following way. With a bubble radius R and an AC voltage V_{AC} supplied to the electrodes, the output current power will be

$$P_{out} = V_{AC}I = V_{AC}^2K = V_{AC}^2gA[1 - (hI/AL)],$$

where A is the electrode area. With a DC voltage of V_{DC} supplied to the electrodes, the input power will be

$$P_{in} = V_{DC}I.$$

Substituting for I gives

$$P_{out} = V_{AC}^2gA[1 - (hP_{in}/V_{DC}AL)],$$

so the gain factor is

$$dP_{out}/dP_{in} = V_{AC}^2gh/V_{DC}L.$$

This equation tells us that we can expect a higher gain factor if we use a highly conductive solution and a high applied AC voltage, and a lower gain factor if our bubbles are big, if we increase the voltage to drive the DC electrolysis input, or if the rate of loss of bubbles from the electrodes is high.

And Finally . . . A Cascade of Electrolystors

You can connect up a cascade of electrolystors. You will need to add a rectifier diode to the output of the first (input) electrolystor and then feed this rectified current to the DC electrodes of the second electrolystor. You can even connect the second one to another and on and on. When trying a cascade, make sure that none of the electrolystors is "saturated"; that is, make sure that the current flowing through it is not so high that further current simply causes the evolution of more gas instead of further modifying the AC conductance. One of the problems with the cascade scheme may well be the standing current at the output of the electrolystor, as the devices cannot actually shut off the output current, which must not saturate the next device. You will therefore need to balance the electrode areas of the different devices: a smaller input electrolystor must be followed by a larger output electrolystor. If you don't do this, then a small electrolystor in the output stage may be saturated by DC from the input stage.

The use of a different salt in the electrolytic bath, or different electrode materials, might also be worth exploring. An obvious choice is an alkali or an acid, both of which have higher ionic conductance, although there may be undesirable electrode reactions. A general chemistry book may suggest alternatives, or perhaps an introductory book specifically on electrochemistry, such as Bryan Hibbert's *Introduction to Electrochemistry,* would be better.

What about an electrolystoronic oscillator? With the basic inverting function of the electrolystor, you cannot simply connect the output of one to its own input: if the input current goes up slightly, the output goes down a moment later and then the input current goes down slightly to correct itself. We would have, rather than an oscillator, a negative-feedback amplifier. (This would have a lower gain factor than an unmodified amplifier, but might, like its transistor negative-feedback cousins, have better characteristics, as discussed with transformer transistors.) To get an oscillator, we need to feed back the output of electrolystor A to the input of electrolystor B whose output is input to A. The delays intrinsic to the electrolystor will give some characteristic (maximum) oscillation frequency, although this could be modified, for example, by including resistors R between the connections and large electrolytic capacitors C from input to ground in the circuit. The capacitors, if large enough, may slow the circuit down so that it oscillates with a period T, where $T \sim RC$.

REFERENCE

Hibbert, D. Bryan. *Introduction to Electrochemistry.* London: Macmillan, 1993.

Vibrations, Rotations, and Chance

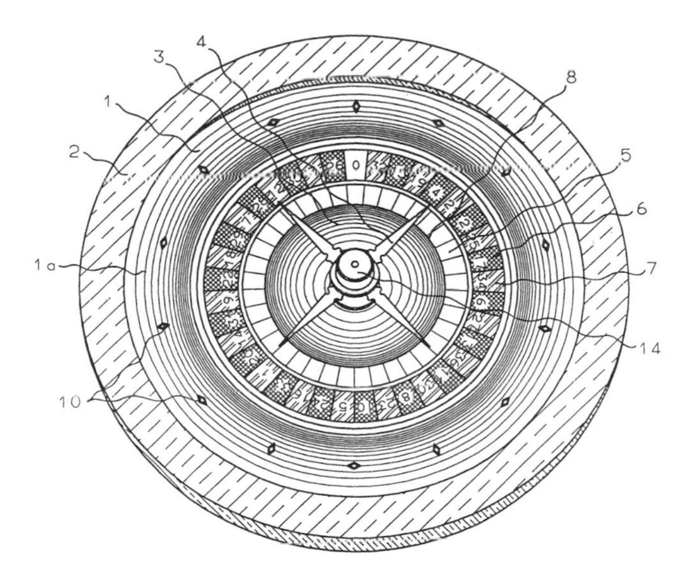

Schrödinger's cat's a mystery cat, he illustrates
the laws;
The complicated things he does have no
apparent cause;
He baffles the determinist, and drives him to
despair
For when they try to pin him down—the
quantum cat's not there!
Schrödinger's cat's a mystery cat, he's given to
random decisions;
His mass is slightly altered by a cloud of
virtual kittens;
The vacuum fluctuations print his traces in
the air
But if you try to find him, the quantum cat's
not there!

—John Lowell's quantum mechanical variation on
T. S. Eliot's "Macavity: The Mystery Cat"

Albert Einstein's work on the photoelectric effect provided the first evidence for the quantization of light, which led to the development of a new field of study called quantum mechanics. Despite his part in its development, Einstein did not think much of quantum mechanics, and he hoped that a better theory, a deterministic theory, would come along to replace it. With a deterministic system, you can, in principle at least, predict its behavior perfectly if you have perfect knowledge of the system to start with, something that quantum mechanics does not allow. Quantum mechanics does allow prediction, but only of probabilities, not hard certainties. Its intrinsically probabilistic nature makes it especially difficult to grasp: the concepts and the mathematics of probability are difficult to understand and widely misunderstood, even among professional scientists. The two simple projects in this part may improve our understanding of random probabilistic processes. One project appears to be governed by chance but is perfectly deterministic, and the other, though it looks as if it ought to be deterministic, is in fact probabilistic.

20 Waltzing Tube

Fortuna imperatrix mundi
O Fortuna, velut luna statu variabilis
semper crescis
et decrescis . . .
Fortuna Rota volvitur
descendo minoratus
alter in altum tollitur.

Fortune, empress of the world,
Fortune like the moon ever changing
Ever waxing
Ever waning . . .
The Wheel of Fortune turns
I go down
Another is raised up.

—From Carl Orff's *Carmina Burana*

Although the waltzing tube is a device that calls to mind a "wheel of fortune," it is not random or probabilistic but rather perfectly deterministic. It is ludicrously simple, so simple, in fact, that no one should read past this chapter without writing some symbols on a piece of tubing and spinning it around. Once you know its secret, perhaps you can convince someone else that it is a probabilistic device like a wheel of fortune and then win some money from them!

What You Need

- ❑ A piece of rigid cylindrical tubing whose diameter is an exact multiple of its length (e.g., a length of 25-mm drain pipe or a piece of small-diameter cardboard tubing)
- ❑ That's it!

What You Do

Cut your tube; a piece with a length that is three times the diameter is a good place to start. (You can either measure the diameter directly, or you can measure the circumference and then calculate the diameter, recalling that the circumference of a circle is π times its diameter.) Make sure that the tube is fairly close to an ideal cylinder; that is, make sure it has no burrs and so on, and cut the ends square. Using a permanent marker, draw a bold symbol on the outside of each end of the tube, for example, an *X* on one end and an *O* on the other.

Now spin the tube. By placing a finger on one end of the tube and pressing down while pulling backwards sharply, you should be able to cause the tube to rotate both around the horizontal axis running through the empty middle of the tube and simultaneously around a vertical axis. After a second or two, the tube will begin to spin regularly, and then, magically, one of the marks will reveal itself in almost perfect clarity as it flashes briefly into view, momentarily stationary, and the other symbol will remain invisible, blurred by the motion of the tube. The number of times that the visible symbol appears depends on the geometry of the tube.

What does a symbol written inside the tube do? Try writing a symbol inside an otherwise blank tube. Now spin the tube and view it from the side so that you can see the symbol as the tube end passes you. Is the inside symbol also visible one, two, or more times per revolution? With an opaque tube, you won't be able to see more than one. Maybe a tube made of clear plastic would allow you to see more.

How It Works

The key to understanding this trick is to realize that once the tube is rotating in a stable fashion, it is actually rolling around in a circle rather than simply spinning about a point underneath in the center. Furthermore, the rolling is taking place at an end—the end opposite to the end that you flicked. The image on the rolling end is no more visible than a patch of paint on the peripheral tread of a rolling tire. However, the opposite end of the tube is actually rotating backwards relative to the direction of axis motion at exactly the same speed, so the symbol is momentarily stationary relative to a stationary observer. The symbol actually follows a nonretrograde cycloid path. From the observer's reference frame, the symbol appears to be sloping to the left about two-thirds of a tube diameter from the ground, rising up and flattening out and then beginning to slope rightwards and descending—all in fraction of a second. The image will appear in this way twice per tube revolution if the tube's diameter is twice its length, three times if the diameter is three times the length, and so on.

The symbol written on the inside of the tube is simpler to explain. The surface of any rolling wheel is momentarily stationary as the wheel rolls along. If it were not, then the wheel would, logically, be skidding, and most wheels don't usually skid. The view inside the tube is something like the view you would get if you rolled a wheel with marked tread over a glass table and observed it from underneath the table.

This simple gadget, you won't be surprised to learn, has been around for many years, quite possibly centuries. Certainly it was known in Victorian times. (For a recent analysis, see Karl Mamola's article titled "A Rotational Dynamics Demonstration.") Once lightweight tubes were available, perhaps made from parchment, kids would, I am sure, have discovered the effect and adjusted the tube size to make it work nicely. Although I can't prove it, I suspect that kids in Pharaonic Egypt played with papyrus waltzing tubes at the feet of the newly built pyramids.

The Waltzing Tube Way to Wealth

Forget for a moment everything you've just learned about the waltzing tube. Imagine that I approach you with a chance to make a great deal of money. Displaying the tube, I explain how it shows either one symbol or the other, just as a coin can land on either of its two sides. I then say that the probability of either of the symbols appearing is not exactly fifty-fifty. After carefully building up seemingly random sequences in which one symbol or the other wins out over a few throws, I offer to take bets. I lose a few small wagers but make sure that I clean up on most of the big ones. Finally, I lose a biggish one and bow out while I'm still comfortably ahead. I laugh all the way to the bank (once you are out of earshot!). All I did, though, was to appear to flick the tube at both ends while in fact ensuring that only one end was propelled.

If the length of the waltzing tube is an exact multiple of the diameter, then the image that appears on the flicked end will be in the same place with each revolution of the tube—stationary. If you don't have an exact multiple, the image will appear to rotate. Suppose for example that a tube of length L and diameter $2R$ is rotating with angular speed ω_h around a horizontal axis and with angular speed ω_v around a vertical axis. If the tube is rolling on the surface at one end, then the connection between the two angular speeds is given by

$$\omega_h L = \omega_v R.$$

The bottom of the rolling end of the tube has zero speed. At angular speed ω_h, the top of the tube at the rolling end has a speed given by

$$(\omega_h L + \omega_v R) = 2\omega_h L.$$

Once the tube is waltzing, one image will appear to be blurred by twice the apparent speed of the tube end, while the other will appear to be stationary. The bottom of the skidding end of the tube has speed $2\omega_h L$, and the top has zero speed.

For every turn of the tube, the rotating tube end will travel $2\pi L$. The rolling end of the tube will travel that same distance and, in the process of rotating around, will show the symbol N times, where

$$N = 2\pi L/2\pi R = L/R.$$

So you don't need to know the value of π. You just need to cut the tubing to a length equal to N times its diameter. If the tube circumference is a half-integer multiple of the length, then two sets of alternately flashing images will appear. If N equals 3.5, then the symbol will appear seven times in two circuits of the tube.

The tube has to be set into rotation about the hollow center and about the vertical axis by the finger's sliding and pulling backwards. If this is perfectly done, the tube will immediately commence waltzing correctly. However, the launch speed is always at least slightly wrong, which means that for the first few seconds, the tube is skidding at both ends. However, once the end that is nearly rolling is slowed or speeded until it actually is rolling, and the other end is rotating freely, waltzing will begin.

The tube waltzes at a very slight angle to the horizontal: the freely rotating end of the tube has to be clear of the surface. What keeps the tube from lying flat and refusing to waltz deserves some investigation. The tube does indeed stop waltzing once it is going slow enough—it grips the surface at both ends and goes rolling off in a straight line. At higher speeds, perhaps gyroscopic forces help: that is, perhaps the tube behaves like a horizontally precessing gyroscope. Or maybe aerodynamic forces are responsible? The tube is light in weight after all, and it will be subject to the Magnus effect. In the early nineteenth century, German physicist Heinrich G. Magnus showed that a cylinder rotating counterclockwise in a horizontal fluid flow directed from the left to the right will tend to lift upward. The Magnus force, L, is (given certain assumptions that I won't go into) proportional to the fluid velocity, v_0, the fluid density, ρ, the cylinder's circumference, $2\pi R$, and the velocity, v_c, of a point on its surface:

$$L = \rho v_0 2\pi R v_c.$$

When the skidding end of the waltzing cylinder is rotating underneath against the direction of motion of the cylinder, there should be a strong lifting force, L. However, because the Magnus force equation holds for a cylinder in a large uninterrupted stream and because our cylinder is very close to a stationary surface, the numerical value given by the equation is incorrect for our system. The effect of a stationary surface is often referred to as the "ground effect," and it usually results in much larger lift values for an airfoil, a rotating cylinder in this case.

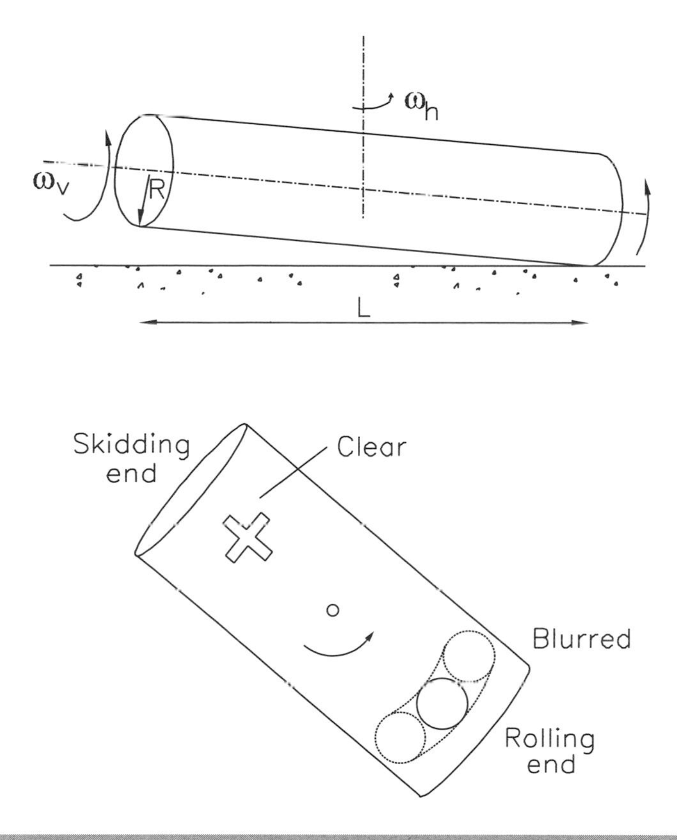

And Finally . . . Quickstepping and Foxtrotting Tubes

What happens if one end of the tube has a different diameter than the other—that is, if the tube is conical—or if it has a ridge at one end? Can you make a tube that flashes up three symbols when spun one way and four symbols in the other? Is a hollow tube better than a solid cylinder? Does the lower moment of inertia of the solid cylinder, which means that it has a lower resistance to changes of rotation speed, put it at a disadvantage relative to the tube? Do the heavier weight and the limited lift force, whether aerodynamic or gyroscopic, also militate against a solid cylinder? Can you fit illuminated symbols inside a waltzing tube?

REFERENCE

Mamola, Karl C. "A Rotational Dynamics Demonstration." *The Physics Teacher* 32 (1994): 216–219.

21 Motor Dice

God does not play dice.

—Albert Einstein

The records of insurance transactions at Lloyd's of London reveal that businesspeople have been calculating probabilities accurately since at least 1688. Insurance is basically about exchanging a small risk of a large financial outlay for a small but certain financial outlay, and probabilistic calculations show what the exchange rate should be. Mathematicians interested in the theory of games of chance and betting had laid the foundations for the study of probability, which then progressed rapidly, perhaps spurred to some extent by interest from the world of commerce. In 1718, Abraham de Moivre published the first in a sequence of masterful books on probability.

Although most of probability theory has been known for hundreds of years, it is still widely misunderstood. People believe, for example, that a coin that has landed heads 5 times in a row is much more likely to land tails on the next toss. However, if the coin is perfectly balanced and fair, it is just as likely to land head as tails after 5, 55, or even 555 heads in a row. Even if the coin is not fair, you still shouldn't bet that it will land tails: it might be weighted such that it lands on heads every time.* Even though heads is the only logical bet after 555 heads in a row, you will still find people—lots of people—who prefer to bet on tails!

*I can't see how one could design a coin that was more than a few percent unfair without its being seen as obviously unfair, even upon the most cursory inspection. I tried making a loaded die and found it difficult, and a loaded coin is probably even trickier to make.

Of course, such misunderstandings make probabilistic games more interesting. People who thoroughly understand probability theory may find such games boring, but profitable. Probability games depend upon a fair but random device such as a coin, a roulette wheel, or a die. Here we devise an electric, sixfold random-choice machine that can supplant a die. We can use statistics to determine how fair it is.

What You Need

- ❑ Small electric motor with two permanent magnets, a three-pole armature (1.5–3 V)
- ❑ Battery
- ❑ Rotor disk
- ❑ Plastic disk or upturned dish (e.g., saucer sold to go underneath a potted plant)
- ❑ Push-button momentary switch normally off ("roll"; e.g., microswitch)
- ❑ Push-button momentary changeover switch ("nudge"; e.g., microswitch)
- ❑ Capacitor (e.g., 500 μF)
- ❑ Transformer 240 V/110 V mains input, low-voltage output
- ❑ Resistor to limit current

What You Do

Mount the plastic disk or upturned dish on the motor with its shaft projecting vertically upward. Draw an arrow on the rotor disk and push fit it on the shaft. If your dish is deep enough, it can house the battery and motor underneath, as well as provide a convenient surface on which to mount the roll and nudge switches.

When the motor is activated and then stopped, the rotor will usually stop at one of six stable positions. Spin the motor a few dozen times, and get ready to mark these positions on the dish. With the rotor in one of its stable positions, write the numeral 1 on the dish in the sector to which the rotor arrow points, and then push the rotor to its next stable position. Mark this position with the numeral 2, and continue until all six stable positions are marked.

Once you've assembled the components as in the diagram, check that the circuit is correctly wired. When you press the roll button, the rotor should spin freely. When you press the nudge button, the rotor should jig just a fraction of millimeter in one direction; and when you release the button, the rotor should move slightly in the opposite direction. Carefully pull the rotor around until it is in "limbo," that is, until the arrow is pointing between, say, 1 and 2. Now press nudge a couple of times; the rotor should flick from limbo to either 1 or 2 and

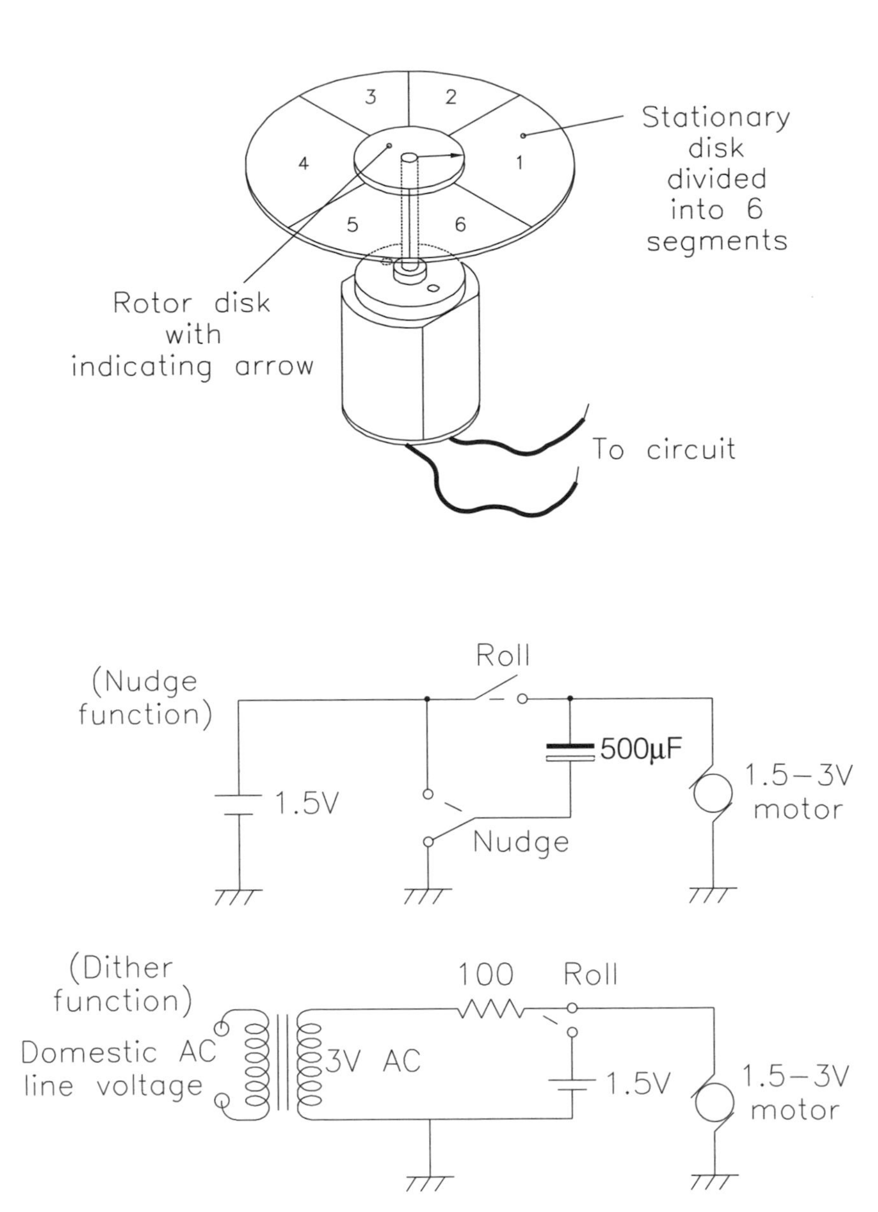

then stay there. If the nudges are so big that they allow the rotor to escape from a stable position, reduce the size of the capacitor. Conversely, if the nudges are too small and don't nudge the rotor from limbo to any of the six stable positions, then increase the size of the capacitor.

Now it's time to test the die for fairness. Make a column diagram with one column for each of the six positions. Spin the die. If it lands squarely on a number (which it should about 90 percent of the time), record that in the appropriate column of the diagram. If it lands somewhere between two numbers, press the nudge button two or three times to kick the rotor forward or backwards so that it points squarely at one of the six numbers. If you continue in this way, patterns like the following (for 600 throws) will begin to emerge: 110, 108, 95, 101, 88, 98. The number of hits for each position will be within about 10 of the average (100 in this case). A computer simulation arranged to simulate a perfectly fair die yields the following patterns (the parenthetical numbers are the approximate deviations from average):

600 throws: 94, 110, 109, 100, 103, 84 (10)

6,000 throws: 951, 1,036, 1,029, 994, 1,014, 976 (30)

60,000 throws: 9,978, 10,046, 10,096, 9,986, 9,913, 9,981 (100).

How It Works

A motor spins so fast that without the aid of high-speed instrumentation, a human operator cannot predict where it will stop. It is therefore a good facsimile of a random-number generator. Conveniently, the motor does not stop at just any rotational position, however. The iron core of the rather open armature (the electromagnetic rotor) of the small electric motor suggested has the effect of giving it several stable positions. When you spin a three-pole armature within the field of its permanent magnets, you will typically find six positions in which the motor prefers to stop. The friction between the armature and the sliding contacts (brushes) of the commutator prevents this effect from being perfect, and therefore the motor will occasionally stop between stable positions. However, the effects of this friction can be corrected for by the application of a "nudge," a small current pulse, first negative in direction and then positive. The nudge is designed so that it is too small to move the motor out of a preferred position but just right to send the motor into one of the stable positions if it has stopped between two such positions.

THE SCIENCE AND THE MATH

The motor die we have constructed is a mechanical random-number generator. Random-number generators are used in many fields of science and industry—not just for generating lottery winners. They are used, for example, to investigate the probability of failure in dangerous machines or systems, where precise mathematical analysis is not possible. (This technique is often called *Monte Carlo analysis,* after the famous casino in the south of France.) Each component in the system—which could be an airplane or a power station—is assigned a probability distribution, and failures are generated with random numbers to match that probability distribution. The scenario is repeated thousands of times, and the computer records the statistics on how many times the system reaches a dangerous condition. This process allows logically sound design decisions to be made.

All computers come with a random-number generator, typically a program that generates a long sequence of numbers that appear to be unrelated in any way. Because the sequence eventually repeats itself, the numbers are not truly random. However, a thirty-two-bit shift register sequence, for example, provides more than a billion random tosses of a "coin." Other "random" sequences can be obtained from multiplications using modular arithmetic: multiply a large number, A, by another large number, B; then divide by a third number, the modulus, and use the remainder as the new A for the next cycle.*

Of course, you can always convert binary sequences into the equivalent of die rolls. (You can make a "die" out of three coins: mark them 0/1, 0/2, and 0/4, for example. As you toss them, add up the totals and disregard any tosses that total 0 or 7.) Another common way of providing a die function is to connect a high-speed oscillator to a counter that runs from 1 to 6. The oscillator is arranged to run only when the user is pressing a switch. If the oscillator runs at, say, 60 kHz, then it will run the counter from 1 through 6 a thousand times in 100 ms, which is about the minimum duration of a typical button press. Since no one can control how long his or her finger stays on a button to within fraction of a millisecond, the number displayed on the counter is essentially random. This setup is a true electronic analog of our motor die. Our die has a similar mode of operation, although the speed of the rotor is more limited. To be equivalent to the high-speed oscillator, our motor die would have to spin at 600,000 rpm!

In trying to estimate whether our motor die is fair, you have to spin it many times, as you can see by inspecting the simulation I described earlier. The variation in the height, N, of each column will vary approximately as $\sqrt{N}$. When N equals 100, for example (600 spins of the motor die), most columns in your column diagram will be within 10 of the average, and only a few columns will be more than 2×10 higher or lower than the average. With $N = 10{,}000$, most columns will be within 100 of the average. A variation that is more than a few times $\sqrt{N}$ larger or smaller than average indicates a genuine bias on the part of the die.

You can use a computer to generate simulated dice rolls for comparison with real dice rolls by using the QBasic function RND,† which gives a random number between 0 and 1 when it is invoked:

```
DIM N(6)
INPUT IX
FOR J = 1 TO 6
N(J) = 0
NEXT J
```

*Try modular arithmetic on a pocket calculator or write a Basic program for your computer. With 334 and 127 and a modulus of 227, I got a sequence that hopped between 0 and 227 in a fairly satisfactory way. Be aware, however, that there are many pitfalls for the unwary and that quite a few math experts have been foxed by these pitfalls. The sequence with 334 and 227, for example, repeats itself over a long enough time.

†QBasic is perhaps obsolete now, but it used to come with computers as part of the operating system. It is still widely available. Note that you can access a version of Basic—Visual Basic or VBA—via an extension of the Microsoft Excel spreadsheet program. QBasic programs will work with VBA. There are many other programs you could use to run simple calculation routines like this.

```
FOR I = 1 TO IX
Y = 1 + 6*RND
IY = INT(Y)
N(IY) = N(IY) + 1
NEXT I
PRINT N(1), N(2), N(3), N(4), N(5), N(6)
INPUT X
STOP
END
```

As we have seen, our random-number generator works because the iron core of the armature has stable positions within the field of its permanent magnets, owing to the fact that iron is easily magnetizable. The poles of the permanent magnets induce opposite poles in the armature and attract them. As I mentioned, however, friction between the armature and the brushes of the commutator, as well as friction in the bearings, will occasionally cause the motor to stop between stable positions. This happens only when, by chance, the armature comes to rest almost exactly halfway between the poles. The nudge switch solves this problem by applying two pulses, one positive, the other negative (the circuit diagram illustrates why this is so). When the nudge switch is moved from the upper to the lower position, a 3-V positive pulse is applied to the motor, the charge transferred being proportional to the value of the capacitor. When the nudge switch is allowed to spring back, a 3-V negative pulse is applied to the motor. The two opposite-polarity pulses must be big enough to jog the motor from its unstable halfway position but small enough not to spin it to any pole other than one of the two adjacent poles. The mechanical impulse, I_m, given by the nudge is the energy stored in the capacitor, less some allowance for inefficiencies:

$$I_m \sim < CV^2/2,$$

where C is the capacitance and V the battery voltage. With a 1,000-µF capacitor charged to 1.5 V, we have a nudge impulse of 1.1 mJ. That isn't much. If you applied that nudge to a 27-g weight, you could move it up against gravity by only 4 mm, even at 100 percent efficiency.

The nudge switch is not the only way to avoid in-between results. Applying a small continuous alternating current to the motor causes it to dither—that is, make small movements to and fro—which is a neat alternative for minimizing the effect of friction.* When a little dither is applied, the effects of friction are again overcome, and the motor halts only at one of its preferred pole positions. Dither helps the motor to decide which pole position to take.

The easiest way to apply dither is to use low-voltage AC from a mains transformer via a resistor to give a controlled dither of a few tens of milliamps AC. If a completely portable motor die is desired, a battery-driven oscillator can be used to give the necessary dithering current. Further improvements could be envisaged: for example, a device to ensure that the motor is running at a good speed before the "roll" button is released.

Is our motor die a quantum device? Are its answers deterministic, or are they determined by Schrödinger's shadowy probability equations? The die is certainly not deterministic in the way that a mathematical pseudo-random-number sequence is,

*Dither is still used in various contexts in industry. Some years ago I found I could get more-accurate control of a solenoid valve in a gas supply system by injecting, along with the DC control signal, a small 60-Hz AC signal. Without the dither, the valve tended to stick a little too open or a little too closed. When the system was adjusting its flow rate, the valve would stick; the control signal would grow bigger and bigger in response to the error until it suddenly came unstuck, and the control system would give a big pulse of gas and then readjust correctly. These days, dithering current is incorporated in electronic controls for some solenoid valves, although a higher-frequency AC signal, from 150 to 400 Hz, is often used. In purely electronic systems, dither is encountered in devices like digital frequency synthesizers, where it allows systems to avoid interference bands, for example, or to increase the frequency resolution in cellular (mobile) phones and radio sets.

but it may be deterministic in that careful and accurate measurements of the device's parameters on a particular trial may enable you to predict the outcome. Thomas Bass's *The Newtonian Casino* is a gripping thriller about a real-life attempt to make just this sort of prediction. A bunch of math and computer-science students tried to win money from Las Vegas casinos by predicting the outcome of spins of the roulette wheel.

You might be able to test for determinism by running a motor die from a fixed starting position at a constant temperature and supplying a constant voltage for a constant, accurately metered time. If this procedure consistently gave the same answer over a number of trials, then I suppose you could say that the motor die was not quantum mechanical. If, on the contrary, you found that the answers were still random, you might be able to say that random variations in the conditions of and inputs to the trials—variations that are fundamentally quantum mechanical—caused the random answers.

And Finally . . . Cheating at Dice

You may be curious to know how fair a regular cubic die is. I tried to load one by putting heavy lead solder underneath its spots. Surprisingly, perhaps, even quite heavy loadings failed to produce much bias. Even when I loaded a die so that it produced a 50 percent increase in the probability of rolling a six, the people I tested it on could easily tell which die was loaded just by handling it for a moment. Its center of gravity was clearly nowhere near the center of the die, and it did not roll as freely as a regular die. When I tried making dice that were rather oblong and thus most likely not to land on six or one, I discovered again that only dice that were obviously noncubic produced a noticeable bias. But gamblers who play dice should still beware: methods that involve magnets or special rolling techniques are probably more effective and much more difficult to detect.

To provide for games that require two dice, you could make advanced motor die units that included two motors operated by common roll and nudge buttons. You could use two vertical plane rotors facing each other with numbers around their edges; this setup would give the double dice readout in neat two-digit form.

REFERENCE

Bass, Thomas A. *The Newtonian Casino*. London: Penguin, 1991. Originally published as *The Eudaemonic Pie*. Boston: Houghton Mifflin, 1985.

Maverick Measurement

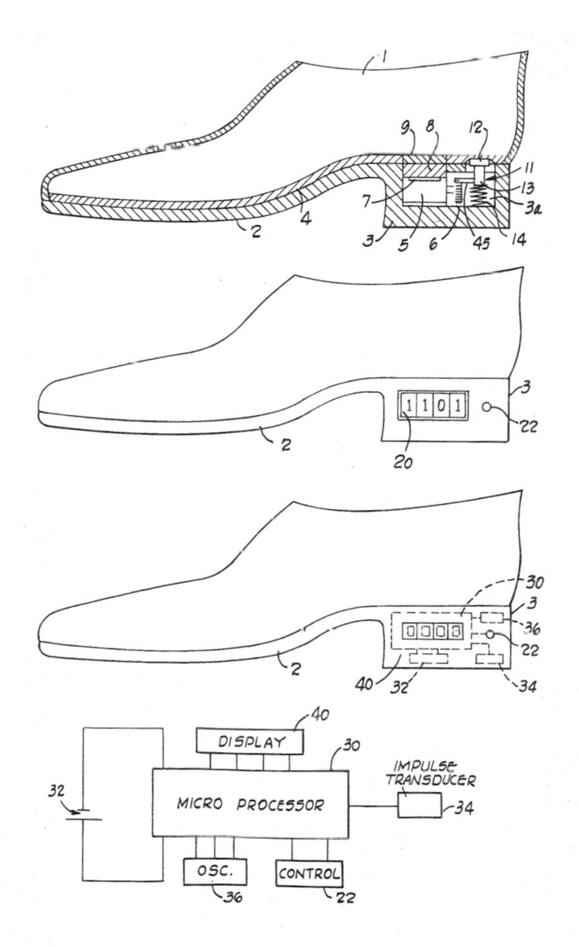

Neither the naked hand nor the understanding left to itself can effect much. It is by instruments and helps that the work is done, which are as much wanted for the understanding as for the hand.

—Francis Bacon, *Novum Organum*

Instruments are the senses of science: they expand our human senses to encompass phenomena of which we are normally unaware, and they allow us to record and quantify our observations of the world around us. The advance of science has depended upon the continuous development of improved instruments, development that has paralleled the progress of science itself. Human senses cover only narrow ranges: electromagnetic radiation from violet to red, sound from 100 Hz to 10,000 Hz, sweet, sour, salty, and bitter tastes, and some dozens of distinct smells. Of the instruments in this part of the book, all but one measure things we have difficulty estimating with our senses:

- Gas flow (To humans, gas is invisible and more or less intangible.)
- Force (Human touch is qualitative only—ask ten people how much a 1-kg object weighs and see how many different answers you get.)
- Angle (We are relatively good at judging some angles, e.g., 90 degrees.)
- Light intensity (Our response to light is logarithmic rather than linear.)
- Electric currents in a gas (Our hair stands on end when we touch a static charge, but that's hardly quantitative.)

Sound frequency is one of the few phenomenon we can sense quantitatively. However, the sounds we will measure in one of our projects have fundamental frequencies that are too low to be sensed directly. There are people with so-called perfect pitch, people who have a surprisingly acute sense of the absolute frequency of a sound, down to an accuracy of a musical semitone (6 percent) or less. However, the fundamental frequencies of engine vibrations in ships and the like are too low to be sensed directly. We hear only higher-frequency sounds modulated at the low frequency of a slow engine.

REFERENCES

There are many books on instrumentation, which has become a specialty in its own right, with numerous periodicals and degree programs in many universities today. Here are some textbooks you might try for an introduction.

Bolton, W. *Measurement and Instrumentation Systems.* Oxford, U.K.: Newnes/Butterworth Heinemann, 1996.

Hankins, Thomas L., and Robert J. Silverman. *Instruments and the Imagination.* Princeton, N.J.: Princeton University Press, 1995. A history of the development of instrumentation from ancient times up to the nineteenth century.

Jones, Barry E., ed. *Instrument Science and Technology.* 2 vols. Bristol, U.K.: Adam Hilger, 1993.

22 Coffee-Cup Revolution Counter

> The principle of generating small amounts of FINITE improbability by simply hooking the logic circuits of a Bambleweeny 57 Sub-Meson Brain to an atomic vector plotter suspended in a strong Brownian Motion producer (say a nice hot cup of tea) were of course well understood.
>
> —Douglas Adams, *The Hitchhiker's Guide to the Galaxy*

There is a certain fascination with things that go on in a cup of tea, as acknowledged by Douglas Adams in his beautifully written sci-fi spoof, *The Hitchhiker's Guide to the Galaxy.* I am always fascinated by tea phenomena: Why, for example, do the tea leaves end up in the middle of the cup instead of at the edge (where you would expect centrifugal force to put them)? I'm not the only one. Reading the future in the dregs of coffee or tea is a popular pastime, especially in Eastern European countries such the Czech and Slovak Republics.

Fortunetellers may soon be joined in their intense study of beverages by mariners. If you look carefully into your cup the next time you travel by ship, you will see a faint but distinct set of waves, often in the form of concentric circles. Cinema aficionados will recognize the phenomenon from scenes in the movie *Jurassic Park,* in which such ripples signaled the approach of a distinctly unfriendly *Tyrannosaurus rex*. The vibrations of a ship's engine are usually transmitted faintly through the entire hull and will be visible as slight undulations in the surface of the liquid in your cup. (Things get more complicated if there are multiple engines working at different speeds.) With a little science you can figure out how fast the engine is going without calling up Scotty in the engine room. Readers who lack the required ocean liner but are in possession of a cup can simulate the phenomenon and calibrate their revolution counter with the aid of a small electric motor.

What You Need

- ❑ Ship (You can use this item for other things.)
- ❑ Mug or cup of liquid (I used a cup, 73-mm diameter, containing coffee with milk.)
- ❑ Electric motor
- ❑ Wheel with an eccentric weight and small projecting vanes
- ❑ Variable-speed power supply (e.g., a battery with resistance wire in series or a pulse-width controller arrangement)
- ❑ Resistor
- ❑ Photodiode
- ❑ Multimeter with frequency readout
- ❑ Tape
- ❑ Reading lamp

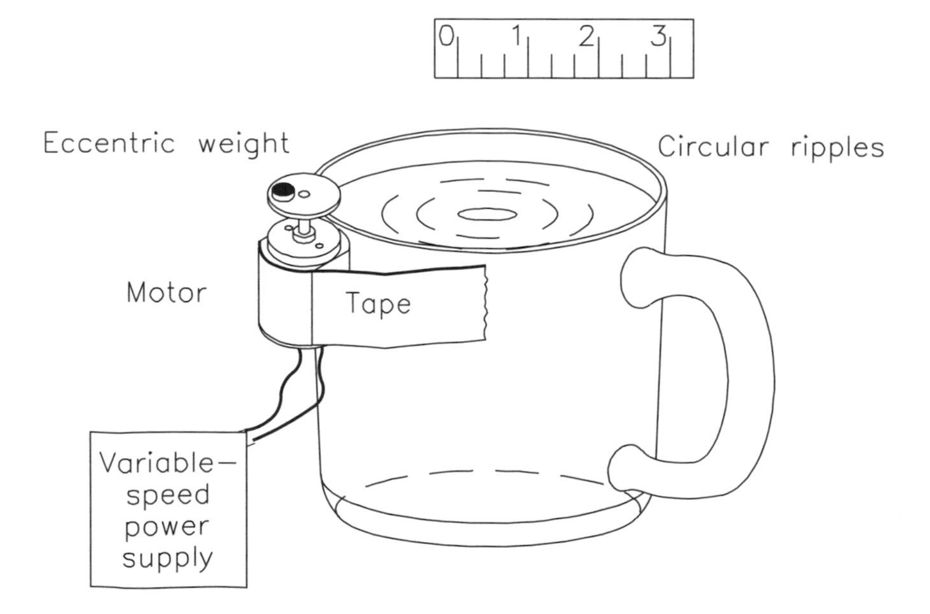

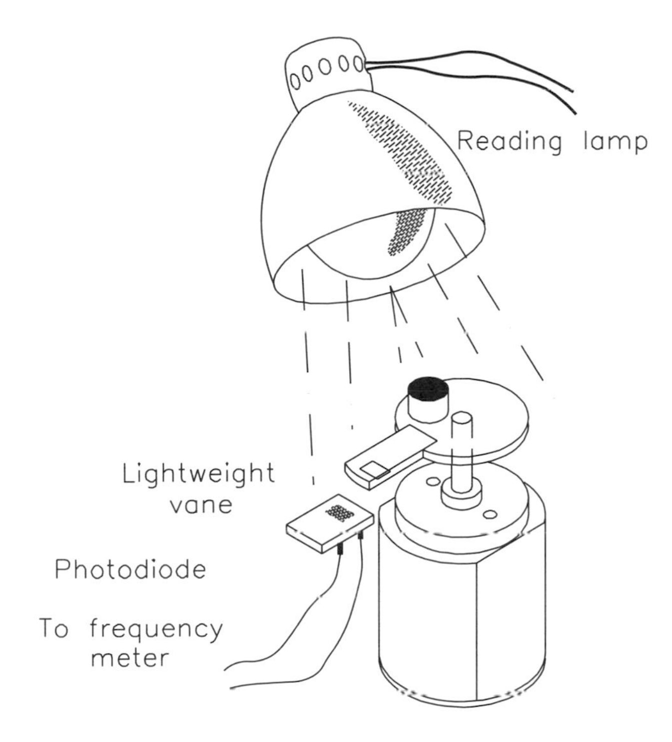

What You Do

Tape the motor firmly to the coffee cup. Set the motor going and start calibrating your cup. Set the motor speed to give a good set of concentric circles. (I measured the motor speed by arranging the eccentric on the motor so that one or more small projecting vanes passed over a photodiode and blocked the light from a reading lamp on each turn of the motor. I used the resulting pulses of current through the resistor connected across the photodiode to operate the frequency-counter mode of a multimeter.) Count how many circles you have, and measure the revolutions per second of the motor on the multimeter (dividing the reading by the number of vanes on the eccentric). You should be able to calibrate your coffee cup up to about six or seven concentric ripples—up to a few hundred revs per minute—a suitable range for measuring the speed of a ship's engine. So the next time you go out on the water, take your coffee cup with you.

How It Works

The concentric circles formed on the liquid surface by the motor start as tiny ripples traveling from the cup's walls toward its center, where they are reflected back to reinforce new ripples just starting out. At certain motor speeds, the superposition of the incoming and outgoing ripples produces a standing wave. A standing wave consists of a certain number of stationary ripples that simply oscillate between positive and negative deviations from the flat surface without traveling. At low motor-rotation speeds, there will be just one ripple between the edge and middle of the cup. The motor sends out, say, three ripples per second, which travel to the center of the cup in one-third of a second and so form one crest. At successively higher speeds, two to six ripples will form because the motor sends out two to six ripple crests in the time it takes the ripples to travel to the center. In this way, the number of ripples observed is a precise analog of the rotation speed of the motor—whether the motor is taped to the cup or is 50 m away in the engine room of a ship.

THE SCIENCE AND THE MATH

The wavelength of the concentric waves that form in the coffee-cup revolution counter is determined largely by surface tension. At this scale, the waves go faster as they get smaller. Larger waves have wavelengths that are for the most part determined by gravity, and these larger waves travel faster than small ones. At a certain frequency, f, and wavelength, λ, the speed of the waves, ω, is at a minimum:

$$\omega^2 = gk + (\sigma/\rho)k^3,$$

where $\omega = 2\pi f$, $k = 2\pi/\lambda$, g is the acceleration due to gravity, σ is the surface tension, and ρ is the mass density of the liquid. If the wavelength is long enough, then surface tension can be neglected, and the waves are called "gravity waves," since the force restoring the surface to flatness is gravity. For short wavelengths, the restoring force is surface tension, and the waves, which we might in ordinary speech call ripples, are known scientifically as "capillary waves." The waves in our rev counter are capillary waves, so

$$\omega \sim \sqrt{[(\sigma/\rho)k^3]}.$$

Using this formula, we can quickly estimate the correlation between the frequency and the number of ripples in the cup:

Ripples	f (rpm)	f (Hz)
1	192	3.2
2	544	9.1
3	999	16.6
4	1,537	25.6

The frequency calibration should be roughly the same for all cups of coffee, but the rpm value refers to a single-cylinder engine, or at least an engine with a dominant vibration frequency of once per crankshaft revolution.

Under the correct lighting, ripples only a fraction of a millimeter in height are readily visible. Their surprisingly high visibility stems, in most cases, from their ability to reflect and focus light from backgrounds of variable brightness or color; and as they progress, the ripples reflect different parts of the background. Even a small ripple gives rise to a substantial deflection of the imaged light because moving a mirror through angle α results in a light beam deflected through angle 2α. (To demonstrate this, just draw, on paper, a beam coming in at a constant angle and reflecting off a mirror, with the angle of reflection equal to the angle of incidence. Then redraw the diagram with the mirror rotated by some angle, and you will see, if you superimpose the two drawings, that the reflected beam is deflected by twice as much.)

And Finally . . . Right-Angle Ripples

At higher rpm values for the motor, the surface of the coffee will show radial divisions or even more complex patterns. How can standing-wave ripples form at right angles to the direction of the wave travel? Some of these patterns bear comparison to the complex patterns, known as Chladni figures, seen when dust is scattered on a vibrating plate or drum skin. Further calibration could no doubt be done on our coffee-cup patterns. Maybe you can check out the engines on your next jet ride with these higher-frequency patterns. More-violent vibrations result in other effects, including water fountains and the strange radial ripples first described by Michael Faraday nearly two hundred years ago. If you see these, you may want to ask the pilot to let you off the plane!

23 Coulter's Bubbles

I'm forever blowing bubbles,
Pretty bubbles in the air.
They fly so high,
Nearly reach the sky,
Then like my dreams
They fade and die.
Fortune's always hiding,
I've looked everywhere,
I'm forever blowing bubbles,
Pretty bubbles in the air.

—Jaan Kenbrovin, "I'm Forever Blowing Bubbles"

If there is a proliferation of devices to do a particular job, then you can be fairly sure that all of them have problems. Usually, if there is one good way to accomplish a task, the other ways of doing it disappear for the most part. Consider mussel fishery, for example. Look inside the equipment shed at a cultivated-mussel fishery, and you will see a cornucopia of different machines and devices, from rubber boots to boats with hydraulic wheels and crawler tractors. The mussel's habitat in the shallow water and soft sand and mud of the tidal zone means that there is no completely satisfactory means of access.

The measurement of gas flow is another task that is not completely straightforward. You can buy rotating-vane (dry) meters, wet-vane meters, oscillating-diaphragm meters, ultrasonic meters, thermal mass flow meters, Coriolis flow meters. The list is long. An old-fashioned alternative is the soap-film flow meter, which is an upright tube with a soap-solution reservoir (a rubber bulb) at the bottom and a gas inlet near the bottom. While air or gas is pushed in at the bottom, the bulb is squeezed; a soap film forms across the tube and is carried up by the flow toward the exit. By measuring the soap-film transit time you can measure gas flow with surprising accuracy. Here we look another way of using bubbles to measure flow.

What You Need

- ❑ Flow cell (e.g., a bowl)
- ❑ Small chamber, with an open conical top, for the sensor (e.g., plastic reflector assembly from a flashlight or a homemade plastic cone on top of a glass jar)
- ❑ Rigid insulated copper wire to make electrodes
- ❑ Rigid wire to position tubing
- ❑ Connecting wires
- ❑ Multimeter
- ❑ Pump or other air source (e.g., aquarium pump)
- ❑ Tubing
- ❑ Smoothing capacitor
- ❑ Power source (either a 1.2-V cell and a diode or an AC source; see diagram)

What You Do

The flow cell consists of two compartments, each containing an electrode, that are separated by a cone of insulating material; the cone should have a hole in the top (the narrow end). I found that the most convenient object that satisfied these criteria was the reflector and window assembly of a small plastic flashlight. The assembly is plastic, has a conical part (the reflector) with a hole of about 10 mm in diameter, and is sealed at the bottom by the clear window through which the light shines. To make the sensing-zone hole, saw off the part where the lamp screws in and then make two holes, one on either side of the assembly, that are a reasonably good fit for the tubing and one of the electrodes. The easiest thing to use for the electrodes is rigid copper wire, bared at the ends, which you can pull out of household mains wiring. One electrode is clipped to the side of the bowl, and the other must pass, insulated, through into the cone, where its bare copper is exposed. Once you've completed the sensor, place it in the bowl and add

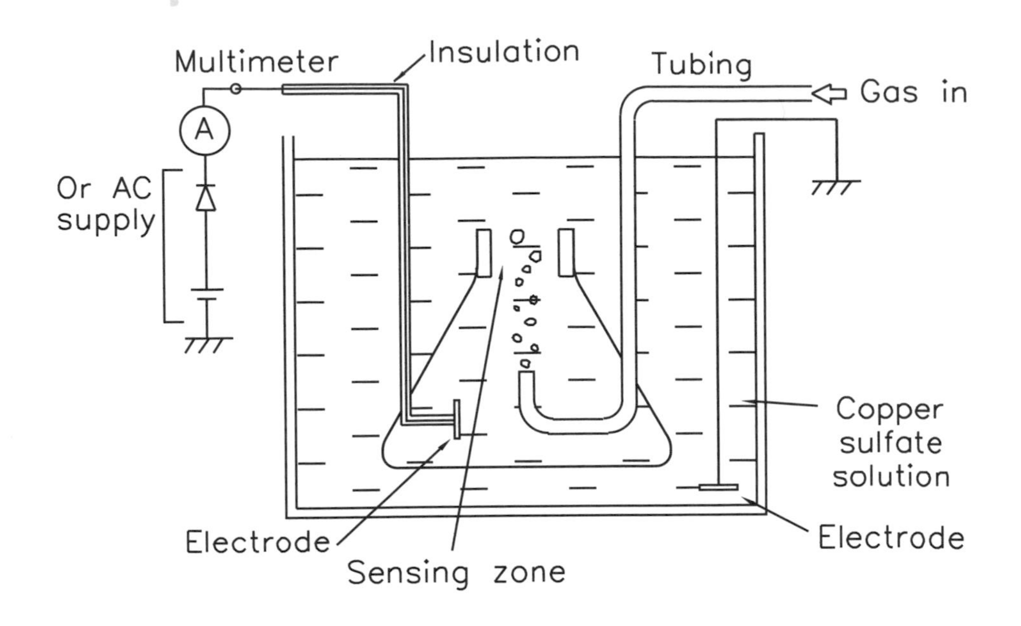

enough salt water to cover the top of the sensing-zone hole by at least 25 mm. What should you use for the salt? If you are using a DC power supply, then try copper sulfate with copper electrodes. If AC, then the nature of the salt doesn't matter much. It is best not to use too strong a salt solution—just a teaspoonful in a half liter—but the exact concentration is not important.

To get an adequate flow of bubbles passing through the sensing zone, make a small hole in the side of the tubing, perhaps with a hot needle or small heated nail. Block the end of the aquarium pump tubing, with a small plug, for example. The use of a small piece of rigid wire inside the tubing will allow you to position the stream of gas bubbles right below the sensing zone.

As the bubbles pass through the sensing zone, the current measured goes up and down, but the average falls significantly, which clearly indicates the effect of the bubbles passing through. I found that when I used a small aquarium pump with a 10-mm sensing-zone hole, the conductance of my dilute sulfate cell decreased by up to 50 percent, from about 1 mA on 0.6 V to about 0.5 mA. Modulating the flow rate clearly showed that the system gave a flow readout.

You will probably find that the readings from an electronic multimeter vary by 5 percent or so. This variation, which limits the accuracy of the system, occurs because the meter tends to give a snapshot of the conductance—that is, an average over a fraction of a second rather than an average over several seconds, which is what is really required. To get around this inaccuracy, you could take multiple readings and average them, or you could try using a smoothing capacitor, which is actually more convenient. The capacitor needs to be a substantial one—several thousand microfarads. An old-fashioned analog meter is another possibility: try a 0- to 3-mA meter with a resistor in series (a few hundred ohms) instead of the diode, to get the standing current within the range of the meter. You will find it easier to get the average reading using an analog meter. A capacitor of 5,000–10,000 μF will also help steady the reading from the analog meter.

How It Works

Our bubble flow meter works because bubbles do not conduct electricity, whereas the salt solution (electrolyte) does. Each bubble that passes through the exit funnel momentarily increases the resistance of the circuit. The electric resistance that the circuit shows is diminished in proportion to the volume, but irrespective of the shape, of any blocking object placed in the sensing zone. The net loss of charge passed through the electrolyte is proportional to the volume of the bubble multiplied by the time it spends in the funnel. If most of the bubbles take a similar time to drift up through the funnel, then each bubble will contribute a loss of charge proportional to its volume—even if the bubbles have different shapes, and even if multiple bubbles are present in the funnel at any particular moment. The charge loss per unit time, which is the decrease in the electric current flowing, is thus proportional to the volume flow rate of bubbles, that is, the gas flow rate.

The change in electric conduction across an orifice when particles pass through it is also the principle behind the Coulter counter, an instrument used in biology and medicine for counting and sizing small particles such as blood cells and organelles.*

An AC power source is preferred for the system. Although the use of DC rather than AC is convenient, it creates inaccuracy. The accidental formation of electrochemical cells in the apparatus will lead to spurious currents and drifts in the zero (no bubbles) signal. Surface contamination of the electrodes almost invariably leads to such zero drifts. This is the reason that chemists always measure electrical conductivity with AC current.

*The principle behind the instrument was discovered by Wallace Coulter, an electrical engineer from Little Rock, Arkansas. The Coulter counter used in medicine consists of a glassware apparatus with a very small sensing hole in it. The counter can detect human cells that are a fraction of a micron in size, and it can count them at a rate of thousands per second. The pulses from the sensing zone are put through a hundred-channel pulse-size analyzer, which displays a particle-size spectrum. Nowadays there is a whole science, flow cytometry, of cell counting and sizing, which makes use of other techniques, such as laser scattering, as well as the Coulter principle. (For more information on the subject, see Michael Ormerod's *Flow Cytometry*.)

The principle behind the Coulter counter is that any object with a nonpathological shape (for example, a long tube that is closed at one end) will decrease the electric conductance through a resistive fluid by an amount proportional to the object's volume. The resistance, R, of a tube of fluid with resistivity ρ, cross-sectional area A, and length L_t is given by

$$R = \rho L_t / A.$$

If a bubble with zero conductivity, cross-sectional area α, and length λ is placed in the middle of the tube, the new resistance, R', will be

$$R' = [\rho(L_t - \lambda)/A] + [\rho\lambda/(A - \alpha)].$$

For small α/A,

$$R' \sim (\rho L_t/A)[1 - \lambda/L_t + \lambda/L_t - \alpha\lambda/(L_t A)]$$

or

$$R' \sim (\rho L_t/A)(1 - V/V_0),$$

where V is the volume $(\alpha\lambda)$ of the bubble, and V_0 $(L_t A)$ is a constant for the sensor, a constant that might be described as the sensing volume or the sensing zone. Provided that α/A is small, it does not matter whether the blocking object is long and thin or short and fat: what matters is its volume. The shape of the cross section of the object also does not matter, provided that it is prismatic (i.e., has a constant cross section). Nonconducting objects with other shapes can also be analyzed in a similar way; they do not have to be prismatic. They also don't have to be oriented parallel to the sensing volume axis. The resistance of the sensor as a bubble passes through will be modulated by an amount that depends on the volume of the bubble. By using a hole rather than a long tube for the sensing zone, we can ensure that there will be only one bubble at a time in the sensing zone, and our analysis indicates that there is no advantage to making L_t overly large, since doing so will decrease sensitivity.

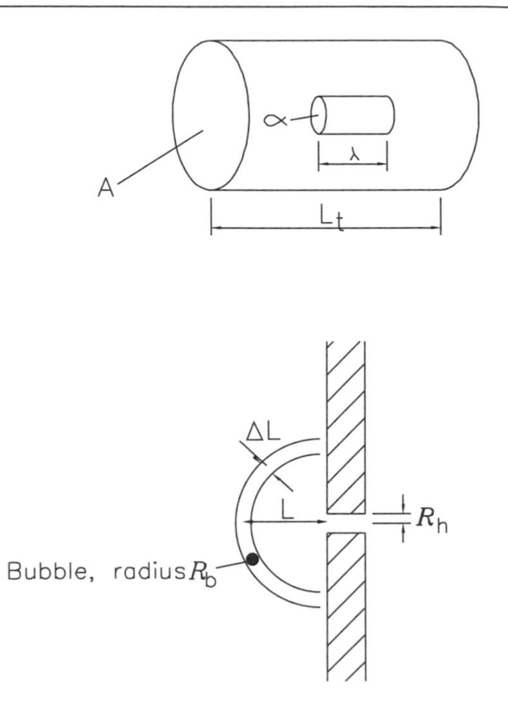

Our analysis tells us only the maximum change in resistance, that is, the change seen when the bubble is in the middle of the sensing zone. Further analysis can tell us how the resistance seen by the sensor varies as the bubble passes through the zone. If we assume that the pattern of current flow is approximately radial through the sensing zone, then we can consider the resistance to be made up of concentric spherical shells, whose area will increase as L^2, where L is the distance of the bubble from the center of the sensing zone, at least when L is large compared to the hole radius, R_h. (We might expect to see a slight difference in sensitivity to bubbles on each side of the cone, since the cone angle restricts the size of the spherical resistive shells on the upstream side. But let's ignore this for the moment.)

The resistance, R_s, of a spherical shell with thickness ΔL (equal to the bubble radius, R_b), and radius L (where the bubble is located) is given by

$R_s = \rho R_b / 2\pi L^2$.

The relative change, $\Delta R_s / R_s$, in the resistance due to the bubble is approximately equal to $4\pi R_b^2/(2\pi L^2)$. Taking these two formulas together gives us

$$\Delta R_s = 2\rho R_b^3 / 4\pi L^4,$$

which gives the resistance change at large distances L from the sensing zone. So we expect the change in resistance due to the bubble to be proportional to the bubble volume, as in our previous calculation, but the resistance should change rapidly with distance from the sensing zone. Once the bubble is in the sensing zone, the current flow will be roughly parallel to the sensing zone walls, so we expect to see a constant ΔR_s with change in L, for $L << $ hole size R_h. In fact, if we reinterpret the original formula, $R = \rho L_t / A$, then we might expect the following equation to hold for small L:

$$\Delta R_s = \rho R_b^3 / (4\pi R_h^4).$$

We could provide one formula to deal with both this situation and that in which the bubble is far away from the sensing zone by imagining a geometry in which L is replaced by $\sqrt{(L^2 + R_h^2)}$; perhaps we could justify this replacement by looking at the geometry of the annuli of resistance and current flow around the bubble as it approaches the hole:

$$\Delta R_s = \rho R_b^3 / [4\pi(L^2 + R_h^2)^2].$$

The plot of this function at least qualitatively reproduces the actual trace shape, which is interesting: it has a peak with a "skirt" on either side. The top of the peak represents the maximum resistance, which occurs as the bubble passes through the plane of the orifice. The skirts represent the resistance measured when the bubble is moving into and out of position.

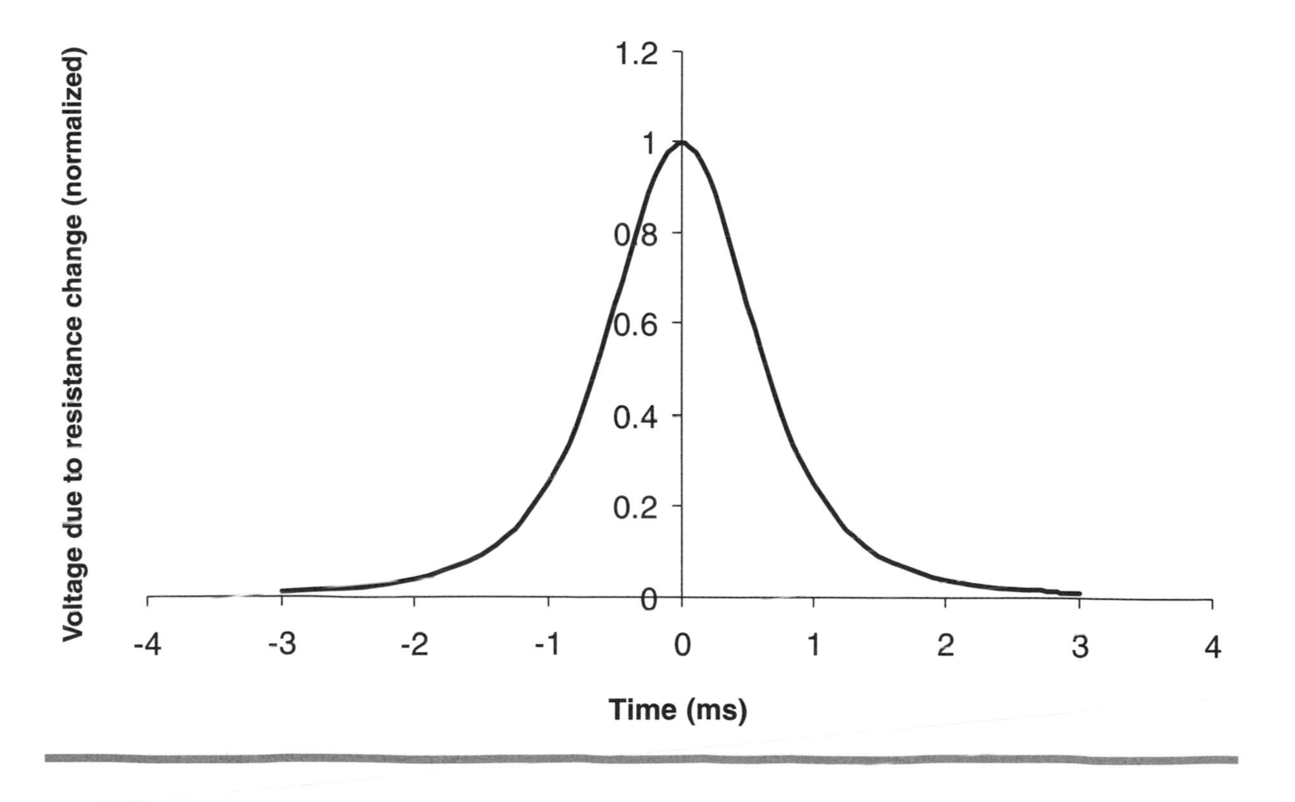

And Finally . . . Oscillographs of the Flowing Bubbles

By sizing the sensing-zone hole differently, you can make systems that sense larger or smaller flows. The hole needs to be big enough to pass all possible bubbles, but for maximum sensitivity and maximum freedom from zero drift, the hole should be no larger than necessary.

Try connecting the system to an oscilloscope. You will see pulse heights and shapes that reflect the passage of bubbles through the sensing zone. When the system is used in flow-metering mode, the pulses pile up to form mountain-range shapes on the oscilloscope, and the trace never returns to the zero. If we adopted

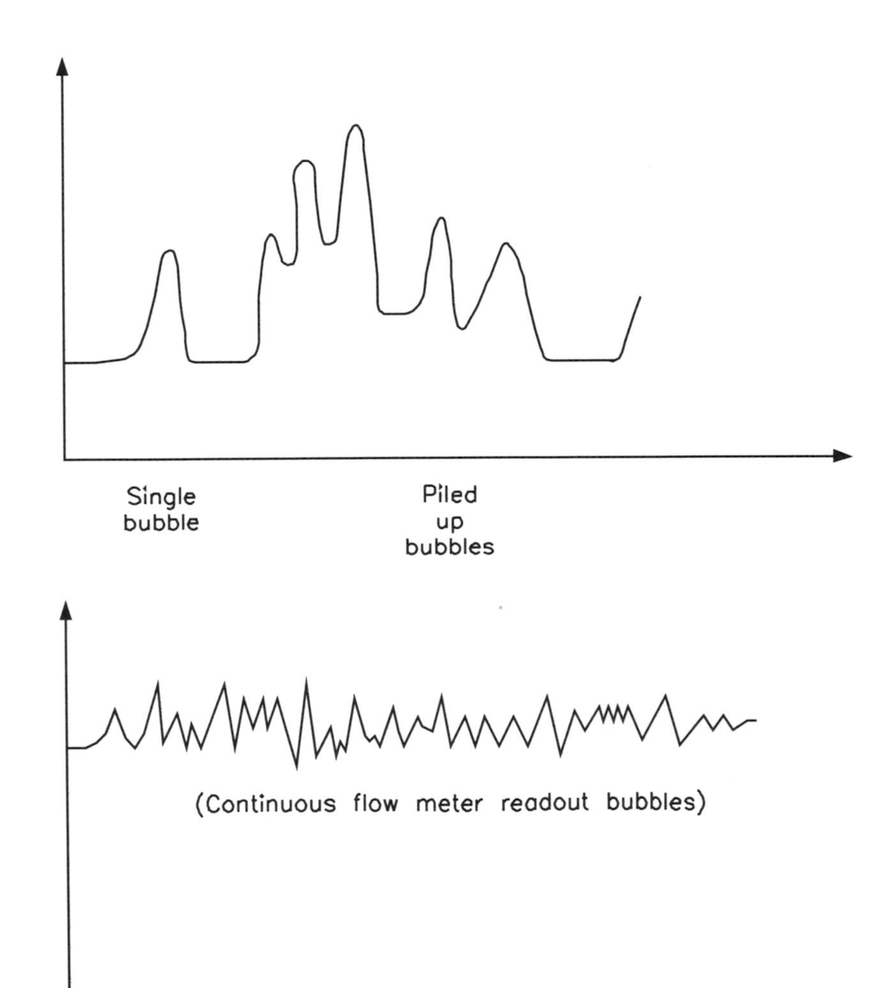

a long, cylindrical sensing zone, then the peaks would have flatter tops. If you connected these signals to a pulse-height analyzer (as used in nuclear physics) and most of the bubbles came as singles, you would have a full Coulter counter for bubbles. But I suppose most people don't have a pulse-height analyzer just lying around.

REFERENCE

Ormerod, Michael G., ed. *Flow Cytometry: A Practical Approach.* 2d ed. Oxford, U.K.: IRL Press at Oxford University Press, 1994.

24 Electronic Elastic

To see a world in a grain of sand
and a heaven in a wild flower,
Hold infinity in the palm of your hand
And eternity in an hour.

—William Blake, "Auguries of Innocence"

Poet William Blake was fascinated by the magic of everything around him, and although I don't have his poetical imagination, I can appreciate the magic in the physical idea of time and the mathematical notion of infinity. The wild flower of Blake's "Auguries of Innocence" delves into the depths of science, from the chemistry of the flower's color and scent and the ecology of the insects that pollinate or parasitize it, to the biochemistry of its growth and the genetics of its DNA. Even the humble and apparently simple grain of sand holds something of interest. Quartz crystal, for that is what the sand grain is, possesses the curious property of piezoelectricity: when you squeeze it, a static electric charge appears, and, conversely, when you apply an electric charge, the crystal contracts slightly. This property allows us to fabricate quartz oscillators, which are the basis of most timepieces today, including wristwatches and computer clocks.

When you hold a rubber ball in the palm of your hand, you also have something magical. Try bouncing something else on the floor, and you will generally find that it doesn't—bounce, that is. Rubbery properties are confined to a rather small number of mostly hydrocarbon polymers that are generically known as elastomers. Elastomers have unusual thermal properties too. Try stretching a big elastic band tight in front of your face, and then gently touch it to your lips. It will be hot. Elastomers heat up when you stretch them. Keep the band stretched and allow it to cool back toward room temperature for a few seconds, then let it relax and feel its temperature again. You will find that another curious property of elastomers is revealed, but I won't spoil the surprise by telling you what it is!

The story of how elastomers were made from latex, the milky sap of the rubber tree, is a complex one. Building on the ancient know-how of native Amazonians who pasted latex onto cloth, pioneers like chemist Charles Macintosh in Scotland used solvent solutions of rubber to coat cloth and make waterproof clothing. It must have been nice to have clothes that kept heavy rain out, but apparently early macintosh raincoats were hopelessly stiff and prone to crack in winter, and they went sticky in summer. I imagine it must have been rather like wearing a coat covered with chewed gum.

Charles Goodyear worked on vulcanization processes from his base in mid-nineteenth-century Connecticut. These processes stabilize elastomers by cross-linking the polymer chains, typically with sulfur. Vulcanization enormously expanded the utility of rubber and made it the important industrial material we know today. Nevertheless, there were many secrets still to be learned: for example, synthesizing rubber from its monomer, isoprene, was difficult because the polymer comes in different isomeric forms, one of which is not elastic. Protecting rubber from boiling heat, atmospheric oxidation, and ultraviolet damage; coloring it successfully; compounding it with strengthening fibers; molding it into large items like tires—the process of overcoming all these challenges was littered with surprises. (Tom French describes many of the problems with tires in his book *Tyre Technology.*) Elastomers are complex and have other unusual properties, and in this project we investigate the curiosities of their light-transmission properties.

What You Need

- ❑ Rubber band (e.g., 6–9 mm wide, 0.7 mm thick, natural color)
- ❑ Ultrabright light-emitting diode (LED; I used green.)
- ❑ Current-limiting resistor for LED (50 ohms)
- ❑ Multimeter with a 2-mA or 200-μA scale
- ❑ Photodiode

- ❑ Transistor
- ❑ Batteries, battery boxes
- ❑ Black tape
- ❑ Ruler (for measuring length to which the rubber band is stretched)

What You Do

The circuit is so simple that you can simply solder the parts to one another, but you may find a piece of matrix circuit board convenient. Set up your circuit so that the LED shines brightly on the photodiode, and the rubber band is guided between the two, perhaps by means of loops of wire on the circuit board, if you have used one. If the rubber band is short, cut it so that you have a good length to stretch out along the ruler. Alternatively, leave it as a band, but pass only one strand through the LED-photodiode pair. You can minimize the influence of ambient light falling on the photodiode by surrounding the arrangement with a little black tape, with cutouts for the LED and the photodiode. The transistor multiplies the diode current approximately two hundred times, so that it can be measured easily on the multimeter. You should see current values in the range of 100 μA or so with the rubber in its relaxed state. When you stretch the rubber band, putting increasing strain on the band, note how the current changes.

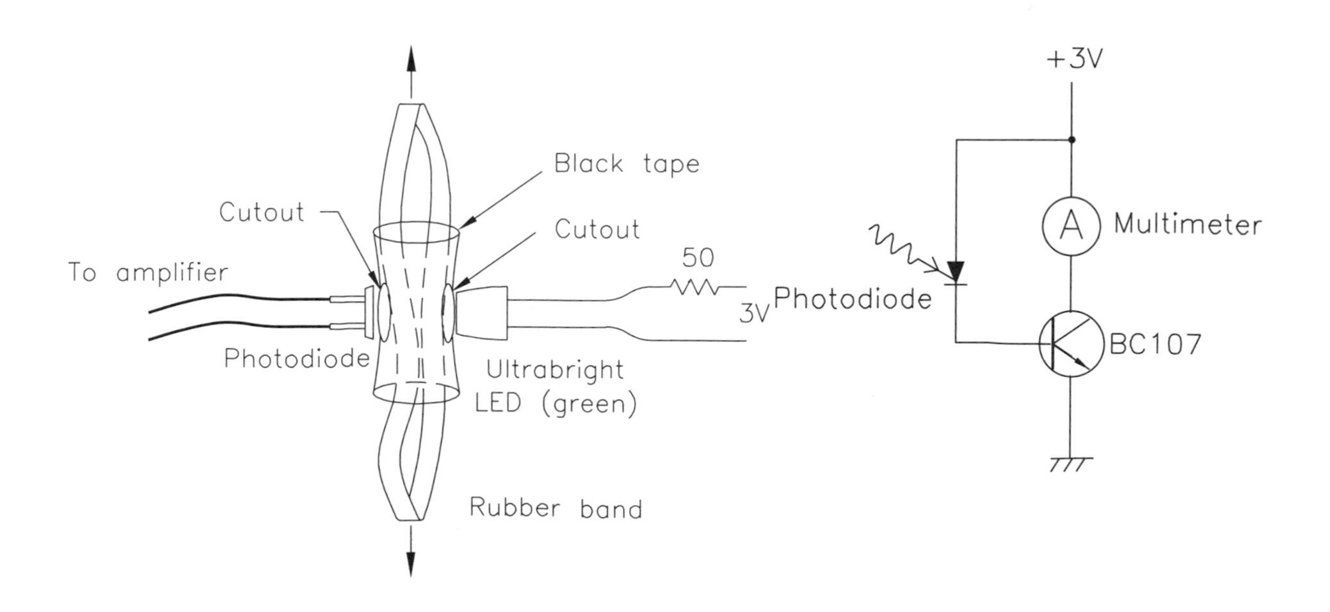

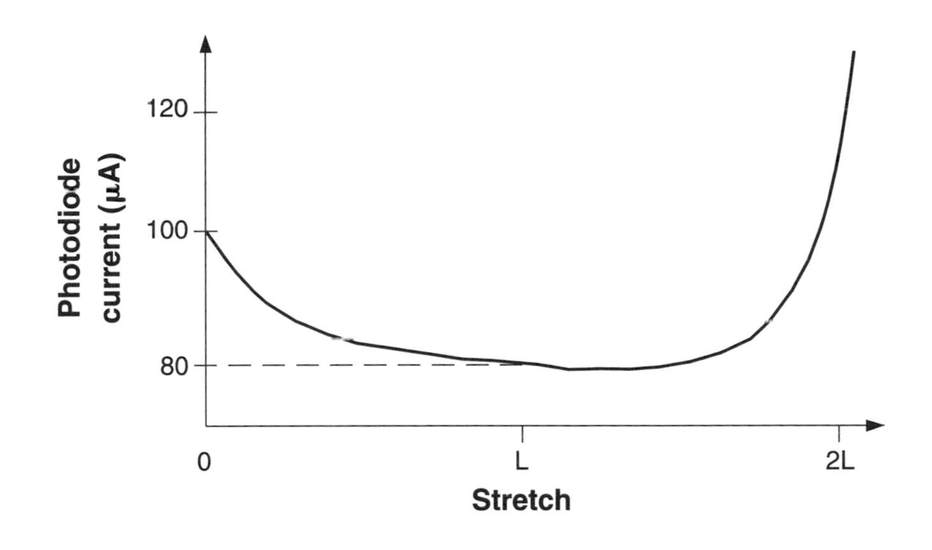

How It Works

When you stretch the rubber band, it gets thinner and allows more light to pass from the LED to the photodiode. The more you stretch the rubber band, the larger the signal, right? Wrong! Actually, you will find that for moderate extensions, the signal will go down as you stretch the band—the rubber gets optically thicker as it stretches. This behavior could be due to the Poisson's ratio of the rubber band (see The Science and the Math), but I think that it's the result of increases in light scattering in the stretched rubber. Many plastics become more opaque when they are bent through a sharp angle. For example, if you sharply fold a reasonably thick (0.5–1 mm) piece of rigid clear or colored plastic, you will, in many cases, find that the creased area will turn opaque white. The increase in light scattering in our rubber band may stem from a similar phenomenon. You will probably measure a decrease in light transmission of approximately 20 percent for all stretch values below about twice the original length.

THE SCIENCE AND THE MATH

Poisson's ratio, ν, is defined as the lateral strain $(\Delta D/D)$ divided by the longitudinal strain $(\Delta L/L)$ in a stretched piece of material of length L and diameter D:

$$\nu = -(\Delta D/D)/(\Delta L/L),$$

where ΔD is the decrease in the material's diameter, and ΔL is the increase in its length. Poisson's ratios generally fall in the range between 0.17 (quartz) and 0.44 (gold). A solid that can be stretched without any shrinkage in the lateral direction (think of the outside surface of a bundle of horsehairs in a violin bow) has a Poisson's ratio of 0. A solid whose volume remains constant upon stretching will have Poisson's ratio of 0.5, as you can see from the following analysis. To make the math easier, we will consider only the case of a cylindrical shape. The total volume, V, of a cylindrical solid is

$$V = \pi D^2 L/4,$$

and the change in volume upon stretching is given by

$$\Delta V = \pi(2DL\Delta D + D^2\Delta L)/4.$$

If $\Delta V = 0$, then

$$2DL\Delta D = -D^2\Delta L,$$

$$2L\Delta D = -D\Delta L,$$

and

$$-(\Delta D/D)/(\Delta L/L) = \nu = 0.5.$$

If the Poisson's ratio of our rubber band was between 0 and 0.5, the band would thin when stretched, which does not match with our optical transmission observations. (Some elastomers do have Poisson's ratios close to 0.5.) So the Poisson's ratio of the rubber is not the whole story.

At high extensions, you will see that the rubber band has shrunk in width, but the change in thickness is not so obvious. The fact that the band's cross section is not circular but rectangular, with one side much longer than the other, may be part of the story. If you consider the band to be made up of a set of square-cross-section bands lying side by side, you might expect to see the same percentage thinning of the band in both directions during stretching. However, there may be some second-order effects of that thinning, since those side-by-side bands may pull at each other laterally.

As I mentioned earlier, many elastomers whiten and turn slightly opaque when fully stretched. This also happens to many colored plastics when you bend them. This effect was the basis for the old Dymo label-making system. When a letter pattern was impressed into a sheet of colored plastic, the letter showed up as a white strained area on the colored background. I don't know exactly why this should be, but since the white plastic is still quite strong, at least for a certain degree of bend, I don't think the effect is due to microcracking. It seems more likely that the strained polymer chains don't fit together as well, which results in microvariations in density and hence refractive index. The fact that variation in refractive index on a microscopic scale leads to opacity is well known to those who make salad dressing or mayonnaise. Mix olive oil and water, both of which are clear but have different refractive indexes, in a blender, and you soon get a creamy opaque mixture. This kind of effect probably accounts for the optical behavior seen in this project.

And Finally . . . High-Speed Electronic Elastic

You may want to test just how fast your electronic elastomer can react. The speed of sound in the material is probably comparable to the speed in air. If you pull on a 34-mm piece of rubber band, a sound wave will travel along the strip, and the whole strip will stretch in less than 1.0×10^{-4} s ($34 \times 10^{-3}/340$), or 0.1 ms.

Perhaps you could use the strip as a "microphone": you could try attaching a slightly stretched band with a sensor to a diaphragm, say, the bottom of a disposable plastic (but not foam plastic) coffee cup, and then use the sensor to give an input to an audio amplifier.

REFERENCES

Cowie, J.M.G. *Polymers: Chemistry and Physics of Modern Materials.* Glasgow: Blackie; London: Chapman and Hall, 1991. A polymer text that provides insights into elastomers.

French, Tom. *Tyre Technology.* Bristol, U.K.: Adam Hilger/IOP Publishing, 1989. An excellent discussion of the peculiarities of rubber.

25 Light Tunnels

> Through the opening she saw a tunnel that curved slightly, so that its goal was not visible. . . .
>
> She was frightened of the tunnel: she had not seen it since her last child was born. It curved—but not quite as she remembered; it was brilliant—but not quite as brilliant as a lecturer had suggested. Vashti was seized with the terrors of direct experience. She shrank back into the room, and the wall closed up again.
>
> —E. M. Forster, "The Machine Stops"

A number of today's sensors depend upon the measurement of the amount of light transmitted through an object and subsequent correlation of that amount to the measurement needed. Medical applications often use this principle. For example, blood sugar tests used by diabetics employ a small light-transmission meter together with a piece of paper soaked with a reagent that reacts with sugar in the blood. The most common medical application of this principle is probably the pulse oximeter, the gadget that clips onto your finger or earlobe and gives both your pulse and the level of oxygenation in your blood, both of which are vital parameters in medical treatment. The pulse oximeter is actually quite a complex device, and many years of trial and error, starting in the 1950s, were required before reliable instruments became available in the 1970s (see John Webster's *Design of Pulse Oximeters*). The device uses two different light-emitting diodes (LEDs) so that the transmission of both infrared and visible light can be measured. The pulse oximeter measures the change of light transmission with time rather than the absolute transmission. The amount of transmitted light changes from moment to moment because blood flows in pulses driven by the pumping of the heart. By measuring only the changes in transmission with time, the instrument can ignore the effects of ambient light, skin color and thickness, and other static effects, so that the instrument reads only the blood oxygen level.

The light tunnel sensor in this project is another application of the principle of measurement via light transmission. This application—although not used widely, if at all, today—actually works particularly well. We've all used the expression about seeing the light at the end of the tunnel, and many of us have lived it, stumbling along in the damp and dark, worrying about bats, creeping insects, and spider webs, perhaps even suffering from the simple fear of being in an unfamiliar and unattractive place. When a little glimmer of light appears as you round a corner in the tunnel, a wave of relief sweeps over you. In this project we use the light at the end of the tunnel to produce an optical or electrical output that allows us to sense angle.

What You Need

- ❑ Opaque tubing (e.g., 40–100 mm of black tubing, 5-mm internal diameter)
- ❑ LED or IRED (infrared emitting diode)
- ❑ Glue (preferably opaque)
- ❑ Resistor to limit current to LED (e.g., 30 ohms)
- ❑ Photodiode, possibly with an infrared filter (or another photosensitive LED or IRED)
- ❑ Batteries, battery holders (e.g., 3 1.5-V AA)
- ❑ Multimeter
- ❑ Wood and hinges (or a plastic adjustable drafter's square or protractor)

For light tunnel servo

- ❑ All the above items
- ❑ Operational amplifier, µA741
- ❑ Transistors (see circuit diagram)
- ❑ Resistors
- ❑ Motor with gearbox (e.g., low-current type sold as a "solar motor" capable of 6 rpm on 0.5-V input)
- ❑ Batteries, battery holders (e.g., 8 1.5-V AA)
- ❑ Potentiometer
- ❑ Wires

What You Do

Glue the LED and the photodiode to the ends of the tubing. I found that black hot-melt glue was perfect, but if you can't find that, try stirring a little black toner powder (from a photocopier) or some other black coloring material into epoxy resin. If you use clear tubing or clear glue, cover the entire assembly with black tape.

Now apply current to the LED and the photodiode and adjust the multimeter to a microamp scale. As you bend the tubing, the photodiode output should fall progressively as the angle of bend increases. Fastening the sensor assembly to the wood (or plastic protractor, if that's what you are using) with one end on each side of the hinge and measuring the angle on the hinge (or protractor) will enable you to measure the readout quantitatively.

How It Works

The sensor works by largely preventing the reception of direct light by the photodiode, that is, by forcing the light to bounce at least once. The light transmission

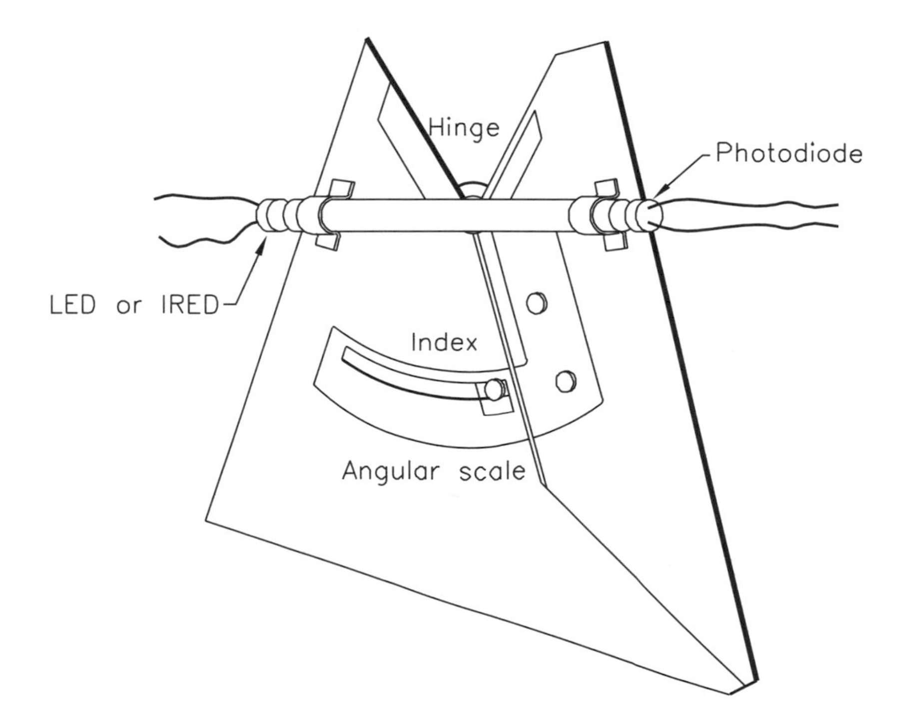

decreases with the number of bounces required for the light from the LED to reach the other end of the tube. You may find that the system also works if you use another IRED or LED instead of a photodiode as a light detector. Many LEDs are light sensitive, and since they are generally packaged in round-nosed, bullet-shaped plastic capsules, they can be conveniently plugged into the end of the tubing. You should ensure that the "bandgap" of the emitter—the photon energy that it emits—is larger than the minimum photon energy required for the receiver. Photon energy is highest at the blue end of the spectrum and lowest at the red and then infrared end. This means that you can't expect to receive infrared light with a red-sensitive detector, or red light with a green-sensitive detector.

Because the current provided by the light tunnel sensor varies smoothly and is stable and predictable relative to the deflection applied, the sensor can be used as a feedback sensor for a servo motor. A servo motor, or servo-actuator (servo for short), is a device for providing a motion output under the command of an input that has a feedback sensor element. The feedback sensor ensures that the servo's output precisely imitates the input drive, by sensing the difference between a feedback of the servo's output and the input and by driving the motor to minimize that difference.

Control systems, like that of the servo, that use feedback are sometimes described as a *closed loop* because a loop connects the servo output back to the

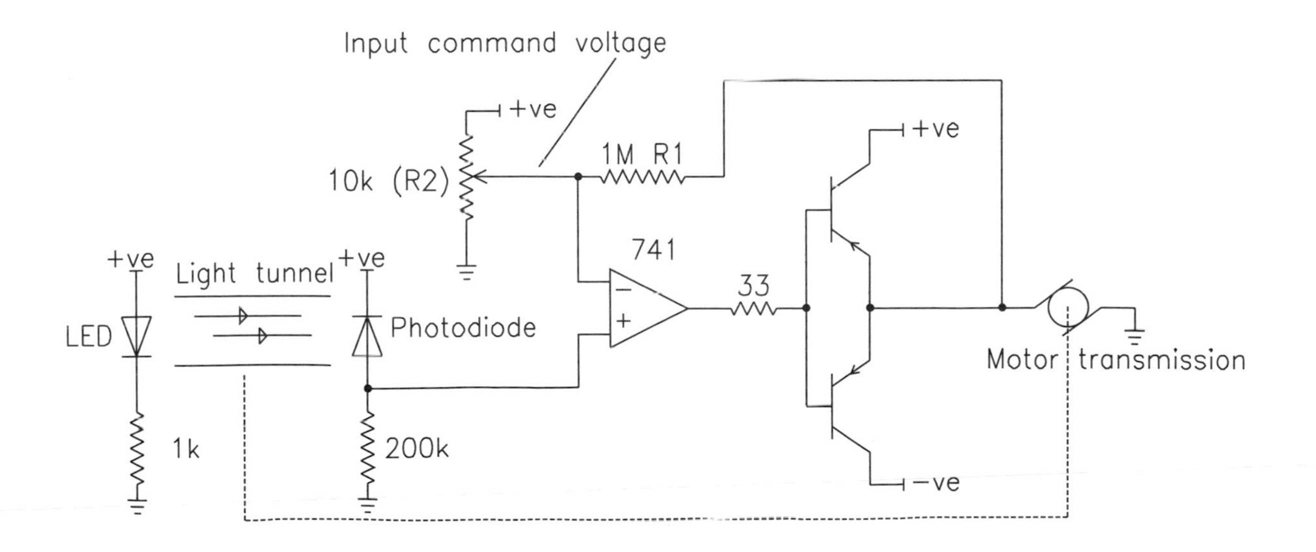

input. (The circuit diagram makes this clear.) The normal sort of control, whereby you amplify the input by a constant factor, M, and then rely on the output to follow it, is called an *open loop*. The needle on an analog meter is an open-loop system, whereas the model control servo in the servo telegraph project in this book is a closed-loop system.

The light-tunnel servo circuit is designed to amplify the difference between the command input and the feedback input by a factor governed roughly by the ratio between the resistance of resistor R_1 and half the resistance of potentiometer R_2. The light tunnel on the servo feeds back the position of the output arm to the operational amplifier (op amp). The two transistors on the output are "emitter followers" and are there simply to boost the relatively small current provided by the op amp to a suitable level.

The easiest way to demonstrate how the servo works is to use another light tunnel as an input. By connecting the op amp inputs to the input light tunnel and the feedback light tunnel, you should find, if the polarity of the various connections is correct, that the servo tries to copy the input light tunnel. If you deflect the input down, the output will move down, and vice versa. The motor is being activated whenever the output feedback from the light tunnel is different from the input.

The advantage of using a closed-loop control system will become apparent if you try to move the output when it is sitting stationary with a constant input. The servo, although not apparently doing anything, will immediately wake up and apply motor power to push back against your efforts if you try to move the output arm away from its proper position. The servo is applying power to correct the error in its output. If you try that with an analog meter, there is no such correcting action: even if the analog meter were a large powerful mechanism, it would not do anything different if you pushed on its arm: it would not apply any special correcting force.

THE SCIENCE AND THE MATH

The light tunnel sensor works by preventing reception of direct light and thus forcing the light to bounce at least once. The light transmission decreases smoothly with the average number of times the light bounces before reaching the other end of the tube. One way to quantify the operation of the sensor is to estimate the number of bounces needed to get from one end of the tube to the other. The maximum length, L_{max}, between bounces is given by Pythagoras' theorem:

$$(L_{max}/2)^2 + (R - d)^2 = R^2,$$

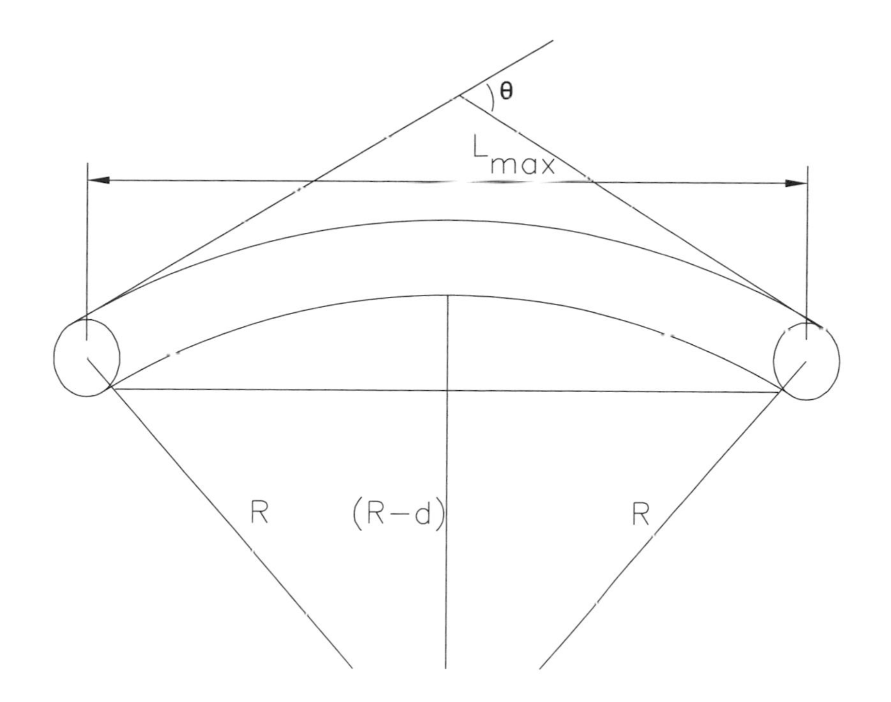

or

$$L_{max} = 2\sqrt{(2Rd - d^2)},$$

where R is the tube bending radius and d is the tube diameter. For more typical bounce lengths, L_{typ}, we might take the same formula with $d/2$ as the effective tube diameter:

$$L_{typ} = 2\sqrt{(Rd - d^2/4)}.$$

The deflection angle, θ, for a tube with length L_{tube} is given by

$$\theta = L_{tube}/R,$$

so

$$R = L_{tube}/\theta,$$

and

$$L_{typ} = 2\sqrt{(dL_{tube}/\theta - d^2/4)}.$$

If we estimate the transmission as the number of bounces, B, times the reflection coefficient of the tubing, taken to be approximately constant at $1 - \alpha$, α being the absorption coefficient per bounce, then

$$B = L_{tube}/L_{typ}.$$

The transmission coefficient, E, is given by

$$E = (1 - \alpha)^B.$$

For small numbers of bounces and small values of α—that is, when most of the light is reflected—we can use a slightly simpler formula:

$$E \sim 1 - \alpha B.$$

This formula is correct within a couple of percentage points for values of E greater than 0.75. Of course, for large numbers of bounces and small values of α, we will see an exponential decrease of light transmission with distance along the tube:

$$E = \exp(k/L_{typ}),$$

where k is a constant. This equation takes the same mathematical form as the Beer-Lambert law for the

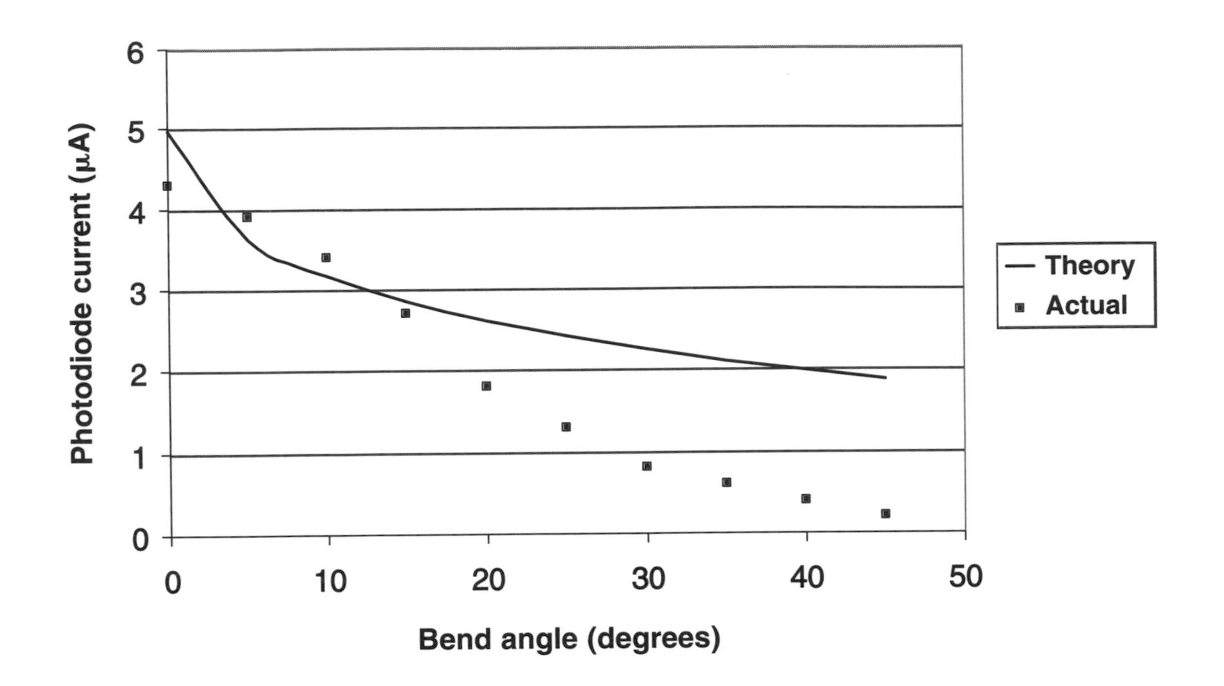

absorption of light as it passes through a homogenous absorbing medium, the law used for calculating, for example, light transmission in medical pulse oximeters.

Our analysis so far has ignored the direction of light as it enters the sensor from the LED. It also ignores the change in reflection coefficient with angle (the coefficient will typically be large for small angles—which is why you can get such a good reflection from a piece of black glass). The analysis also sweeps noncircular bending of the tube and many other things under the carpet. Taking all these things into account would be very complicated. Could a slightly more sophisticated analysis give a simpler formula? Perhaps we need some computerized "ray-tracing" to come up with a better estimate of *E*.

In any case, comparison of a graph of the exponential function with output-current figures from a real sensor shows a reasonable degree of agreement, at least for medium and large values of the deflection angle, θ. The agreement is poor for small θ because we have failed to correctly account for what happens with direct light transmission when the tube is nearly straight. If you peer down a short length of tube and bend it, the initially circular disk of direct light will change to a leaf shape and gradually disappear as you continue to bend the tube, a bit like an eclipse of the sun in reverse. To account for the small-angle behavior of the light-sensor tunnel, we actually need something like the "eclipse equation." The equation gives the relationship between the area of overlap area, A, of two similar-size disks (e.g., the Sun and the Moon) of radius r and the distance, $2s$, between their centers:

$$A = 2r^2 \cos^{-1}(s/r) - 2s\sqrt{(r^2 - s^2)}.$$

Actually, the area of the LED that the sensor can see directly, A_{vis}, is given by $\pi r^2 - A$, that is, by

$$A_{vis} = \pi r^2 - 2r^2 \cos^{-1}(s/r) + 2s\sqrt{(r^2 - s^2)}.$$

Because the sensor has a number of lines of sight, not just an axial one, this equation will not describe

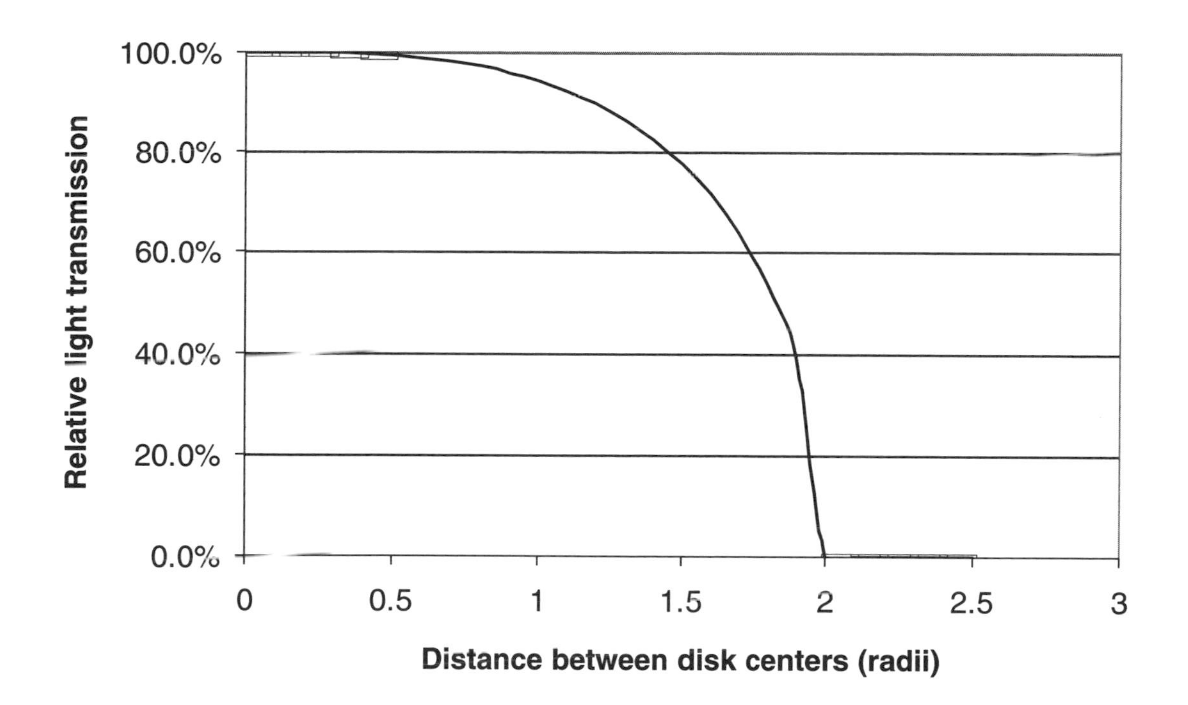

its behavior precisely, but a graph of the function at least shows the right kind of shape. Adding this shape to the shape obtained from the multibounce formula gives us quite a good model of the sensor.

And Finally . . . Double Bending Light Tunnels

Various sensors that use some variation on the principle of the light tunnel have been proposed, although none seems to have caught on. The most popular application for tube-based sensors has been for monitoring the angle of human limbs and fingers. The Sayre glove was a system built for tracking hand movements in experimental arrangements. In 1985, Thomas Zimmerman patented what today is called a "data glove," an "optical flex sensor" that uses flexible sensors to allow hand movements to be input to a computer in a virtual reality system. Zimmerman even proposed complex arrangements of colored LEDs and multiple, or multiplexed, detectors that he claimed would be able to sense movement in different directions of bend. Maybe you can devise a simpler system for sensing the degree of bend in two planes, say, theta and phi deflection angles, using a single tube.

Alternatively, you could just try to make our simple sensor work better. The mechanical characteristics of elastomer tubes change slowly with time. Natural rubber, for example, will lose strength and elasticity with age unless it is heavily protected with an antioxidant. Synthetic rubbers may be more durable. Plastic tubing might work well, but plastics will undergo "creep" to a much greater extent than do elastomers: that is, they will tend to slowly mold themselves into their new positions, adopting a curled shape if they are left for a long time in a curled posture, or adopting a straight shape if they are left straight long enough. Paying attention to the mounting of the tube will help to mitigate the effects of creep but will not entirely eliminate it.

It is not necessary to use an electrically driven light source at both ends of the tube. You could use ambient light, or you could pipe light in with an optical fiber. And the photodiode could be replaced by an optical fiber pickup. When both the sensor and the emitter are optical fibers, we have the opportunity to mount the sensor in a relatively hostile environment since, within limits at least, pressure, temperature, corrosive liquids, vibration, and electromagnetic interference will not affect the optical fibers.

REFERENCES

Webster, John G. *Design of Pulse Oximeters*. Bristol, U.K.: Institute of Physics Publishing, 1997.

Zimmerman, Thomas G. "Optical Flex Sensor." U.S. Patent no. 4,542,291, September 17, 1985.

26 Reverse Electric Lamp Solar Tracker

> You drop a cup and it breaks, and you can sit there a long time waiting for the pieces to come together and jump back into your hand.
>
> —Richard Feynman, *The Character of Physical Law*

Our world appears to be a place of irreversible processes. If you drop a cup on the floor, the shards of pottery will certainly never leap up and join seamlessly together to re-form into a cup. After the break in a game of pool scatters the pyramidal group of balls over the table, they don't come back together to re-form the pyramid. But although many of the world's processes are irreversible, certain physical phenomena are reversible, or nearly so. In his BBC lectures, physicist Richard Feynman gives a number of examples of apparently irreversible phenomena and explains that they are irreversible only at the macroscopic level. The collisions of atoms and electrons are completely reversible. Many physical sensors are actually transducers—that is, they can convert one form of energy into another and, to some extent, back again. Loudspeakers, for example, are also microphones: they can convert electricity into sound, or sound into electricity. Others are not so reversible, however. Electric filament lamps normally accept electricity as an input and produce light as an output: when electricity passes through the piece of tungsten wire, it heats up to white-hot and emits light. Nothing reversible there, you might think. Nevertheless, the process can indeed be reversed. Here we see how you can put light back into a lamp and get electricity (or at least an electrical sensor output) from it.

What You Need

For the reverse electric lamp

- ❑ Electric lamp (e.g., auto 12-V, 5-W type)
- ❑ Magnifying lens (e.g., 3 or 4 diopter reading lens, 90 mm in diameter)
- ❑ Multimeter with a low ohm range
- ❑ Connecting wires

For the solar tracker

- ❑ Bicycle wheel
- ❑ Base on which to mount the wheel
- ❑ Two electric lamps
- ❑ Relay
- ❑ Operational amplifier or comparator amplifier
- ❑ Transistor
- ❑ Resistors (100 kohm, 10 kohm, 100 ohm)
- ❑ Matrix circuit board to mount components
- ❑ Batteries
- ❑ Electric motor with very slow drive output (A battery-operated barbeque spit is ideal.)

What You Do: Reverse Electric Lamp

Connect the multimeter and the lamp, and set the multimeter to its lowest resistance range, normally 0–200 ohms. Now arrange the lens so that it focuses an image of the sun on the lamp filament. Illumination of the filament causes an increase in resistance that will register on the multimeter. By measuring the maximum change in resistance, you can measure the power of the sun. The maximum resistance change will depend on the sun's height in the sky and on the amount of cloud cover and haze. On a completely or partially clear day, you should have no trouble getting a substantial signal.

CAUTION

Never stare directly at the sun; you could easily damage your eyes. Human eyes are not designed to withstand direct sunlight for more than a fraction of a second. When we are outside in the sunshine, our normal reflexes cause our eyes to shift direction frequently to avoid the light. We ignore those reflexes at our peril.

What You Do: Solar Tracker

The reverse electric lamp can be used as an instrument for monitoring the intensity of the sun, that is, as a kind of solar gauge, if you like. By using a focusing lens, you can be assured that you are measuring only the sun's output, not any other effect. A nonfocusing solar gauge would be affected by other variables; for example, a gauge that used a thermistor would be affected by the ambient temperature. You can also use the reverse electric lamp to construct a solar tracker. As the sun sweeps the sky, the tracker will try to follow it around. A tracker is a useful gadget if you want to generate solar power: you can use it to ensure that you get the most out of a photovoltaic solar panel, for example. In Egypt, thousands of years before electric light was available, hired boys would track the sun with a mirror, reflecting its light to fixed mirrors in the tunnels of the pyramids and the tombs of the ancient pharaohs, carrying light deep into the structures. The boys had to move the mirror at just half the angular speed of the sun to keep the output light traveling in a constant direction.

To track the sun's path across the sky, you generally need two tracking systems, one for azimuth and one for altitude. However, if we mount our tracker

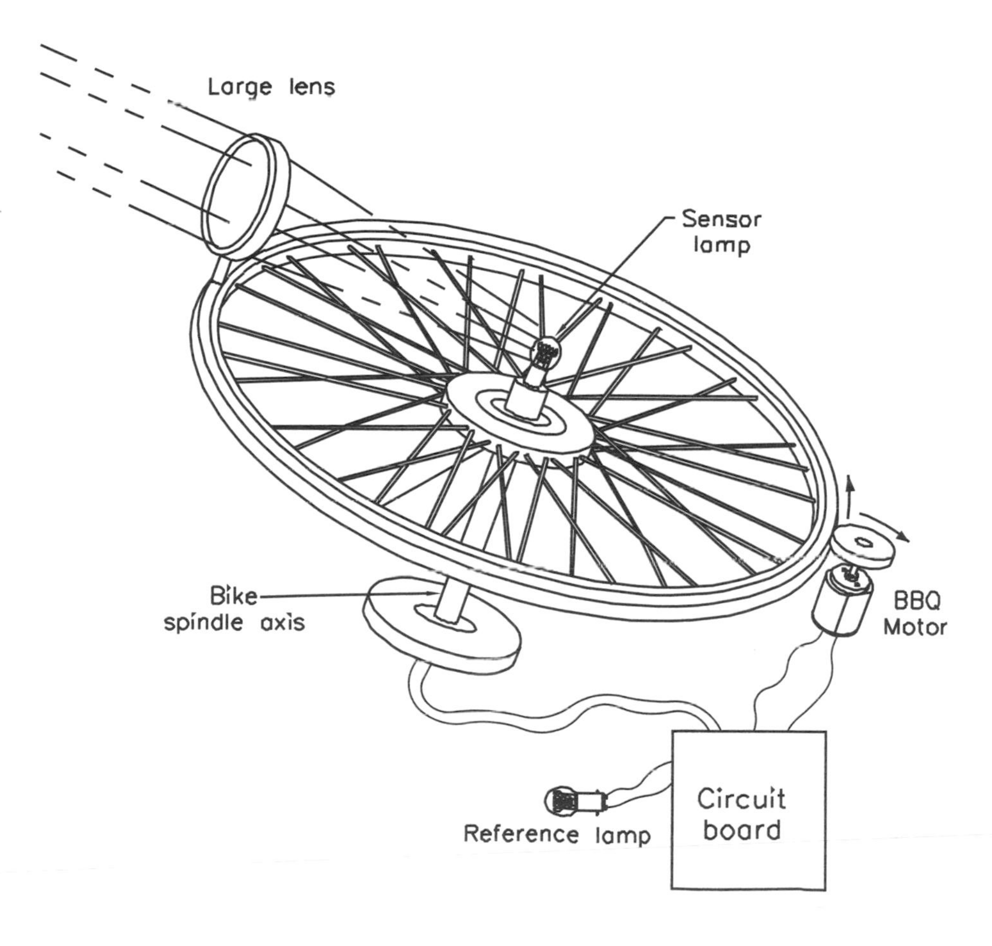

equatorially—that is, if we arrange it at an angle to the ground such that it sweeps the same arc as the sun—then a simple rotation in one plane is sufficient.

Our tracking system is based on a bicycle wheel driven by a barbeque-spit motor. We mount the lens and a lamp on the wheel, feeding the output from the lamp to the circuit, as illustrated. Another lamp is mounted on the circuit board to act as a reference resistance. The wheel is bolted via its spindle to a strong wood board, which is then propped up at the angle of latitude of where we happen to be, adjusted for the season of the year, pointing due south. The relay connects the barbeque motor to a battery, which causes the bicycle wheel to swing slowly around when the motor is switched on. The resilience of the tire and its grip on the wheel are a convenient way of transferring power. I used a lens with a focal length almost equal to the wheel's radius, meaning that the sensor bulb could remain stationary at the axle position; however, other focal lengths will work.

How It Works

When the sun illuminates the filament, its resistance increases relative to the resistance of the circuit board–mounted lamp. This increase is detected by the circuit, which causes the motor to run, rotating the bicycle wheel until the image of the sun moves just a little off the filament. The filament then cools, and the motor stops. The tracker thus moves around in small distinct movements, following the sun.

The resistivity of a tungsten filament varies by almost a factor of five between room temperature and red-hot (700 °C) and by a factor of eight between room temperature and yellow-hot (1,200 °C). By contrast, the alloy commonly used in electrical heaters, Nichrome, varies by only a few percentage points from 0 to 800 °C. (Using a completely heatproof sensor element—tungsten does not melt until 3,400 °C—is a wise precaution in a device that could, at least in principle, become white-hot.) Even with crude focusing, the filament resistance should change by 50 percent, which represents a temperature change of at least 100 °C.

The circuit I built is a bridge-type circuit that incorporates two lamp filaments, one in the sunbeam, the other not. When the sun illuminates one lamp, the change in the balance on the bridge causes the op amp to swing its output and switch the relay. A separate low-voltage single-cell battery is used to activate the bridge, since the low resistance of the lamp filaments would otherwise be a problem in that too much power would be consumed.

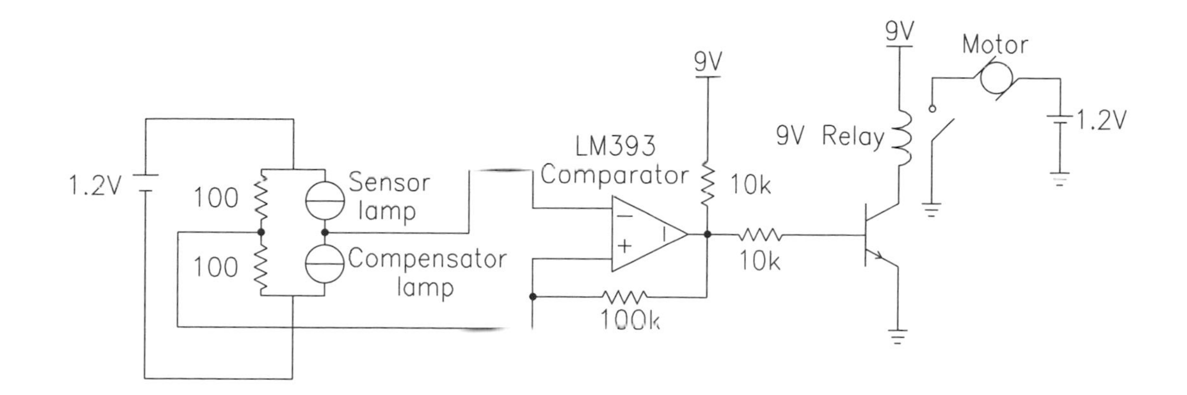

Our tracker won't work on an overcast day, but it might also run into problems under partly cloudy conditions: the tracker could lose the sun during a cloud transit and then be unable to recover if the sun went past the sensor entirely.

THE SCIENCE AND THE MATH

Where is the irreversibility in the emission of light from a filament lamp? First, let's consider what happens in the "forward" direction. Negatively charged electrons in the metal filament are pushed through the wire under the influence of an applied voltage. They collide with irregularities in the regular array of ions that make up the metal's solid matrix. In these collisions, the energy of the electrons is transferred into the vibrational energy of the metal ions, and the metal heats up. When the metal is hot, the vibration of the charged electrons and ions emits little packets of energy called photons. (Any accelerating charge radiates electromagnetic energy, quantized in photons. The electric charge moving up and down in a radio antenna is what makes a mobile phone work, for example.) The emitted photons are mainly in the infrared region of the electromagnetic spectrum, but as the temperature rises to 2,000 °C and higher, a proportion of the photons are in the visible region: the filament becomes white-hot and visible.

Now let's think about the backwards process. Photons from the sun, focused onto the wire filament, are absorbed and cause the ions and electrons in the metal to vibrate. So far, we've got the reverse of the forward process. However, the next step is where irreversibility comes in: the vibrating ions and electrons do not conspire together to push electrons along in the "reverse" direction or, indeed, in any particular direction at all. Focusing sunshine on the filament does not create a DC current flowing along the wires to the lamp terminals. Nevertheless, we can still use the part of the backwards process that is reversible: the temperature rise in the filament. We can detect that rise by means of the resulting increase in the electrical resistivity of the filament. The temperature that the filament will reach when illuminated by the sun will depend upon the temperature of the sun, which is about 5,000 °C at the surface.

The filament temperature will also be affected by whether the lamp bulb is filled with gas; if the surface of the bulb remains near room temperature, some heat is lost to convection. Some small lamps are still made with vacuum bulbs, and if you can find one of these, then you will have a instrument that is more sensitive and that, perhaps, has a simpler response, that is, a response nearer to simple proportionality.*

*There used to be a type of pyrometer, the disappearing filament pyrometer, based on a glowing filament. The radiation from the target surface was focused on the filament by means of a lens, and then a carefully calibrated variable current was applied to warm the filament. To measure the temperature of a furnace, for example, the user looked through the pyrometer's viewfinder at the red-hot inside of the furnace and then adjusted the current until the filament was neither dimmer nor brighter than the furnace image. The current control knob could be calibrated directly with temperature. I actually used one of these antiques quite recently to test some burners in which I had melted several expensive thermocouples; I eventually convinced myself that a noncontact thermometer of some kind was called for. A more modern electronic instrument later showed that the temperature given by the disappearing filament pyrometer had been substantially correct.

And Finally . . . The Bolometer

You could build a more sensitive version of the solar gauge, called a bolometer, using the bridge sensor from the tracker. The bolometer, invented in 1880 by airplane pioneer Samuel P. Langley, is just an optimized version of our lamp filament, but it works with infrared photons and measures sources with much smaller temperature differences. A bolometer could indeed be considerably more sensitive than our solar gauge if a bundle of fine wire or a thin-film element—the fine, enamel-insulated copper wire used for coils in transformers might be a good choice—was used as the sensing element. By squashing a large random tangle of such wire into a small ball, you'll get an assembly that will absorb much of the incident light by multiple reflection, although the copper wire will increase in resistance by only 50 percent with a 100 °C rise in temperature. Studio lamps would surely be powerful enough to give you a good signal, but perhaps if you optimized your detector, you could pick up an ordinary lamp, or even sunshine. Or a polar bear?

> O Langley devised the bolometer:
> It's really a kind of thermometer
> Which measures the heat
> From a polar bear's feet
> At a distance of half a kilometer.
>
> —Anon.

REFERENCES

Feynman, Richard. *The Character of Physical Law.* London: BBC, 1965. A series of lectures broadcast by the BBC in the summer of 1965 and later issued in book form.

Weber, Robert L., ed. *A Random Walk in Science.* London: IoP, 1973. The bolometer poem is quoted on p. 84.

27 Gravity Diode

"You know," said he, "what progress artillery science has made during the last few years, and what a degree of perfection firearms of every kind have reached. Moreover, you are well aware that, in general terms, the resisting power of cannon and the expansive force of gunpowder are practically unlimited. Well! Starting from this principle, I ask myself whether, supposing sufficient apparatus could be obtained constructed upon the conditions of ascertained resistance, it might not be possible to project a shot up to the moon?

". . . by calculations I find that a projectile endowed with an initial velocity of 12,000 yards per second, and aimed at the moon, must necessarily reach it. I have the honor, my brave colleagues, to propose a trial of this little experiment."

—Jules Verne, *From the Earth to the Moon*

Frenchman Jules Verne's scientific imagination gave the world pioneering epics of science fiction such as *Twenty Thousand Leagues under the Sea* and *Around the World in Eighty Days.* Until recently, scientists have pooh-poohed his idea for a gargantuan gun for launching satellites, proposed in his book *From the Earth to the Moon.* It is certainly true that rockets provide the most versatile and practical technology for reaching escape velocity and hence outer space. However, calculations have actually proved that Verne's vision of a gun could—but only just—be made to work. Multiple breeches to "top up" the gas pressure in the barrel would be needed, but the technology to accomplish this is known from the German V-3 weapon as well as from later research. Some even more advanced technology, say, high-pressure hydrogen gas, would also be needed. But, with a NASA-level budget, it just might be possible. The gun might ultimately even be a practical launch system for unmanned spacecraft, but it wouldn't be easy.

It is a curious fact that accelerating an electron to 12,000 yards per second is really easy: all you need is a little vacuum technology and an electric potential —a mere 0.4 mV, one-thousandth the voltage you get from a small battery, will suffice. However, this fact serves as a reminder of the extreme weakness of the gravitational force and the high strength of the electric force. You might be forgiven, then, for thinking that the idea of using electrons to detect gravity is therefore a bit of a nonstarter: surely, you might think, the gravitational force would be completely swamped by the huge electric force in every conceivable experiment. Actually, it is perfectly possible to detect gravity using nothing more sophisticated than an automotive electric lamp.

The type of electric lamp with a white-hot piece of tungsten wire inside a bulb of thin glass filled with inert argon gas is still the most common type. It's not much different from the lamps that Thomas Edison and Joseph Swan made in the late nineteenth century, the lamps that kick-started the whole electricity industry. Actually, in those days the filament was more commonly made from trickier carbon, and the glass bulb was evacuated rather than filled with gas. A few years after this type of lamp was invented, physicist John Fleming came up with a kind of one-way valve for electrons, what we now call a vacuum tube diode or a thermionic diode.* This diode had a filament suspended in a vacuum with a third electrode. In addition to emitting heat and light (electromagnetic radiation), the hot filament also emitted electrons.

*The diode effect was actually observed by Edison and was in fact called the Edison effect. Somehow, however, the great inventor failed to see a use for it, and it was Fleming and the radio entrepreneur Marconi, for whom Fleming acted as a consultant, who actually used the effect to develop a workable device. They put it to use in early radio receivers.

What You Need

- ❑ Dual-filament lamp (The lamps from auto tail lights are best—they have 5-W and 20-W filaments.)
- ❑ 1 or 2 PP3 batteries
- ❑ 2 car batteries or car-battery chargers (1 for the experiment, 1 for blowing one of the lamp filaments)

- ❑ Multimeter with low current range (e.g., 0–100 μA or 0–50 μA)
- ❑ Transistor (optional)

What You Do

Prepare the lamp by blowing the larger filament. I did this by approximately doubling the normal voltage, to 24 V: I wired up two auto batteries in series and connected them to the lamp. Within a few minutes the lamp (which was impressively bright, by the way) blew.* Although the filaments sometimes break right in the middle, they most often break nearer to one end or the other. I don't know whether it matters much, but I suspect that the diode will give a greater current if the sensor electrode has a piece of the blown filament sticking to it, as it will then collect more charge and be more sensitive. If you blow several lamps, you can choose the best.

After you've blown the larger filament, connect the lamp to the battery charger and the multimeter, as shown in the diagram. The filament that remains, the smaller one, needs to be heated to bright yellow—say, at least 12 V for a normal auto lamp designed for about 14 V. Now rotate the lamp around the axis of the filament and monitor the current measured. (Don't touch any of the bare

*Filament lamps have a lifetime that depends strongly on the applied voltage and the temperature of operation. As temperatures rise, tungsten evaporates, particularly from notches in the wire filament. Because a slight notch has high resistance, it becomes hotter and loses metal by evaporation, thus becoming a bigger notch: an accelerating vicious circle is set up that soon destroys the filament. Measurements of lamp lifetime at the typical operating temperature of 2,500 °C vary as the thirty-third power of the temperature! A mere 50 °C (2 percent) rise in temperature will shorten a typical filament's life by a factor of two. Manufacturers of incandescent filament lamps typically give a formula for the lifetime that varies as the twelfth-power of the ratio between the applied voltage, V, and the rated voltage, V_0:

$T = T_0(V_0/V)^{12}$,

where T_0 is the rated lifetime. So by applying 24 V to a nominally 14-V auto lamp, I reduced its lifetime to 0.15 percent of normal. With a normal lifetime of, say, 500 hours, we might expect it to last 45 minutes; in practice the lamps I tried blew in just 10 minutes or so.

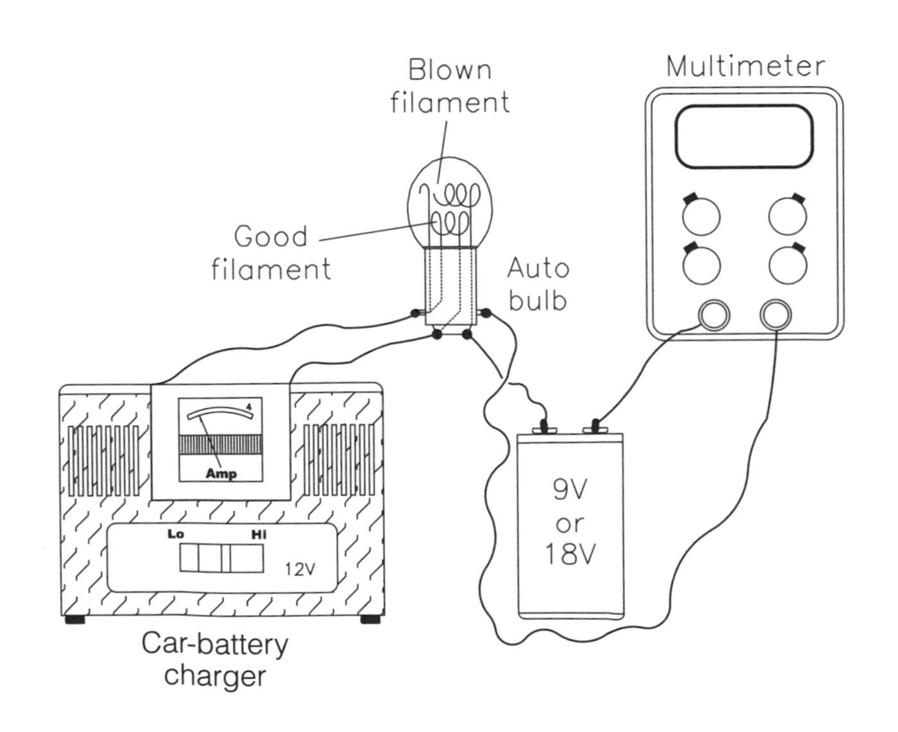

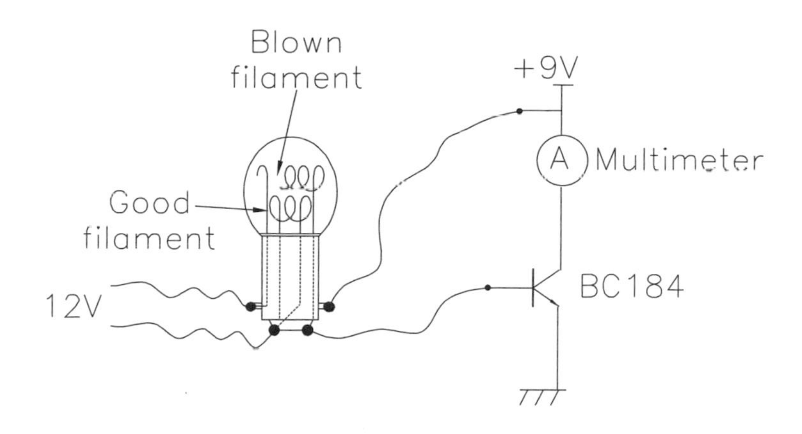

wires—they are not dangerous, but the tiny currents, a few microamps, that you are measuring will flow through your skin and you won't get sensible results.) You should find that the current, rather than being constant—which you might expect if you thought of the device as a simple thermionic diode—peaks sharply when the sensor electrode is above the filament. With 18 V applied (two PP3 batteries), you should get 1–2 μA with the sensor electrode above the hot filament. With the sensor electrode below the filament, the current should decrease by a factor of two or so, to 0.5–1 μA. This diode, unlike any other electron device, is sensitive to gravity!

You can try boosting the current output from the device by using a simple transistor arrangement. By putting the input current through a transistor, you can multiply the output current several hundred-fold. If you test your transistor for gain with a multimeter in advance, you can divide the observed current by the gain factor. Be aware that the hfe gain factor of the transistor varies with temperature—with the transistor I used, a BC184, hfe changed from 490 to 460 when I moved from the house to the garage, which was only 10 °C colder.

How It Works

How can we explain the current variation we observe with the gravity diode? Is it possible that the variation has nothing to do with gravity and is instead a result of the photoelectric effect? Perhaps our device is sensitive to light, which tends to come more from above than from below. Could it be that light from the filament hits the sensor electrode and causes electrons to be emitted? You could test this effect on the device with the aid of a powerful lamp or just sunlight from a window. But the answer is no anyway. The sensor electrode is positive, so it will attract electrons rather than emit them.

Remember that the lamp contains argon gas. Could the sensor electrode be producing positive gas ions by collecting electrons from the argon atoms as they migrate toward the hot filament? Perhaps, but I think there is a better explanation, one that involves thermal convection currents, which occur only in a gravitational field. When a fluid, liquid or gas, is heated, it almost invariably expands. Because the expanded portion of fluid is lower in density than the neighboring fluid, buoyancy forces exert a pressure differential more than equivalent to the heated fluid's weight, and the heated portion is thus accelerated upward relative to gravity, creating a convective gas flow. In our gravity diode, the hot filament creates a convective gas flow in the low-pressure argon, a flow that carries the electrons emitted by the filament upward, probably as negative argon ions, toward the sensor electrode, which is positively charged and can then collect some of them and generate the electric current that we measure. As you orient the device relative to gravity, you change the direction in which most of the electrons are carried by the convective flow, and you change the electric current picked up by the broken filament charge collector, just as we observe. The current is highest when the diode is oriented with the collector electrode above the hot filament. With other orientations, depending upon the convective flow pattern inside the lamp, you will generally pick up less charge.

THE SCIENCE AND THE MATH

When a modest voltage is applied to the gravity diode, the charge collected depends strongly on orientation. However, when a higher voltage is applied to the diode, it becomes insensitive to gravity. If you apply at least 40 V to the sensor electrode, you will find that the current increases to 5 μA or so and then remains essentially constant irrespective of the lamp's orientation. At 40 V, there is enough electrostatic pull on the negative ions and electrons to pull them over the sensor electrode against the convective flow when the electrode is below the hot filament.

We can estimate the speed of a negative ion under an electrostatic field to see if this explanation is reasonable. Let's say that the current, *I*, is 1 μA ($10^{-6}/1.6 \times 10^{-19} = 6 \times 10^{13}$ electrons/s) and the distance traveled is 4 mm. The drift velocity, *U*, of an ion in a gas is given by the following formula:

$$U = (dE/dx)DNe/P_0,$$

where P_0 is atmospheric pressure, *N* is the number of molecules in 1 m^3 of air at atmospheric pressure, *e* is the charge on the electron, *D* is the diffusion coefficient, and dE/dx is the electric field in volts per meter. Lord Rutherford, the physicist who first "split the atom"—that is, split the atomic nucleus—measured some of these drift velocities in 1897. He found a value of 3 cm/s for air gases using a dE/dx value of 100 V/m. Here we are using 4,500 V/m, so if we assume Rutherford's value for drift speed, we should expect the ion velocity to be 1.3 m/s.

This may seem quite fast—no wonder the convection current is defeated by the voltage we apply—but it is actually very slow compared to the typical speed of electrons in ordinary electric fields. The voltage required to speed up an electron to escape velocity is remarkably small. An electron accelerated through a voltage, V, will achieve a velocity, U, given by

$$U = \sqrt{(2Ve/m)} = 593{,}000\sqrt{V} \text{ m/s},$$

where e and m are the electric charge and mass of an electron. So for an electron to achieve escape velocity (12,000 m/s; roughly 12,000 yards/s), it needs a voltage of only about 0.4 mV. But for Einstein's theory of relativity, the electron would get to the speed of light (3×10^8 m/s) at only 250,000 V, which can easily be achieved with specialized power supplies.

And Finally . . . Gravity Diodes to Replace Bubble Levels

You could compare the characteristics of our gas-filled gravity diode with those of a thermionic vacuum tube diode.* Some types, such as the ECC83 double triode, are now readily available and, curiously, are probably cheaper in real terms than they ever have been. If you can't find a diode, by the way, you can just use a triode by connecting all the nonfilament electrodes together. Could you make a bubble level out of our gravity diode? To increase the sensitivity to angle, you could maybe use a pair of them, aligned not quite parallel, with a readout of the voltage difference between them.

*I am just old enough to remember vacuum tubes. Although they were obsolescent in 1970, they were still being specified for new equipment like televisions and amplifiers, even though transistors had been available for ten years. I used to repair TV sets while I was at school—I am in good company here; so did Richard Feynman—and I was once very familiar with the technology. Thermionic tubes have been making a bit of a comeback in the audio industry in the past ten years, for use in guitar amps as well as in hi-fi systems.

REFERENCE

"Warranty Cost Savings of LED Signal Lights," Application Note 1155-1, Lumileds (a Philips Lighting and Agilent Technologies joint venture), San Jose, California (1998). Available at www.lumileds.com/pdfs/an1155-1.pdf (as of March 10, 2003). This publication gives details about tungsten lamp lifetime and light emission, perhaps with a more open viewpoint than the tungsten lamp manufacturers, who also publish data.

28 AMIPLEX

> Rail travel at high speed is not possible because passengers, unable to breathe, would die of asphyxia.
>
> —Dionysius Lardner

Irish-born Dionysius Lardner was a brilliant student who wrote or edited many important nineteenth-century books, including the well-known *Popular Lectures on the Steam Engine*. However, he had a habit of making predictions that proved to be either inaccurate or just completely wrong. In 1835, Lardner set a theoretical upper limit on the range of the steamboat: he calculated that no ship could carry enough coal to sail more than 2,500 miles. There was more. In 1838 he wrote, "Men might as well project a voyage to the moon as attempt to employ steam navigation against the stormy North Atlantic Ocean." One hundred and thirty-one years later, this particular prediction was proved to be not just wrong but doubly wrong! As Woody Allen once said, "It's tough to make predictions, especially about the future." I won't be making any predictions about AMIPLEX; I'll just say that it's pretty neat.

In research and development work, you often need to fit many sensors and rapidly connect them up and analyze the results that come back. If you have a big budget and lots of time, there is no problem. You laboriously wire up all your sensors to a huge computer interface and get busy with the software. However, if you have to work quickly on a small budget, then you don't want to spend time programming the computer, and you certainly don't want to spend hours on scads of wiring. AMIPLEX will help you.

AMIPLEX evolved from the need for a system that could connect a dozen or two sensors on a simple wiring system to a display or recording unit. The particular requirement that led to the project in this chapter was the need for light sensors; I needed a system that would allow me to assess simply and rapidly how evenly light was falling on an unevenly colored scene. I was going to wire up a couple dozen photodiodes to pairs of wires festooned all over the place and then read the current from each one with a meter, or, if I got time, feed them all into a computer analog-to-digital interface. But I thought there had to be a better way. So I decided to connect all the photodiodes to a large multiway switch and then switch between the sensors to look for similar readouts on the meter. But I still needed a heap of wiring and the multiway switch, and the readout was slow and clumsy. What about a switch on each sensor that I could remotely access in a simple way? Thus was AMIPLEX born.

What You Need

- Dual comparator chips
- Resistors
- Diodes
- Capacitors
- Photoresistors (or other sensors)

What You Do

Build the circuit as shown in the circuit diagrams, paying attention to the daisy wiring. You will see that there are several common wires to each node—ground wire, address command wire, data wire, daisy wire, and power wire—only one of these wires is "daisy-chained." In a system with *N* nodes, the daisy wire is interrupted by a resistor at each node and is connected to ground beyond node *N* and to the address command voltage (or ramp voltage) at the first sensor.

You operate the system by connecting a variable voltage—say, from 0 to 15 V if your system power supply is 15 V—to the address command wire. Connect a resistor, say, 1 kohm, between the data wire and the 15-V power supply wire, and monitor the voltage across this resistor with a multimeter or an oscilloscope. As

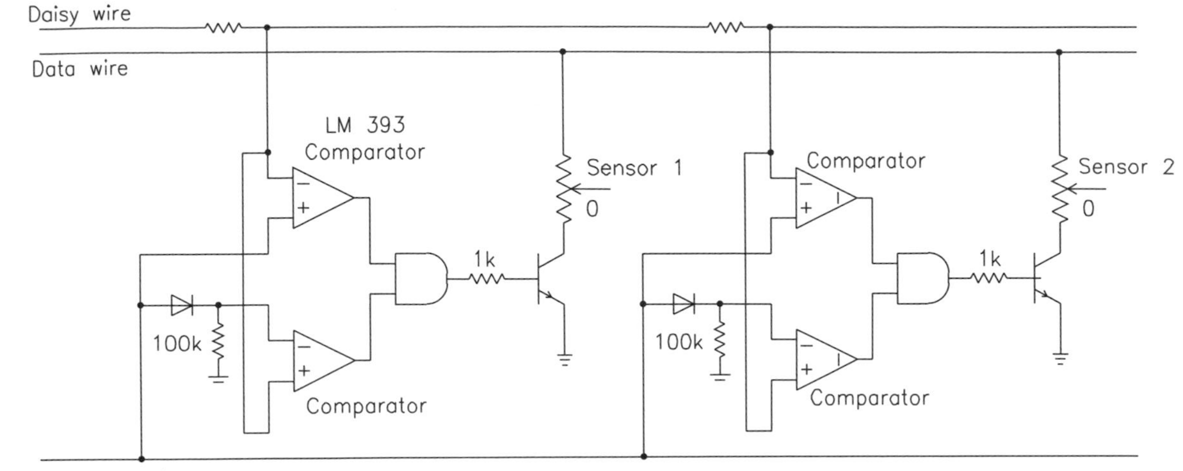

you increase the voltage on the address command wire, you will switch from the sensor at node N to the sensor at node $N - 1$, to the sensor at node $N - 2$, and so on until at the full 15 V you will address node 1; that is, as you apply a ramp voltage to the address command wire, you will cause the data wire to scan over all the node sensors in turn.

How It Works

AMIPLEX allows the connection of a single master station to many simple slave stations (nodes) via a single daisy-chain four-wire cable and minimal node electronics. The nodes are wired in daisy-chain fashion, with a power wire, a ground wire, a daisy wire, an address command wire, and a data wire. The common wire is continuous, but each node interrupts the daisy wire with a resistor. When a current is flowing down the daisy chain, each node sits at a different voltage, and these different voltages can be used as "addresses."

The voltage on the address command wire at each node is used to activate that node individually. When this voltage ($V_{address}$) is equal to the voltage on the daisy wire (V_{daisy}), then the node will switch on and connect its sensor to the data wire; that is, the voltage on the data wire (V_{data}) will be equal to the sensor voltage (V_s). If you connect the ramp voltage from the back of an oscilloscope to the address command wire, then the data wire will display a histogram of all the sensors across the screen of the oscilloscope. The beauty of AMIPLEX is that it doesn't need computers, software, or complicated digital electronics. With just a

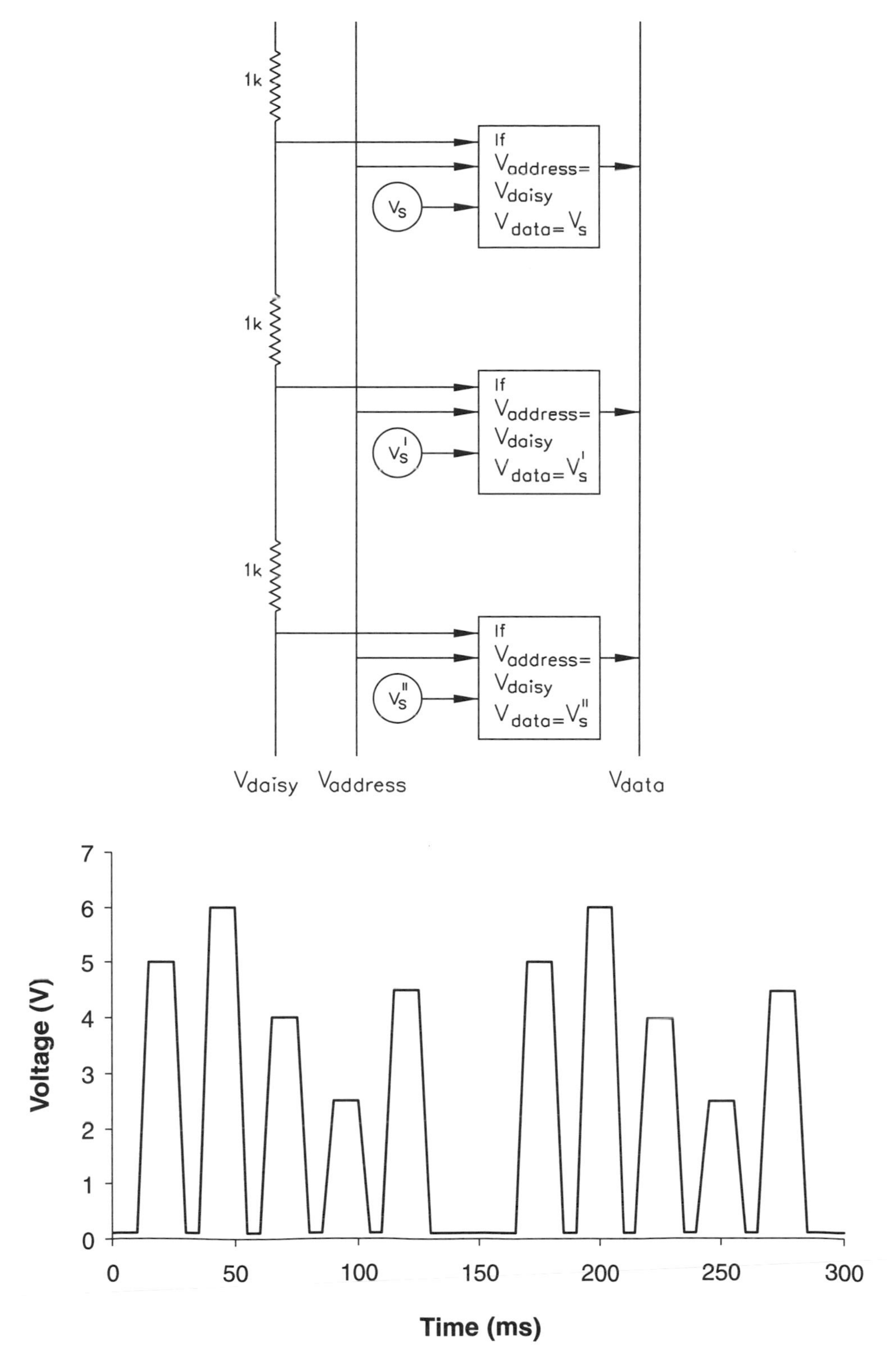
1k
1k
1k
If
Vaddress=
Vdaisy
Vdata=Vs
Vdaisy
Vaddress
Vdata
Voltage (V)
Time (ms)
0
50
100
150
200
250
300
0
1
2
3
4
5
6
7

simple analog comparator (operational amplifier) chip at each node and an oscilloscope as the master, you can scan dozens of sensors into an instant histogram display.

Although photoresistors are among the simplest sensors, you could use other sets of photosensors, arrays of temperature probes, level sensors, reed switches (magnetic reeds are used in the window- and door-sensors of burglar alarms), Hall effect magnetic sensors—there are many possibilities.

THE SCIENCE AND THE MATH

AMIPLEX uses the voltage on the address command wire to activate each node individually via two comparator chips. These chips have two thresholds for each node, equal to the daisy-wire voltage and 0.6 V below that. The address command voltage is compared to the daisy-wire voltage at each node. If the voltages are equal (within a tolerance band given by the upper and lower comparator thresholds), then that particular node's sensor signal is impressed on the data wire. AMIPLEX can provide a full bandwidth connection (from DC upward; in other words, the equivalent of a piece of solid wire) from master to a single slave as an option when required.

Analog Voltage or Current Addressing

The AMIPLEX system relies on a voltage level to address each of the nodes in the daisy chain. Each node's daisy-wire signal is modified by means of a voltage-dropping element (typically a resistor or diode) connected between the daisy-wire input and the daisy-wire output. Each node's "address" is therefore defined by the voltage on the daisy wire. When the address command and daisy-wire voltages are equal (again, within a small defined range), the node electronics is triggered to provide an output to the data wire, by gating the sensor via an open-collector transistor. Here we use two open-collector transistors to provide the "AND" gate function on the upper and lower comparator outputs as well as the sensor-output gating.

The use of a constant current source to feed the nodes via the daisy wire means that the voltage levels are evenly spaced. New nodes add more "stairs" to the staircase waveform needed. The voltage will be NIR at the first of the N nodes, $(N - 1)IR$ at the second node, and so on, until the last daisy wire has only the voltage due to the constant current, I, flowing through the last node resistor of value R.

An alternative version of the system employs a constant voltage source and current drain circuits located in each node. The current flowing in the daisy wire is used as the "address" for each node.

Master Node Function

The master node generates address command levels, typically by using a digital-to-analog converter (DAC). Systems I have tried work adequately with sixteen voltage levels at submillisecond speeds with a reasonable degree of noise immunity in the laboratory even on unshielded cables. Systems with thirty-two or more voltage levels should be possible, although these would require more care in rejecting sources of electrical interference. Shielded cable, for example, might be needed.

The address command voltage can be tailored to provide different functionality. For example, the voltage can be a simple "staircase" waveform that addresses each possible node in turn. Alternatively, it can be a "sparse staircase" in which known missing addresses are avoided (the master having a list of valid addresses). Or it may be a waveform that visits a particular node or nodes more often than it visits other nodes. If you want to hop nodes in this way, you must take care that the momentary replies from other nodes during switching transients are ignored. You can do this easily by requiring that the master read the data from the data wire only after a short delay that allows the address command voltage to settle.

Analog Sensors or Actuators at Nodes

AMIPLEX can be used with analog sensors. The open-collector data-wire transistor and the AND gate that feeds it can be changed for, say, a field effect transistor (FET) analog switch if higher-quality analog signals are needed. Alternatively, the analog signal can be encoded, as pulse width, for example, before connection to the data wire. In either case, the master electronics interprets the received signals to recover the analog data.

A more complex alternative is to use an analog-to-digital converter (ADC). An ADC converts the sensor input into serial digital form, which is sent in a burst of binary bits (encoded, for example, as long and short pulses for 1 and 0) along the data wire when the node is addressed. But I think this would lose some of the simple elegance of the concept.

The system can be used as an addressable actuator bus in a number of ways. If an actuator needs only a single pulse to activate it, then the data wire can be connected to a driver circuit at the master unit. The data wire at an actuator node is then fed via a buffer to the actuator operated by that node. If a continuous actuation command voltage is needed, then a latch circuit can be employed at the node, with status feedback if necessary. AMIPLEX systems can also support both sensor nodes and actuator nodes on the same bus.

Maximum Number of Nodes and Minimum Number of Wires

The limit on the number of nodes on AMIPLEX systems is mainly imposed by noise on the wiring and by the maximum voltage that can be used on the circuits employed. With a noise immunity comparable to that with transistor-transistor logic (TTL) digital systems, up to thirty channels are possible. You can extend AMIPLEX to very large numbers of sensors if needed by using two address command bus wires and two daisy chains.

The basic system employs conductors with five functions: power, address command voltage, daisy wire, data wire, and ground. However, two or more of these functions can be combined so that fewer wires are used. Systems can easily be built with full functionality on three wires plus ground or even, with increasing sophistication of node electronics, two wires plus ground. For three wires plus ground, you need to use diodes to connect nodes and you need to provide an additional resistor down to ground on each node. The daisy wire can then be used to provide address command.

What happens as the voltage on the address/daisy wire rises? With zero voltage applied, all sensor address inputs are at zero. Then, as the voltage begins its rise, the voltage on the first node exceeds the lower threshold and the first sensor is switched onto the data bus by its node. (All the other nodes are held at ground by their resistors.) Then with further voltage rise, the first node rises above its upper threshold and the second node becomes high enough to switch its sensor on, and so on.

And Finally . . . Expanding AMIPLEX

As I mentioned, you can wire up AMIPLEX with fewer wires, but can you add more wires and increase the number of nodes in the system? An additional data wire would allow you to use a dual-beam oscilloscope to display sensors from two AMIPLEX systems: use the ramp output from the scope for the address wire of all, but take the sensor outputs from each data wire to the Y1 beam and the Y2 beam. (Many scopes have a ramp output on the back that you can use for this.)

Could you just add an additional address wire, for example, and include more comparators in the gating arrangement? Could you avoid more wires and just use higher voltages for addressing? The silicon circuits might be blown out by voltages much higher than 30 V, but perhaps the circuits could be protected in some way.

REFERENCES

Downie, Neil A., and Craig G. Sawyers. "Multiplexed Electronic Bus System." U.K. Patent no. GB2,353,448, 2001.

Lardner, Dionysius. *Popular Lectures on the Steam Engine, in Which Its Construction and Operation Are Familiarly Explained; with an Historical Sketch of Its Invention and Progressive Improvement.* New York: Elam Bliss, 1824.

Curious Communications

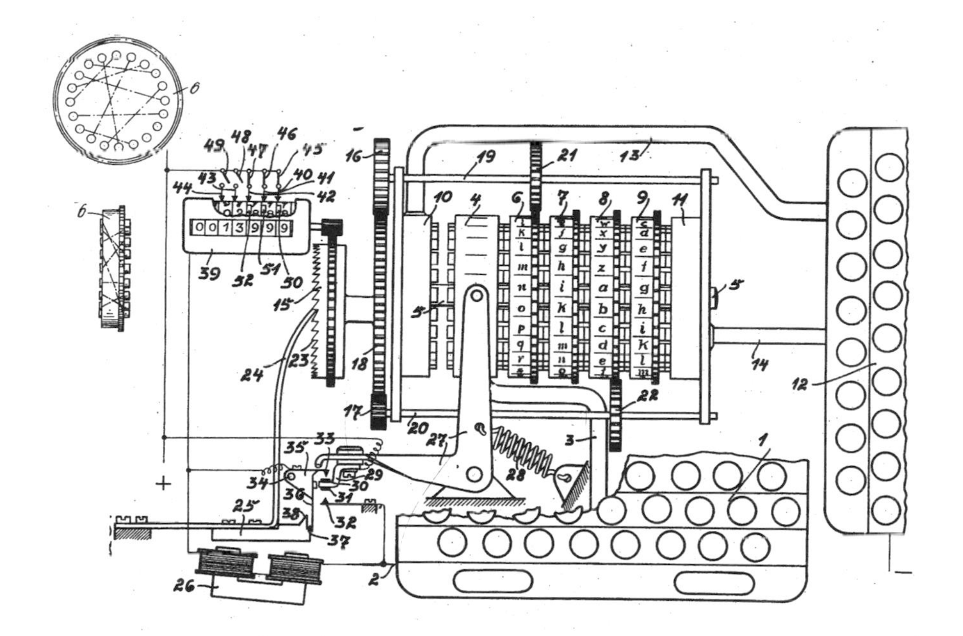

> Mr. Photography is going to marry Miss Wireless, and heaven help everybody when they are married. Life will be very complicated.
>
> —Marcus Adams, society photographer, quoted in the *Observer,* 1925, on the possibilities of television

Up until the early 1900s, the principal preoccupation of Western society, and especially American society, seems to have been transportation. However, by the quarter-century mark, this preoccupation was rivaled by communication. Today, we seem to be more preoccupied than ever with communication, in the form of the Internet. Communication systems need certain characteristics to succeed. They must code efficiently, with few errors, and they must sufficiently utilize the capacity of the communication link. They must offer an easily assimilable display—we don't want our Internet terminals to produce Morse bleeps. And they must offer high capacity and high speed—nobody wants last month's news.

The ticker-tape telegraph news, devised by Thomas Edison, was an early system that satisfied all these criteria, at least to the satisfaction of late-nineteenth-century businesspeople. The ticker tape coded efficiently; its output, a continuous stream of letters, could be instantly read by anyone; it was capable of sending eight characters per second; it was inexpensive (its costs were shared among many users); and, finally, its transmission speed was close to the speed of light.

In this part, we look at various exotic communication systems, considering the problems of low-speed transmission in flowing air and water, and at alternative display systems.

29 Servo Telegraph

What hath God wrought.

—Samuel Morse,
first telegraphic message sent
in Morse code

The Morse telegraph was by no means the first electric telegraph to transmit messages around the world, but it eventually became the dominant system. In its heyday, before the telephone, millions of people used it every day, and the Morse alphabet was still being used for low-bandwidth radio transmissions at the end of the 1990s. The significant feature of Morse's system was that a single wire could transmit any message efficiently with combinations of long and short pulses for the characters. Multiwire systems such as Charles Wheatstone's had used an *x-y* grid system for the characters, but the need for multiple wires was a heavy burden. Previous single-line systems had allowed only a limited number of set messages: "Train in next sector," "Train stuck in tunnel," "Send next train," and so on. Only by using tens of pulses could these early single-line systems send alphabetic messages.

Samuel Morse's basic concept of dots and dashes had been tried before—by pioneers like Joseph Henry, Peter Barlow, and William F. Cooke—in various clumsy ways. Even Wheatstone, who took a keen interest in secret codes, did not come up with a scheme as good as Morse's. A number of inventors proposed a simple "unary" system, one in which, for example, the code for *A* was a single short pulse, that for *B* was two short pulses, and so on until *N* was reached.

Then long pulses were used, one for *N*, two for *O*, and so on to *Z*. People soon realized that you could shorten this code by using a binary system to encode the letters; you could get 32 characters by using a total of 5 dots or dashes ($2^5 = 32$). For example, *A* was dot-dot-dot-dot-dash (00001), *B* was dot-dot-dot-dash-dot (00010), *C* was dot-dot-dot-dash-dash (00011), *D* was dot-dot-dash-dot-dot (00100), and so on. Morse was smart, however, because he realized that you could reduce the number of dots and dashes by using shorter sequences for frequently used letters. In Morse code, the most frequent letter of the alphabet, *E*, is a single dot, *T* is a single dash, and rarely used letters such as *Y* (dash-dot-dash-dash) and *Q* (dash-dash-dot-dash) are longer. Morse's system allowed messages to be sent, for the first time, at a rate of more than a letter per second over a single wire.

There were also automatic pulsed systems. The original "uniselector" dial telephone system actually looked a bit like the arrangement we are going to build in this project. Dial telephones communicated numbers to the telephone exchange by sending a number of pulses that was equal to the number dialed. To send letter *N*, you wound the dial around by *N* units and then allowed it to spring back. The mechanism of the dial was such that the spring-back process took place at a controlled speed, and as the dial returned to its starting position, a pair of contacts was made and broken *N* times by pins on the dial. At the telephone exchange, the pulses pulled a solenoid-and-pawl mechanism that moved a set of rotating arms around through an angle $N\alpha$ inside a set of contacts spaced α apart. The rotating arms carried metal wipers that bridged contact pairs *N* and thus connected the incoming telephone call to outgoing wire *N*. The whole assembly was known as a uniselector. This clever system worked well but was slow, managing only about one character every two seconds, whereas the Morse telegraph could send several characters per second.

Communication systems, and electronic systems in general, can be characterized as being either analog or digital. The standard TV is an analog system in that it converts a physical parameter, light intensity, into a voltage that is a representation, an analog, of the original light intensity. A line of image, with an intensity that varies with position along the line, is translated into a voltage varying with time, the voltage being proportional to the intensity and the time being proportional to the position. The voltage is then transmitted by wire or a radio-frequency (RF) carrier wave by broadcast. Digital systems, by contrast, convert inputs into an abstract set of mathematical characters with discrete values: a digital TV picture, for example, is converted into several million elements, and each element is assigned a number, from 0 to 255, that represents its intensity. The system then transmits the resulting string of numerals. At no point do the transmitted digital signals have a voltage that can be related directly to the light intensity pattern at the camera.

Today, messages are rarely sent via analog signals. They are more subject to noise and interference than digital signals, and corrupted signals cannot easily be restored. Digital signals, by contrast, can be fully restored if the interference levels are below certain limits. It is nevertheless possible to send analog messages, and they are still used in special circumstances. The most important example of analog communication today, what might be dubbed a "constant-current telegraph," is actually used every day in chemical and other process plants all over the world: the current transmitter. Current transmitters minimize noise and other errors almost to the same levels exhibited by digital signals, at least on short lines. The current transmitter method makes use of feedback. An input signal from 0 percent to 100 percent governs the output of a constant-current circuit between 4 mA (representing 0 percent) and 20 mA (representing 100 percent). The constant current is sent down a wire, and there is an ammeter at the other end. The beauty of this arrangement is that the constant-current source automatically corrects for any changes in path resistance and imposed voltages; the received current is continuously corrected to the transmitted value.

The project in this chapter illustrates another analog communication method. The servo telegraph is a chimera, a hybrid of analog and digital. We transmit only two voltage levels—one and zero, if you like—but the length of time that the one-signal is on, the pulse length, is an analog of the input. The rotation of an alphabet wheel at the transmitting end of the device is mimicked by a fairly common modern gadget, a servo motor, at the receiving end.

What You Need

- ❑ Servo motor (A $10 model-airplane servo will do.)
- ❑ Receiver dial
- ❑ Transmitter dial
- ❑ 555 timer integrated circuit
- ❑ Resistors
- ❑ Capacitors
- ❑ Circuit board (matrix board or stripboard)
- ❑ 2 trimmer resistors

- ❑ 1 potentiometer
- ❑ Long wire
- ❑ 2 ground connections (or another long wire)
- ❑ 2 battery packs
- ❑ Switch
- ❑ Bulb
- ❑ Transistor
- ❑ LED

What You Do

First build the oscillator according to the circuit diagram, and check that it emits regular pulses whose width varies as you adjust the potentiometer. Take a ground wire and a signal wire from the oscillator and feed them over to the servo, including the 0.5-μF capacitors to block DC from the LED switch. Mount the indicator dials on the potentiometer at the transmitter end and on the servo, and check that they rotate freely. Finally, connect the LED lamp to the transistor, including the resistors, which limit the current to appropriate values.

When you connect the servo to the oscillator, you should find that turning the potentiometer on the oscillator makes the servo follow your input. Once you

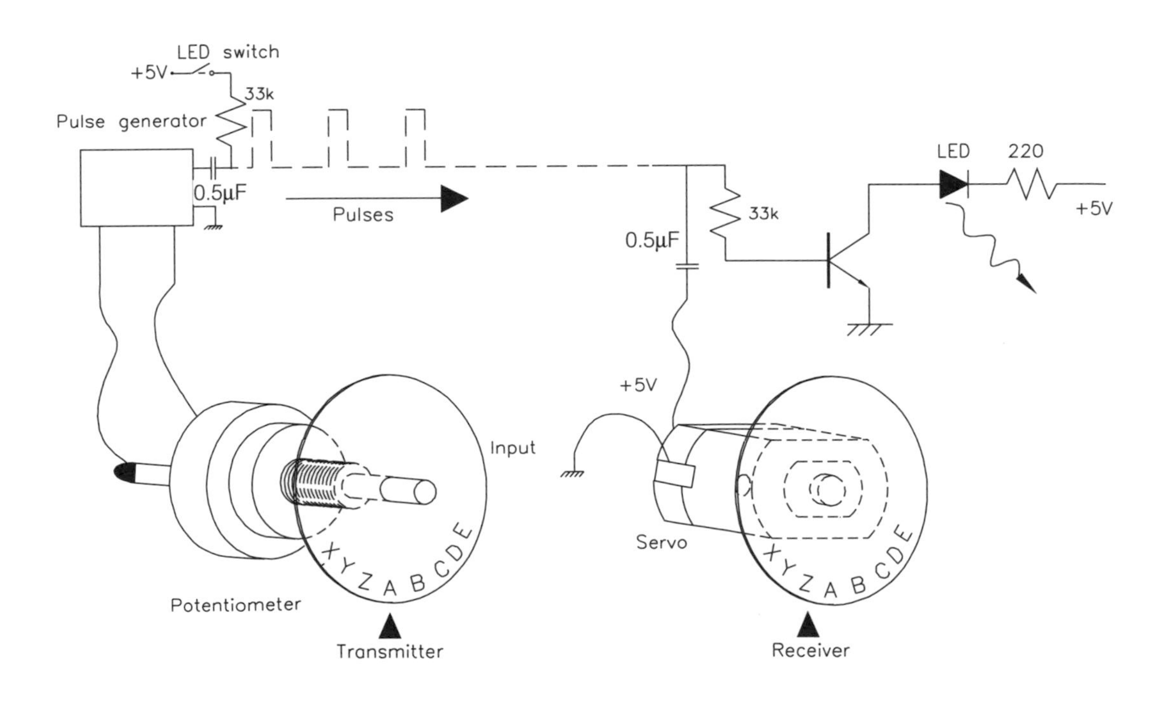

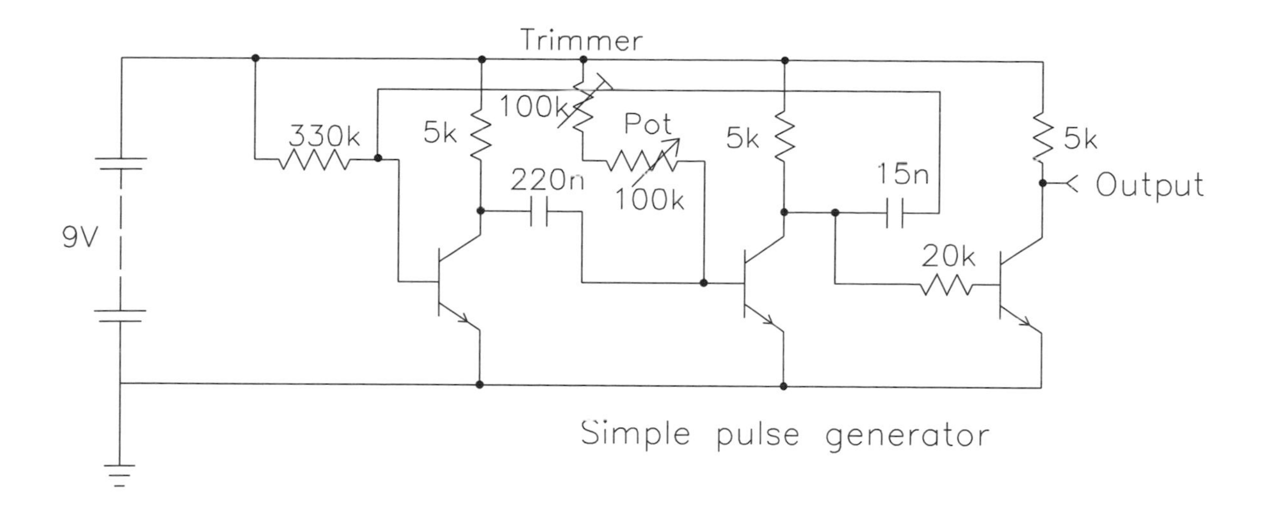

have established communication between the subassemblies, adjust the oscillator trimmer resistors so that the oscillator potentiometer span corresponds roughly to the rotation of the servo and so that the center position on the potentiometer gives the center position on the servo. Now mark the letters of the alphabet around the potentiometer dial, marking the dial on the servo as it turns to match the potentiometer.

When you press the switch at the transmit end, the LED lamp at the servo end should light up, without affecting the servo position. A particularly good arrangement uses a transparent disk on the servo. Connect the switch-circuit LED lamp so that it illuminates the letters on the servo dial from behind.

Now you are ready to send a message. Persuade a colleague to sit at the receiver servo with a message pad. Rotate the transmitter dial to the first letter of your message, and press the switch for a second. Then rotate the dial to the second letter, and press the switch again, and so on. Your colleague simply has to write down the letters that the LED lamp illuminates as you send them.

How It Works

The oscillator in the transmitter converts a rotary position—corresponding to the letter we wish to transmit—into a series of pulses. The duration of these pulses is proportional to the rotary position; the pulses are an analog of the input angle. When the pulses arrive at the receiving end, they are converted in the opposite direction: the pulse length on the input to the servo electronics is converted into the rotary position of the servo. The servo angle is an analog of the pulse length.

The word *servo* is derived from the Latin *servus,* "slave." A servo, or servo motor, is a slave device that produces a mechanical movement that mimics the movement of a remote input device. One typical use for servos is in remote-controlled model airplanes. When you push the control stick over on the radio transmitter, the radio receiver in the airplane picks up the signal and applies it to a servo motor to ensure that, for example, the aileron movement echoes the stick movement. Model-airplane servos swing their mechanical output through about 300 degrees of travel in response to pulses on their input; these pulses are given out by the transmitter and are repeated about twenty to fifty times per second and range from 0.5 to 1.5 ms in length.

The model servo in our telegraph device expects a 1- to 3-ms pulse every 20 ms or so. A signal of 0.5 ms should give the fully anticlockwise position, 1 ms the central position, and 1.5 ms the fully clockwise position, with all possible positions lying between 0.5 and 1.5 ms. The oscillator is designed to satisfy this expectation as the potentiometer is rotated from fully anticlockwise to fully clockwise. If you have an oscilloscope, you can display the waveform from the oscillator as the input potentiometer is moved.

The LED lamp lights up when a small DC current amplified by the transistor is superimposed on the AC signal that commands the servo. The 33-kohm resistors limit the DC current to the transistor, and the 0.5-μF capacitors allow

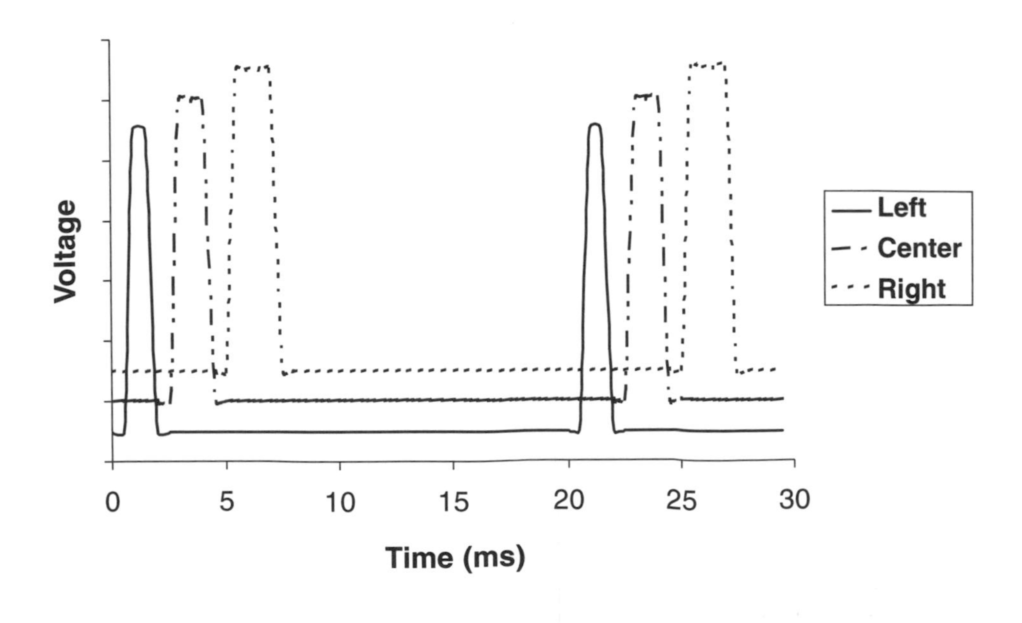

the AC signal to be transmitted to the servo essentially unaffected while blocking any DC (which might cause problems) to the servo amplifier or to the oscillator. The oscillator is a multivibrator circuit, which functions internally by charging and then discharging a capacitor between two voltage limits: the charge-circuit resistance is high so that the charging period is relatively long (roughly 20–30 ms), and the discharge-circuit resistance is low so that discharging period is roughly 1 ms. The oscillator potentiometer varies the latter period by varying the discharge-circuit resistance.

What is the maximum distance over which the servo telegraph can transmit? Well, if you used a telephone connection, you wouldn't find it difficult to telegraph a friend in Australia, in principle at least. In practice, though, you might have at least one limit, an unusual one: the lowest frequency in the band transmitted by the telephone (the highest frequency that can be transmitted is usually the problem). A normal telephone works best between about 400 Hz and 4 kHz. Beyond these frequencies, the line is filtered to eliminate signals for the most part, so that interference due to higher frequencies in other telephone signals does not occur and so that the domestic mains power hum at 50 Hz and its harmonics cannot be picked up in the telephone. The 4-kHz upper limit gives a reasonable pulse shape for our 1- to 3-ms pulses, but loss of the components below 400 Hz in the signal may create a problem.

THE SCIENCE AND THE MATH

The servo telegraph converts an analog quantity—the rotation of the input dial—into digital pulses, and then uses the model control servo to reverse the analog conversion. How exactly does the model servo accomplish the conversion? The necessary sequence of operations is as follows:

1. Amplify any input received with a high-gain AC amplifier, which is followed by a limiter, so that the exact amplitude of the pulse received is not important.
2. Measure the average voltage of the pulse waveform using a resistor-capacitor smoothing circuit: the voltage will be low for short pulses, high for long pulses.
3. Compare the average voltage of the input pulse waveform produced with the voltage on the transmission potentiometer, which is fed with DC.
4. Drive the motor to reduce (if the difference is positive) or increase (if the difference is negative) the potentiometer voltage.

In this way, the output potentiometer voltage, and therefore its position, will mimic the pulse length input to the servo. (For more details on servos, see the light tunnel project in the previous section of this book.)

What is the maximum speed of the servo telegraph? That is a more difficult question. Since the repetition frequency of the system is 20 ms, an

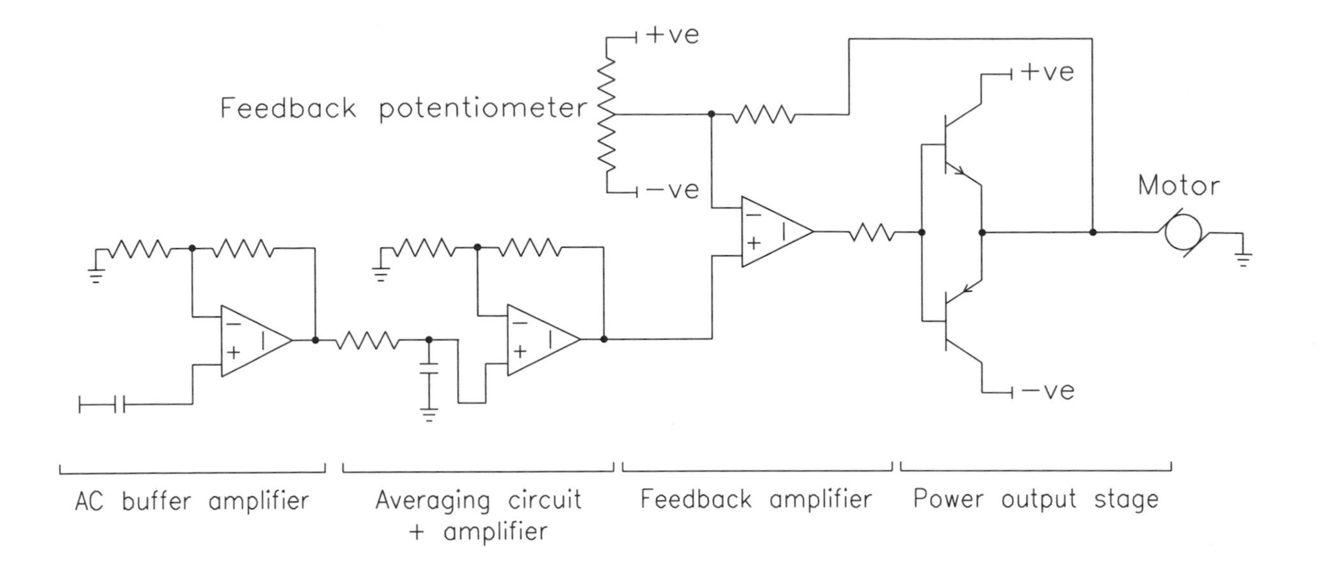

absolute maximum of a few tens of characters per second could be sent. If you could actually get the system to operate at this speed, you might find it difficult to observe: perhaps a Dymo-type printing mechanism would allow you to read the words coming off the printer, like an old-fashioned ticker-tape machine. In fact, the servo speed is limited to fewer than twenty characters per second: mine took more than half a second to go from *A* to *Z*. Although there undoubtedly are faster servos, this may well be the real speed-limiting factor.

We can simulate the effect of filters on the waveform of the servo system using a spreadsheet-based computer model. Suppose that we have one or more filters of the "differentiator" resistor-capacitor (RC) type in series with the signal, limiting transmission below 400 Hz. In this arrangement, the current, *I*, through the resistor is given by

$$I = V_{out}/R = C[d(V_{in} - V_{out})/dt],$$

where V_{out} is the output voltage, V_{in} is the input voltage, R is the resistance across the output and ground, and C is the series capacitance. When R is small, V_{out} is much smaller than V_{in}, so

$$V_{out}/R \sim C(dV_{in}/dt)$$

and

$$V_{out} = RC(dV_{in}/dt).$$

Thus V_{out} is the differential of the input, which is why this kind of filter is called a differentiator.

The effect of the high-pass filter can be numerically simulated with a spreadsheet. The waveform is distorted: its peak slopes, and there are small negative excursions after the pulse has passed. In a similar way, a simple RC circuit can model the low-pass filtering above 4 kHz that we see in telephone connections. In this case, the resistor is in series from the input to the output, with the capacitor in parallel across the output and ground. The current, *I*, is given by

$$I = (V_{in} - V_{out})/R = C(dV_{out}/dt).$$

Solving for V_{out} gives

$$V_{out} = V_{in}[1 - \exp(t/RC)].$$

If RC is very large,

$$V_{out} \sim V_{in}t/RC,$$

that is, the integral with respect to time of V_{in}, which is why this kind of low-pass filter is some-

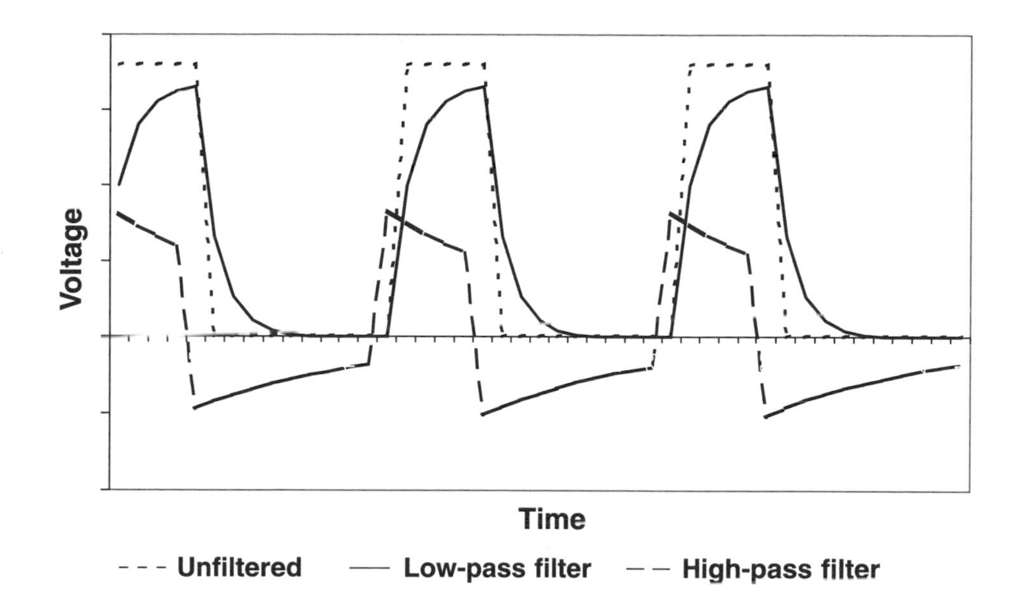

times known as an integrator. Looking at the spreadsheet simulation again, you will see that this kind of filtering also distorts the waveform; in this case, the filter removes the sharp corners of the input waveform and adds in trailing parts. At higher input frequencies, those trailing parts will run together, and the filtered waveform will no longer go from zero to full voltage. The servo telegraph waveform must be slow enough that this low-pass filtering still allows the signal to achieve more or less the full voltage span needed to operate the servo.

And Finally . . . Two-Dimensional Telegraphs

Our servo telegraph is a one-dimensional transmitting system: the single distant servo rotates, mimicking a single input rotation. A two-dimensional system is possible though. Systems in which the receiver station perfectly imitates the movements of a writing hand at the transmitter station have actually been developed, the Teleautograph of Elisha Gray being one of them. (Elisha Gray was nipped at the wire by Alexander Graham Bell, whose telephone patent filing at the U.S. Patent Office beat Gray's by just a few hours.) The U.S. coastal defense guns were operated by the Teleautograph from the late nineteenth century until the 1950s, apparently because telephones were difficult to hear when the guns were actually being fired—the forts were very noisy. Patents on new telewriter derivatives were still being filed in the 1960s and even in 1980 (U.S.

Patent no. 4,289,926). Could a two-channel servo drive like ours make a good new-style Teleautograph?

REFERENCES

Kobayashi, Masahisa. "Transmitter for a Telewriter." U.S. Patent no. 4,289,926 (1981).

Miller, Richard W. *Servomechanisms: Devices and Fundamentals.* Reston, Va.: Reston Publishing, 1977. A useful though older book on servo theory.

Newell, Paul. *Radio Control: A Handbook of Theory and Practice,* chap. 8. Guildford, U.K.: RM Books, 1981. A useful book on model engineering.

Shinksey, F. G. *Process Control Systems.* New York: McGraw-Hill, 1988. A practical and interesting book on control theory.

Thomson, S. *Control Theory.* Harlow, U.K.: Longman, 1989. Describes servo theory, but the description is wrapped up in rather abstruse math.

30 Send Me a Bubblegram

> It is my heart-warmed and world-embracing Christmas hope and aspiration that all of us, the high, the low, the rich, the poor, the admired, the despised, the loved, the hated, the civilized, the savage (every man and brother of us all throughout the whole earth), may eventually be gathered together in a heaven of everlasting rest and peace and bliss, except the inventor of the telephone.
>
> —Mark Twain, Christmas greetings

"Bubbles" of different kinds have a strong track record of leading to useful electronic devices in recent years: bubble memories, bubble-jet printers, and bubble optical switches are but three. The "bubbles" in bubble memories are not air bubbles but small domains of reversed magnetization in thin films of certain magnetic materials. Bubble-jet printers contain tiny heaters that superheat an inky fluid far beyond its boiling point for a period of milliseconds, and the pressure of the resulting steam bubble drives a tiny droplet of ink out of a nozzle toward the paper being printed. Bubble optical switches use a related effect: a heater creates a bubble whose silvery sides reflect light pulses coming down an optical-fiber communication system.

Chemical engineers often need to use bubbles and have long studied them. Engineers use bubbles for dissolving gases in liquids and for facilitating reactions

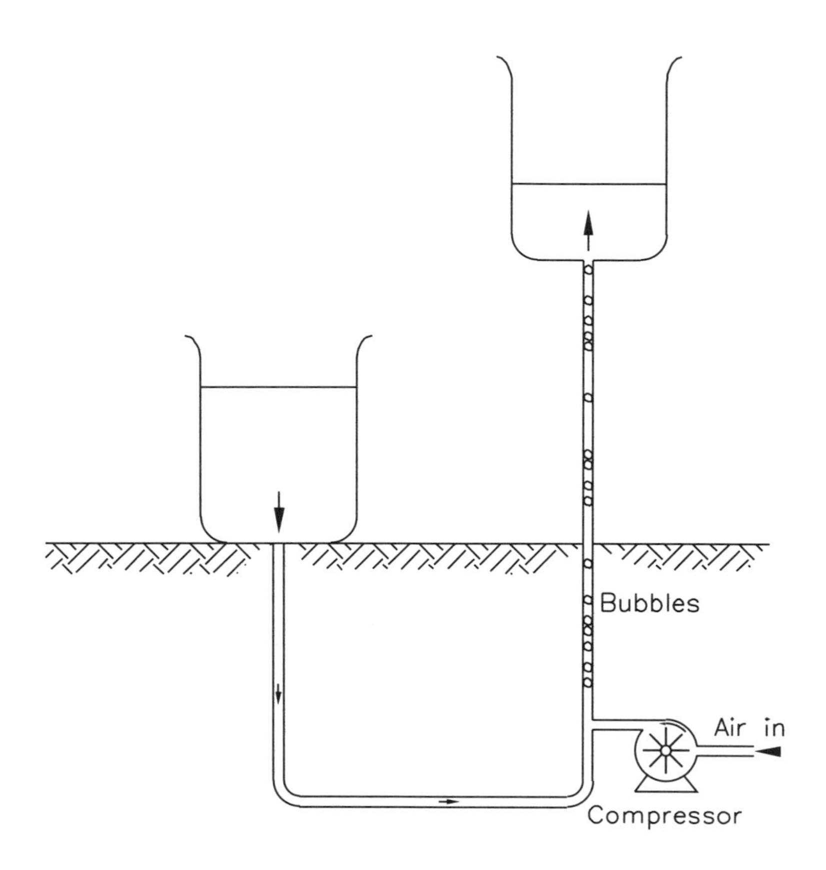

between the two, for agitating baths, for creating foams, and even for pumping liquids (including corrosive liquids) with air-lift pumps. For example, a liquid can be sucked out of an unpressurized supply tank into a higher tank. Air bubbles are injected into the upcoming pipe by a compressor. The bubbles reduce the average density of the fluid in the upcoming pipe relative to that in the pipe coming down from the lower tank, and the result is that liquid is pumped from the lower to the higher tank (see Coulson and Richardson, *Chemical Engineering*, 156–163). The efficiency of the air-lift pump might be almost perfect were it not for the fact that, under the influence of gravity, a thin film of liquid on the sides of the pipe tends to push past the bubbles as they float up. The wider the pipe bore, the worse this "slippage" is. In the horizontal pipes used in this project, slippage should be smaller, but it may still be significant.

The bubblegram in this chapter uses something like a horizontal air-lift pump to transmit and then detect bubbles traveling through a piece of tubing. The bubbles are just like signal pulses along an optical fiber communication system, only many orders magnitude slower.

What You Need

- ❑ Clear plastic tubing, 3 m long
- ❑ Beeper (e.g., multimeter beeper)
- ❑ Stopwatch
- ❑ Water supply (e.g., tap)
- ❑ Tap adaptor (if using tap water)
- ❑ Valve or faucet
- ❑ Air pump
- ❑ T-piece of tubing
- ❑ Photosensor (e.g., photodiode or phototransistor)
- ❑ Light-emitting diode (LED)
- ❑ Resistor (e.g., 200 ohms) to limit current through LED
- ❑ Batteries/battery holder (e.g., 2 AA batteries)
- ❑ Input switch (Morse key or momentary push switch)

What You Do

Set the tap to give a steady flow of water, so that it travels down the tube at just 50–300 mm/s. Keep the tubing as near as possible to horizontal. Connect the air

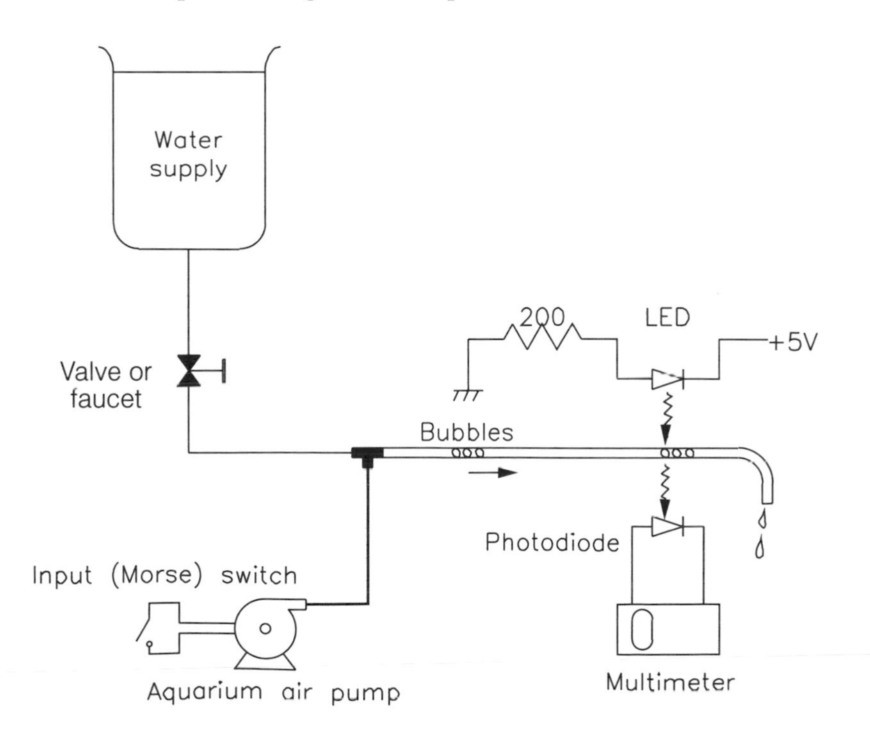

pump to the Morse switch, and inject bubbles into the water flow in the tubing by pulsing the air pump momentarily using the switch.

You are now ready to send your message. If you don't know your Morse code, then you need to convert your message; encode the message first into a series of dots, dashes, and pauses on a piece of paper. Blip the Morse switch on with the series of dots and dashes. Each time you press the Morse switch, a bubble or set of adjacent bubbles is injected into the tube and is followed by water. You should find that the pattern of bubbles and water forms an analog of the code on your message pad and that, furthermore, this pattern is carried along the tube to the receiving end. Surprisingly, perhaps, the traveling bubbles will tend to retain more or less exactly the pattern with which they were injected, even over many meters of tubing. It is worth adjusting the system to run at a suitable speed. Adjusting the pump speed and the faucet should allow you to achieve 50 mm for the length of dots and dashes.

You can simply read the "bubble message" as it comes along, but the bubblegram works rather nicely if you can listen to the message with a receive beeper. To make one, tape a photosensor and an LED to opposite sides of the tubing and cover the whole sensor assembly with black tape to block ambient light. When there is liquid in the part of the tube between the photosensor and the LED, the light from the LED is defocused and misses the photosensor for the

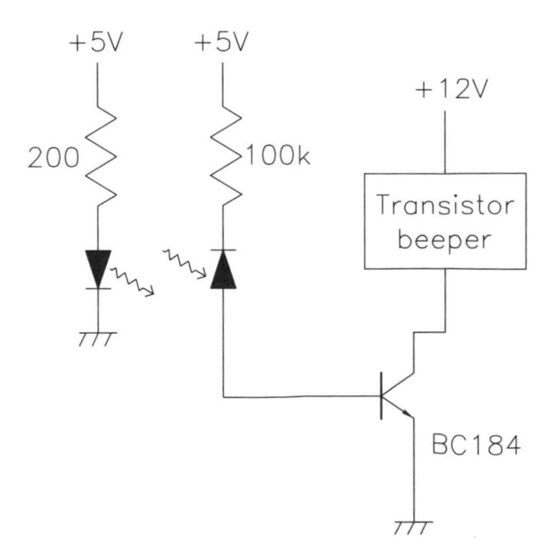

most part, whereas when a bubble is present, light can pass to the photosensor more effectively. You will find that with voltage applied to the photosensor in the correct polarity, the current through the photosensor will vary as a bubble passes through.

The simplest way to set up your receive beeper is to connect the photodiode to a multimeter set to the connection-checker position (this is often combined with the diode-checking position). In this position, the meter beeps when the two probe wires are connected by a conductor. With luck, you will find that the multimeter beeps each time a bubble passes the photodiode, since the illuminated photodiode behaves at least momentarily as a conductor. If you find that this does not work well, try the beeper circuit shown in the diagram.

How It Works

The bubble telegraph works because the air bubbles act almost like solid particles filling the tubing cross section, moving along like loose pistons or plugs. *Plug flow* is the term given to such a flow. Oil pipelines often rely on plug flow. For example, a batch of a few thousand tons of fuel oil might be followed through the pipeline by a batch of diesel oil. Hundreds of miles away, at the receiving end, a sensor tells the pipeline engineers of the arrival of the diesel. The engineers can then divert the pipeline output into another tank. The need for only one pipeline affords enormous savings in construction costs.

Actually, the water flow in our device is not a true plug flow: the water at the walls of the tube is stationary, while the water in the middle of the tube flows at higher than the average speed. However, the bubbles that fill the cross section force plug flow to take place, and this is what allows us to write a message in a sequence of Morse code bubbles and send that message down considerable lengths of tubing without loss of data.

The bubblegram provides a rare opportunity to see a message in transit. If you sent Morse code at one dot or dash per second into outer space using a laser beam, then you could in principle see your message strung out in the same way, with only one difference: scale. The laser beam dots and dashes would be too far apart—300,000 km or so—to see clearly. If the end of a dash you sent was just leaving your transmitter at the earth's surface, then the other end would be most of the way to the moon. With the bubblegram, you get 50-mm spacing instead.

For a more detailed understanding of how the bubblegram works, we need to consider the Reynolds number of the system. The Reynolds number is the ratio between the inertial forces and the viscous forces in a flowing liquid. If a liquid is runny and moving at high speed (that is, if it has a high Reynolds number), its motion will be dominated by inertial forces—the forces needed for acceleration and deceleration—and turbulence is likely. If a liquid is viscous and moving sluggishly (low Reynolds number), then viscous forces will dominate and laminar flow will ensue.

In the bubblegram, we have a liquid flowing in a small-bore tube at low speeds, which means that we should expect largely laminar flow or undeveloped turbulent flow, as indicated by the Reynolds number, R_N:

$$R_N = \rho v L/\mu,$$

where ρ is the density of the fluid, v is its speed, μ is its viscosity, and L is a linear dimension characteristic of the set-up (in this case, we use the diameter of the pipe). If we plug in the appropriate values for our project (ρ = 1,000 kg m^3, v = 0.15 m/s, L = 4 mm, and $\mu = 10^{-3}$ Pa s), we get a Reynolds number of 600, which indicates that some turbulence is possible. However, with the long length and smooth walls of the tubing, we should expect laminar or quasi-laminar flow.

Surface tension also affects the behavior of the bubblegram. The bubbles injected into the tube are held together by the surface tension of the liquid surrounding them. Molecules at a liquid surface are attracted to the neighboring molecules all around them, but this attractive force is counterbalanced by liquid pressure forces. The attractive force manifests itself as a tensioned skin along the surface of the bubbles, and the tension in this surface is constant at a given temperature and is characteristic of the liquid. Surface tension usually causes bubbles to be spherical, but our bubbles, confined as they are by the tube, are formed from two spherical caps on a cylinder whose diameter is nearly identical to the inner diameter of the tube.

For the most part, the bubbles move with the water in the tube. For the bubbles to move relative to the water, the water must squeeze past in a thin film along the cylindrical walls of the bubbles, and this takes place only slowly. Thus bubbles spaced by a reasonable quantity of liquid tend to maintain their spacing even after traveling tens of meters.

What happens if the bubbles vary in length? Will they travel at the same speed, or will there be some dispersion? (*Dispersion* is the technical term for the phenomenon whereby signals at different frequencies travel at different speeds in a medium. For example, light traveling through glass fibers is dispersed because blue light travels slower than red light.) In our case, we might expect small bubbles to slip to a small extent relative to the water flow, whereas the larger bubbles might travel at almost the

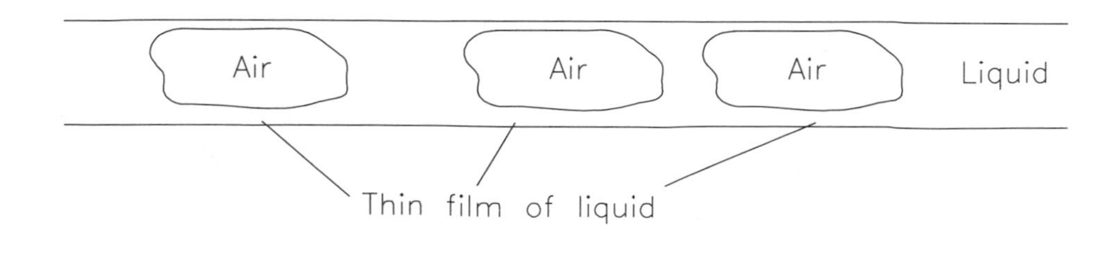

same speed as the water flow. Is this actually what happens? I certainly found that small bubbles—those that were nearly spherical and significantly smaller than the tube diameter—moved slower than larger bubbles and sometimes even stopped and were captured by a larger bubble.

And Finally . . . Sending Bubblegrams Uphill

I tried a double beeper system. If you place one beeper near the beginning of the tubing and another at the end of the tubing and then inject short Morse code bubble sequences, you can hear how well the tubing preserves the Morse. Listen to the input sequence of beeps, and then wait a few seconds for the received beeps; it is easy to judge how well the system is preserving the message integrity.

Try sending bubble messages up a slightly raised tube, so that there is height difference between sender and receiver. (With bubbles running uphill, we have a situation similar to that of the "air-lift" pump mentioned earlier in this section.) Is there an increased tendency for the bubbles to coalesce? Is any information lost in this way? Could you join narrower pipes or slightly wider pipes into parts of the system and still preserve the information? What pipe diameter best preserves information, even at a high flow speeds?

REFERENCE

Coulson, J. M., and J. F. Richardson. *Chemical Engineering.* 1st ed. Oxford, U.K.: Pergamon Press, 1954. Many subsequent editions have been published.

31 Pneumatic Morse

Electric communication will never be a substitute for the face of someone who with their soul encourages another person to be brave and true.

—Charles Dickens

Like Mark Twain, Charles Dickens had a few problems adjusting to the hot new technology of his day. Dickens was not a fan of what he called "electric communication," but perhaps he would have preferred pneumatic telegraphy to the electric sort? In this project, we send a pressure pulse down a long tube to activate a pressure-operated signal. Systems like this are actually in use today: pneumatic control systems in industry often employ long lengths of tubing to send signals.

Signaling using the pressure along a tube is the principle behind the speaking tubes that used to be found in ships and offices. You first had to let your party know that you wanted to call. Each end of the tube was provided with a whistle on a cork, and to call, you removed your whistle and blew down the tube to alert the person on the other end. In the early nineteenth century, before the electric telegraph became established, pressurized tubing was proposed as a long-range method of communication. There was even an extremely ingenious plan to use the coal-gas pipelines of Victorian cities to carry voice or pressure messages. This plan would have been doubly effective: the pipelines were already there, so the network cost was zero; and coal-gas contains a lot of hydrogen, a medium that would have allowed sound to travel much faster. I don't know whether gas-pipeline telegraphy would have interfered with the lighting function of the gas, but since the gas lamps of the day flickered badly anyway, I suspect that no one would have noticed.

What You Need

- ❑ Pressure switch (from a discarded washing machine)
- ❑ Pneumatic tubing (10–30 m, 6 mm in diameter)
- ❑ Beeper (a multimeter with a beeper on its resistance/diode range, or a transistorized 12-V or 5-V electronic beeper unit)
- ❑ Battery
- ❑ Squeeze bulb

What You Do

The setup we use here is a model of simplicity. Connect one end of the tube to the squeeze bulb, and run the other end of the tube to the pressure switch, ensuring that the connections do not leak. Connect the multimeter to the contacts on the pressure switch, or the beeper and battery if you prefer. You may have to experiment before you find a pair of contacts that close when the switch is pressurized—there are often multiple contacts on a washing machine pressure switch. You also have to decide which input to use on the pressure switch—such switches often have two ports, for negative and positive pressure, and sometimes even another port for another switch working at a different pressure. Use the most sensitive input.

First try the system out with a short piece of tubing. Try sending a little Morse code, say SOS (dot-dot-dot, dash-dash-dash, dot-dot-dot). If all works well, put a long piece of tubing in, 10 m or so, and try sending your Morse code again. You will discover that the delay in the system makes it more difficult to send a message successfully. Now put a really long piece of tubing—30 m or so—and try again. With this long piece of tubing, going slow enough to give the system time to react and sound the beeper is a major challenge. I tried out an 80-m piece of tubing and found that I could not transmit more than one dot or dash every 2 s.

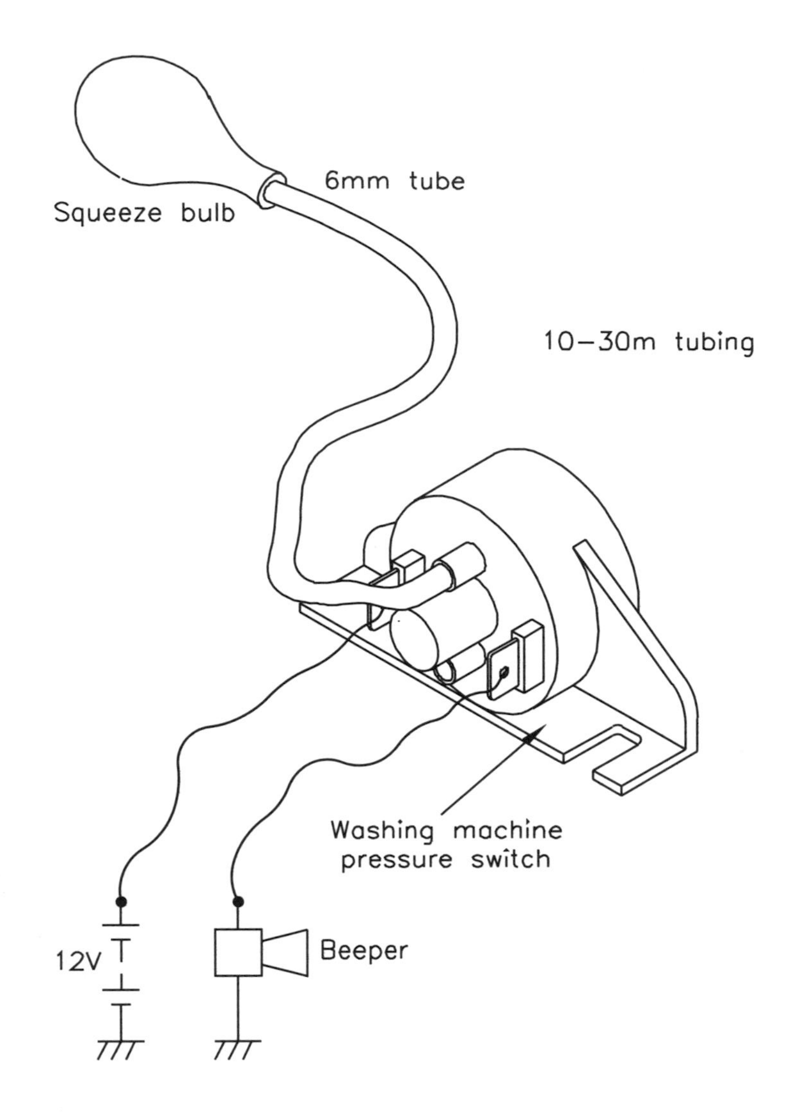

How It Works

What is going on in the pneumatic Morse system? Well, basically, the pressure pulse applied by the squeeze bulb reaches the receive beeper only by virtue of the tubing, whose small diameter forces the air to flow slowly. You can increase the flow speed by using tubing with a larger diameter, but then the volume you are trying to fill with compressed air will be larger, so you'll need more flow to fill it. Overall, you do gain by using larger tubing, but the system is severely limited at long distances. Nevertheless, the clever thing about the system is that it will actually work over an unlimited distance. Unlike a sound wave, which is attenuated with distance, the pressure pulse will eventually transmit the signal, although it may take a very long time indeed unless the pipes are large in diameter.

THE SCIENCE AND THE MATH

Our pneumatic Morse device is limited in its maximum operational frequency by the need for pressurization and depressurization of the tubing. The speed of a pressure pulse in air is governed by the speed of sound: the pulse travels at a constant speed of around 340 m/s. This also holds true for our pneumatic system, at least for short distances in large tubes. However, for longer pipes, and narrower ones, the fluid viscosity begins to have an effect that we can calculate roughly.

Let's assume we have a laminar flow (this assumption is valid for long pieces of tubing). When we apply pressure differential ΔP_0 to a piece of tubing with radius R and length L, the flow rate, Q, through the tubing is given by Poiseuille's equation (once the flow has stabilized):

$$Q = \pi \Delta P_0 R^4 / 8\mu L,$$

where μ is the viscosity of air and P_0 is the ambient pressure. However, what is detected at the far end is not the flow but rather a pressure increase, ΔP. But we can consider the pressure increase to be the result of flow Q filling up a volume, V, that at its very minimum is half the tube volume ($\pi R^2 L$). (The effective volume, V, could also be the detector volume—the volume of the diaphragm chamber in the washing machine pressure switch, for example, if that volume was large compared to the pipe volume—but let's use the pipe volume for the moment.) The rate of pressure rise, $d\Delta P/dt$, at the far end will be given by the flow rate, Q, going into volume V:

$$d\Delta P/dt = -(1/V) P_0 Q,$$

which gives

$$d\Delta P/\Delta P_0 = -P_0 R^2 dt / 8\mu L^2.$$

Solving for ΔP gives

$$\Delta P = \Delta P_0 \exp(-t P_0 R^2 / 8\mu L^2),$$

which is an exponential with time constant τ given by

$$\tau = 8\mu L^2 / P_0 R^2.$$

This equation indicates that there is an exponential *rise* in pressure following the application of the driving pressure, a rise that approaches the input pressure differential, ΔP_0, asymptotically. The maximum data transmission speed is a few times the reciprocal of the time constant, τ. For short distances, the time constant is less than the time taken by a sound wave to traverse the tube, and the time lag will be due to the finite speed of sound. Above a few tens of meters, however, the system will be slower than the speed of sound, at least with small pipes (e.g., those with an outer diameter of 4 mm and an inner diameter of 2 mm):

Radius	Length	Time constant	Signal speed
1 mm	10 m	4 ms (actually limited by speed of sound to 30 ms)	340 m/s
1 mm	1,000 m	40 s	25 m/s
50 mm	6,000,000 m (trans-Atlantic system)	160 h	10 m/s

But before we believe this theory, we need to check the Reynolds number of the flow to determine whether our assumption of laminar flow is valid. To do so, we need the velocity of the flowing fluid (note that this is not the same as the signal speed).* Using our equation for Q and a pressure signal of 10 mbar, we arrive at a velocity of 0.5 m/s. Plugging this velocity, $L = 0.001$ m (pipe radius), $\rho = 1.5$ kg/m^3, and $\mu = 10^{-6}$ into

$$R_N = \rho VL/\mu,$$

gives us a Reynolds number of 750, which indicates that we have laminar flow. However, with larger diameter pipes and larger driving pressures (which will give larger values for Q), the Reynolds number will be higher and flow may be turbulent).

Now that we have checked the Reynolds number, let's go back to our signal transmission time constant equation, which shows the value of larger pipe diameters and high pressure in transmitting data. An demonstration system like ours operating over a distance of 10 m will respond essentially instantly, within tens of milliseconds—although at tube lengths of more than 1,000 m, the response might be slowed to 40 s or more. The equation also shows the difficulty of using longer tubes. However, although it is clear that using large-diameter tubing provides a considerable advantage even with a tube diameter of 100 mm, a trans-Atlantic tube would have a signal delay of 160 hours!

We approximated the volume that the system feeds to be one-half the tube volume, lumping together all the volume in the tube. However, a partial differential equation might more accurately represent the distributed volume in the tube:

$$D\partial^2\Delta P/\partial L^2 = \partial\Delta P/\partial t,$$

where $D = P_0R^2/8\mu$ and L is the position along the pipe. Solving for ΔP gives

$$\Delta P = \left[1/\sqrt{(t)}\right][K \exp(-L^2/Dt)],$$

where K is a constant. This equation gives a bell-shaped pressure-distribution curve that is centered on the input end and whose width increases with time. The time constant for this equation is similar to that obtained with the simpler analysis above for any particular value of L. This spreading-out process, although it does not describe the pneumatic Morse device exactly (which starts with a spike of pressure at time zero near the input end), does give the right kind of behavior.

The partial differential equation is an example of the "diffusion equation" that is often used to describe the diffusive mixing of fluids in the absence of bulk motion and to describe the flow of heat. By thinking of the air flow in the tube as a "plug flow" (that is, as if the flow velocity through a cross section of the tube was the same all over), you can derive the diffusion equation. Accounting for the fact that the flow speed at the edges of the tube will be nearly zero and that the flow speed will be maximal in the middle complicates the picture more, although clearly a more accurate answer would result. The Navier-Stokes equation (discussed on page 32), which allows for rotational (turbulent) flow, would fully describe our system. At low pressure, with irrotational flow and laminar (viscous) behavior, the Navier-Stokes equation reduces to a diffusion equation. Diffusion equations have solutions in which the distance of propagation is proportional not to time but to the square root of time, and our diffusion equation is no exception. This proportionality explains why long tubes are exceptionally slow to respond.

*It may surprise you that the air speed in the tube is not the same as the signal speed. However, this is not unusual: for example, another system in which the bulk speed is not the same as the signal speed is a row of marbles placed close together. With a gap of 1 mm between each 10-mm marble, the signal speed will be about ten times the speed of an individual marble. If each marble has to travel only 1 mm at 1 m/s before it hits the next marble, the signal is transmitted from one marble to the next in 1 ms, which is one-tenth the time (10 ms) it would have taken the first marble to reach the position of the second marble.

The theoretical performance of the pneumatic Morse system looks surprisingly good to me. However, actual systems may well be much slower because they are limited by the volume, V_d, of the detector device. If we use V_d for the volume, we get

$$d\Delta P/\Delta P_0 = [(P_0 \pi R^4)/(4\mu L V_d)]\, dt,$$

which has a time constant of $4\mu L V_d/(P_0 \pi R^4)$. This equation indicates that to minimize time delays, we must minimize V_d, but it also shows a very steep variation with R: it is worth using the maximum bore of tubing possible. However, V_d will eventually become small relative to the tube volume, and then the earlier time constant equation with an R^2 dependence will become operative. The weaker dependence upon L exhibited by this equation also seems like an advantage; but if L is made too large, V_d will again become small relative to the tube volume, and the equation with an L^2 dependence will apply.

Using $R = 1.5$ mm, $L = 30$ m, and $V_d = 50$ cm^2, I get theoretical values for the time constant of 10 ms, which seems rather swift compared with what I see in the actual system and confirms that the real system is very much limited by the speed of sound and the relatively large detector volume, V_d. Perhaps in practice there are also turbulence effects.

With delays of only a fraction of a second, transmitting with the system becomes difficult if the only feedback to the person doing the transmitting is the delayed signal. The difficulties that people experience when they get delayed feedback are well documented. Pipe organists who play in a large church are familiar with the problem. The rank of pipes being played may be 50 m from the keyboard, which gives a delay of more than 100 ms. This delay is surprisingly disconcerting, and organists have to practice diligently to handle it. Psychology researchers in a number of studies have asked volunteer operators to control a robot using a TV screen for feedback. Using a tape-loop delay system, the researchers gave the operators successively longer and longer delays between the time when picture was taken and the time that it was displayed to the operator. With a delay a second or two most people became hopelessly inaccurate with their control.

And Finally . . . Hydraulic Morse

The analysis above is all for laminar flow. What happens if we have turbulent flow, which, as explained, is very likely for larger-bore tube runs over shorter distances and with greater driving pressures? Presumably the energy wasted in the turbulence slows the signal transmission. But might there be some advantages to turbulent flow?

Clearly limitations imposed by the speed of sound can be mitigated by using hydrogen, but would hydrogen actually speed up our system? What if we used water instead of air? The much lower compressibility of water would obviate many of the problems (the hydraulics of auto braking systems take advantage of this lower compressibility). However, we would have to deal with the increased viscosity. Would we need to account for this? On the face of it, repeating the calculations above with the compressibility of water, gives a 20,000-fold advantage in time constant, since the compressibility of water is about 2 GPa, as opposed to 0.1 MPa for air.* So what about a trans-Atlantic hydraulic telegraph?

REFERENCES

Chemical engineering texts are frequently useful when it comes to the practical aspects of pipe work. However, they often don't have pressure-drop calculations, since chemical engineering firms do this sort of calculation by computer these days, using their own proprietary software. But see, for example, the classic *Perry's Chemical Engineer's Handbook* for the published practical material, my *Industrial Gases* for some of the basic theory, and Coulson and Richardson's *Chemical Engineering* for more on this.

Coulson, J. M., and J. F. Richardson. *Chemical Engineering.* 1st ed. Oxford, U.K.: Pergamon Press, 1954. Many subsequent editions have been published.

Downie, Neil A. *Industrial Gases.* London: Chapman and Hall, 1997.

Lee, B. S. "Effect of Delayed Speech Feedback." *Journal of the Acoustical Society of America* 22 (6): 824–826. A discussion of psychological testing of delayed feedback, which can, surprisingly, prevent people from speaking.

Perry, Robert H., Don W. Green, and James O. Maloney, eds. *Perry's Chemical Engineer's Handbook.* New York: McGraw-Hill, 1984.

Smith, K. U., and H. M. Sussman. "Delayed Feedback in Steering during Learning and Transfer of Learning." *Journal of Applied Psychology* 54 (4): 334–342. A discussion of psychological testing of delayed feedback.

*Compressibility, or bulk modulus, is the constant K in the following equation: $\Delta P = -K\Delta V/V$, where ΔP is the change in pressure, ΔV is the change in volume, and V is the volume. For an ideal gas like air, $P = nRT/V$ and $\Delta P = -\Delta V nRT/V^2$, where R is the universal gas constant. Therefore, K for air equals nRT/V, or P.

32 Seven-Segment Telegraph

Bob sells beige shells. Bob's igloo is big. Ellie's is less big. She sells big size shoes. She boils goose eggs.

—A passage that can be written upside down on a calculator, using only the letters *b, E, g, h, I, L, O, S,* and *Z* (that is, the numbers 9, 3, 6, 4, 1, 7, 0, 5, 2)

According to many accounts, Samuel Thomas Soemmering of Munich made the first attempt to send information along wires when he demonstrated a twenty-six-wire system to the emperor of Austria in 1809. His system was based on the principle of sending electricity to separate electrodes in an electrolysis cell where bubbles could be observed. However, this system was very expensive to wire. Even in 1809, Soemmering could have made a superior system using the seven-segment display principle, perhaps by using seven-segment bubbling electrodes instead of our light-emitting diodes (LEDs). But he did not, nor did anyone else for years.

Not until the great electronics revolution of the 1960s did people consider the direct sending of character images made up of separate dots or lines. After that time, a cornucopia of different devices for displaying information arose: starburst displays, dot-matrix displays, and seven-segment displays like the one in this project (see Alan Chappell's *Optoelectronics*).

Seven-segment displays are mostly used to display the numerals 0 to 9, which do not use more than a tiny fraction of the display's capability. The display could show 128 (2^7) different characters, although admittedly, many would not be readily distinguishable to the untrained eye. We actually need only, say, 28 distinct characters to be able to transmit messages telegraphically: the twenty-six letters of the alphabet plus the period and the space.* For example, why not

for the letter *d*? Sometimes the uppercase letter will be more distinctive than the lowercase letter, but I personally find lowercase letters easier to read.

You may disagree with choices for the seven-segment characters. For example, I use a bar above the character to specify a wide letter as opposed to a narrow letter (*w* versus *u* or *m* versus *n*):

and

are *w* and *m*, respectively. But you may be able to devise an alphabet that is, at least for you, more readable than mine. The symbols I have here for awkward letters like *x* (=) or *k* (which looks like an *h* with a additional segment on the top) are also debatable.

What You Need

- ❑ 110–130 diodes (e.g., 1N4004, which cost only a few cents by mail order; the exact number will depend upon your choice of alphabet)
- ❑ Large 7-segment LED display (e.g., 5 V per segment operating voltage)

*Don't forget the space; it is more important to intelligibility than you might think. Consider the water rat's description of the contents of a picnic hamper in Kenneth Grahame's *The Wind in the Willows:* "coldtonguecoldhamcoldbeefpickledgherkinssaladfrenchrollscresssandwidgespottedmeatgingerbeerlemonadesodawater."

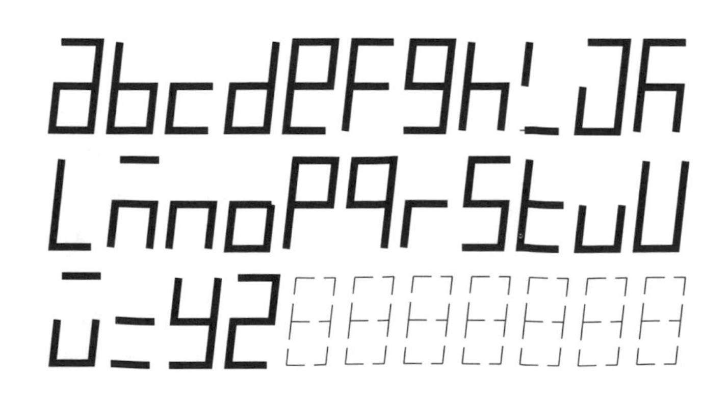

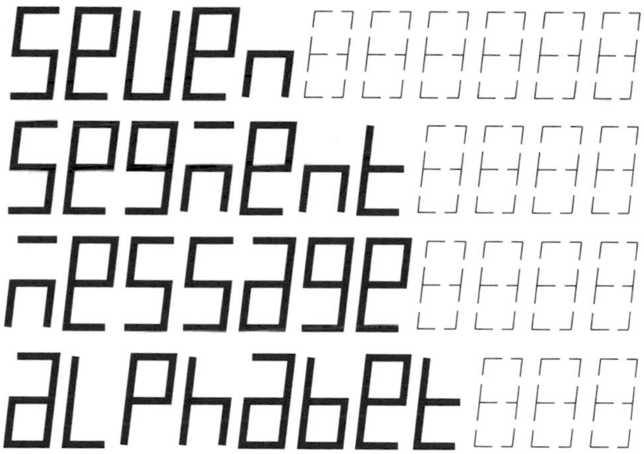

seven
segment
message
alphabet

- Wire
- Mounting board and circuit board for LED display
- 26 brass drawing pins and a brass spacebar strip
- Cheese-head brass bolt to fit into the end of a ballpoint pen body
- Ballpoint pen body, with ink cartridge removed
- Seven current-limiting resistors (e.g., 200 ohm)
- Battery (e.g., 9-V PP3)
- 8-Wire cable (7 wires plus ground)
- Circuit board for diodes/contacts

What You Do

The transmitter of the telegraph requires the construction of a diode array to provide the different segment selections corresponding to the array of brass drawing pin contacts. Depending upon the alphabet you choose, you will need from 110 to 130 diodes. For clarity, the diagram shows only two letters, *c* and *d,* and only 8 of the 114 diodes needed for my alphabet. A convenient way to wire the diodes for the letters is to use a long piece of circuit board, stripboard with the metal strips running lengthwise, and a set of diodes connected to each drawing pin at the bottom running to one of seven "bus" wires running along the top of the circuit board.

The seven wires from the diodes are then routed to the eight-wire cable that leads off to the remote display. Rather than having a separate ground connection or wire, it is easier to use an eight-wire cable: a standard eight-wire telephone

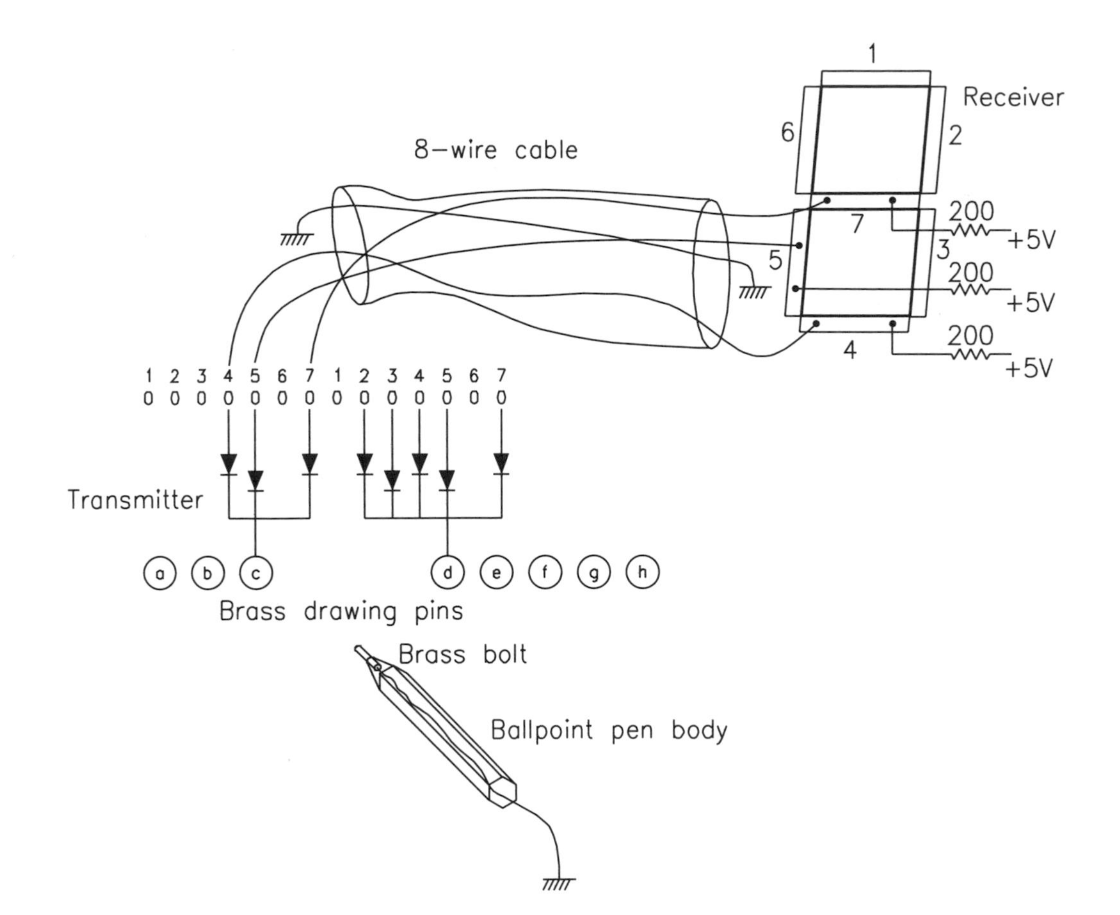

cable is suitable. Each of the segments of the display is connected to the seven "live" wires and a 200-ohm resistor to 5 V (only three resistors are shown).

To make the wand, remove the pen's ink cartridge and replace it with a small brass bolt that is soldered to a wire connected to ground. Transmission is accomplished by touching the wand to a contact, which lights up a pattern of display segments.

How It Works

When you touch one of the brass drawing pins with the wand, electric current is conducted via the diodes to selected segments of the display. The diodes ensure that the connections to another brass pin do not bridge the outputs when that pin is not selected. Without the diodes, the whole display would light up the moment any contact was made. The diode array amounts to a homemade programmable read-only memory (PROM). With a PROM, you select an address using an address input, and the contents of that address are sent to the output wires: the seven-segment display is analogous to the output wires and the wand is analogous to the address input.

And Finally . . . Alternative Indicator Systems

You could try using less-conventional indicator systems. Soemmering's pioneering system of electrolytic bubbles works fairly well, although a current of at least an amp is needed for a quick response. A set of seven electrodes to give the seven-segment indication would not be too difficult to devise, and you would need salt in the water to increase its conductivity. I tried a Soemmering system in the spirit of his original, using common table salt, a copper counter electrode, and twenty-seven tiny copper alphabetic electrodes. It actually worked very well with a modern 12-V battery and was a source of fascination and wonderment to the kids who tried it out. You could try a copper sulfate solution with a copper counter electrode and carbon alphabetic electrodes, which might avoid confusion caused by the bubbles that form on the counter electrode: oxygen gas bubbles are produced on the carbon seven-segment electrodes,

$$4OH^- \rightarrow H_2O + 4e^- + O_2,$$

and copper plating forms on the copper counter electrode,

$$Cu^{++} + 2e^- \rightarrow Cu.$$

A system like this probably should not be operated for long periods: the solution will eventually become weakly acidic (sulfuric acid).

Other candidates for a seven-segment indicator might be simple filament lamps. "Festoon lamps" are linear filament versions of regular lamps. But perhaps you could use simple hot wires. Resistance wire such as NiCr can be safely heated to red-hot in the air, and although it is an inefficient light producer, it is perfectly visible in dimmed lighting. (You would need the very thinnest NiCr—36 gauge [0.19 mm], for example—and a high current, at least 1–2 A.) You could also use a dynamic display like the one outlined in the "Moving Messages" chapter.

REFERENCE

Chappell, Alan, ed. *Optoelectronics: Theory and Practice.* Bedford, U.K.: Texas Instruments Ltd, 1976.

33 Six-Wire Telegraph

O! for a Muse of Fire, that would ascend
The brightest heaven of invention;
A kingdom for a stage, princes to act
And monarchs to behold the swelling scene.
. . . can this cockpit hold
The vasty fields of France? or may we cram
within this wooden O the very casques
That did afright the air at Agincourt?
O pardon! since a crooked figure may
Attest in little place a million;
And let us, ciphers to this great account,
On your imaginary forces work.
Suppose within the girdle of these walls
Are now confined two mighty monarchies
Whose high uproared and abutting fronts
The perilous narrow ocean parts asunder.

—An arrangement of twenty-six letters in different combinations (Shakespeare, *Henry V*)

O! for a Nose of Fire, that woold ascend
The prightest heawen of inwention;
A cingdon for a stage, princes to act
And nonarchs to pehold the swelling scene.

. . . can this coccpit hold
The wasty fields of France? or nay we cran
within this wooden O the wery cascwes
That did afright the air at Agincoort?
O pardon! since a crooced figore nay
Attest in little place a nillion;
And let os, ciphers to this great accoont,
On yoor inaginary forces worc.
Soppose within the girdle of these walls
Are now confined two nighty nonarchies
Whose high oproared and apotting fronts
The periloos narrow ocean parts asonder.

—A different arrangement of twenty-six letters,
with apologies to Shakespeare

The nineteenth-century Morse telegraph system managed with just 2 characters—dot and dash—to code for the English alphabet. What is the minimum number of characters needed for uncoded communication in English? If we include the 26 letters, the space, and the full stop, we come up with the 28 characters used in telegraphic-style communication. If we include the upper- and lowercase letters and the 10 numerals, for a total of 64 characters, we could transmit text in the style used in e-mail messages, for example.

Communication pioneers realized from the outset that the cost of wiring precluded the use of sets of 64 separate wires for telegraphic communication, and they tried strenuously to devise systems with fewer wires. One of the earliest telegraph systems, that of Charles Wheatstone, used just 6 wires to transmit, which is very close to the minimum needed.

With a binary system, you can encode 2^N characters with just N binary signals. This system is the most efficient system for parallel transmission that can be theoretically devised.* Our 64 characters could be transmitted with just 6 signals, or bits, and a cruder alphabet would need just 5 bits, giving 32 characters. In fact, the commercial standard system for character coding, ASCII, uses 8 bits, giving 256 characters, although there is a reduced 7-bit ASCII code that gives just 128 characters. The ASCII code is used for practically all coding purposes in computers and computer communication.

*Information theory predicts that the "information content" of 64 equally probable characters, is 6 bits. However, not all characters are equally probable in real language. The actual information content of the 64 characters in the English language as it is normally used is less than 6 bits per character. Morse code ingeniously uses the differing average probabilities of the letters by using as many as 4 dots and dashes for some characters, such as *Q* and *Y*, and only 1 dot for *E* and 1 dash for *T*. This approach can also be used in parallel transmission systems. Imagine, for example, a system that used just 4 wires to transmit all the commoner characters and only occasionally used 1 or 2 extra wires borrowed from another parallel circuit to transmit the less frequently used characters.

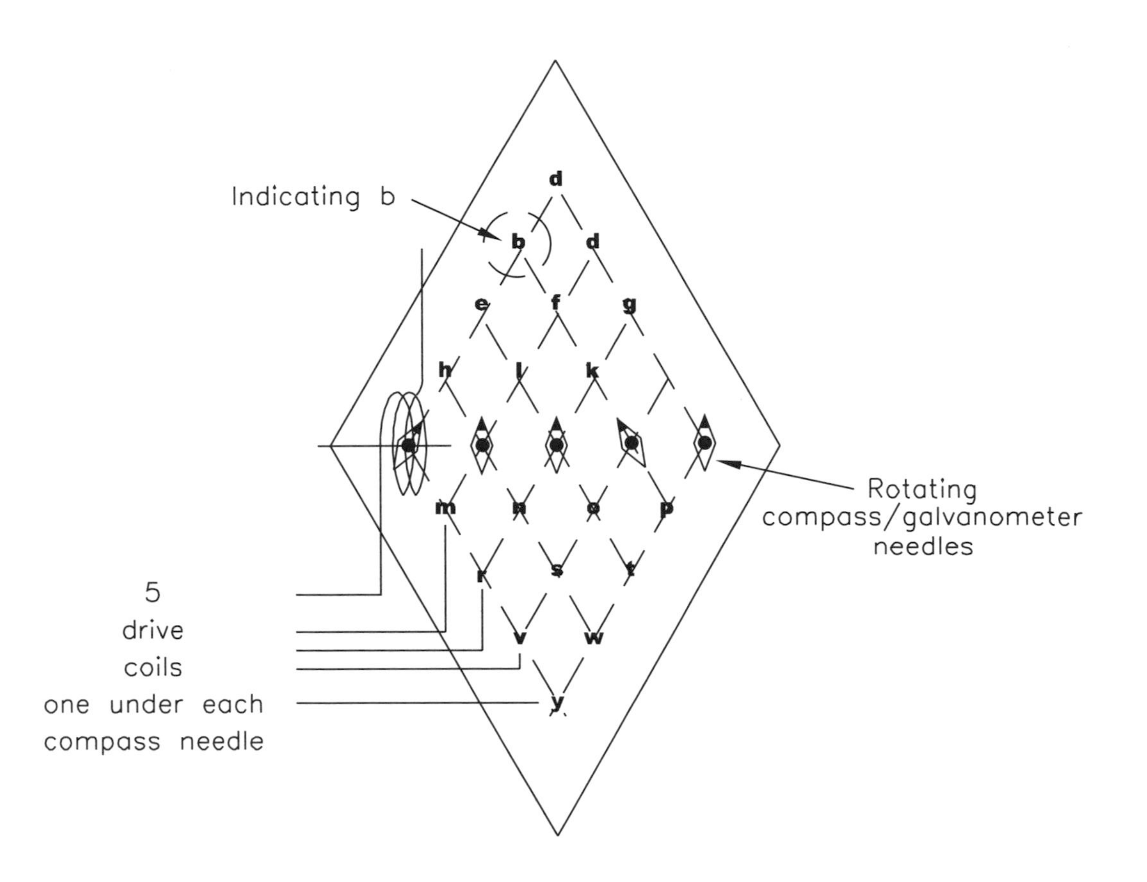

With an 8-wire household telephone cable, we could send 7-bit ASCII (1 wire is used as a ground reference). However, it is difficult to devise a simple binary circuit system that does not involve a considerable number of logic circuits, many-poled switches, or other complexities. For example, consider a binary switching system in which 4 relays are switched by 4 wires from the transmitting station, with an alphabet of 16 possible letters as the output. This looks clumsy and requires multipole relays, including a huge 8-pole relay, which is large and expensive, or a whole bunch of 1- or 2-pole relays wired to simulate the multipoles. Even with modern electronics, you would have to use a number of integrated circuits, including a pair of devices known as 3- to 8-line decoders, or a fairly complex chip known as a 4- to 16-line decoder.

Another possibility is to use combinations of just 2 of the 6 wires. For example, if you have 6 choices for the first wire, which leaves you 5 choices for the second wire, then you have 30 possible combinations of wires. It is easy to see how this is worked by having the wires at the transmitter end powered by wands

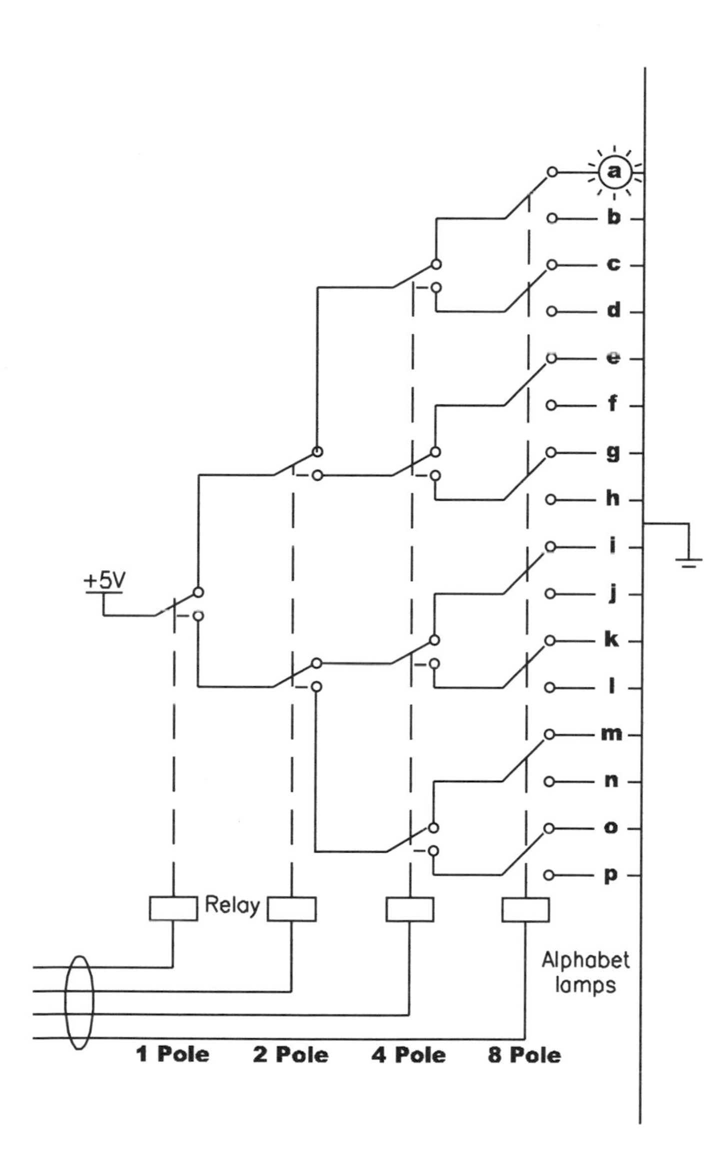

connecting the supply voltage and ground to 2 of the 6 wires, with a suitable decoder on the other end. However, because the system cannot tell the difference between the combinations 1 3 and 3 1 or 5 2 and 2 5, there are really only 15 combinations. In general, with W wires, the number of usable combinations, N, is given by*

$$N = W(W - 1)/2,$$

so that

*In general the number of combinations of r parts chosen from the total number of parts, n, is ${}_nC_r$, where

$${}_nC_r = n!/[(n - r)!r!] = [n(n - 1)(n - 2) \ldots (n - r + 1)]/[r(r - 1) \ldots 3 \times 2 \times 1].$$

Could we use $r = 3$, that is, triplets of wires? What about quadruplets? Quintuplets?

$${}_6C_1 = 6$$

$${}_6C_2 = 6 \times 5/2 = 15$$

$${}_6C_3 = 6 \times 5 \times 4/(3 \times 2) = 20$$

$${}_6C_4 = 6 \times 5 \times 4 \times 3/(4 \times 3 \times 2) = 15$$

$${}_6C_5 = 6 \times 5 \times 4 \times 3 \times 2/(5 \times 4 \times 3 \times 2) = 6$$

This equation indicates that there is some advantage in using triplets, but doing so would be complex electronically, requiring the use of semiconductor triple-input AND gates and a rat's nest of wiring.

2 wires	1 character
3 wires	3 characters
4 wires	6 characters
5 wires	10 characters
6 wires	15 characters
7 wires	21 characters
8 wires	28 characters.

So 8 wires are needed to transmit the full telegraphic character set.

Here are some wiring combinations for pairs from 6 wires if 1 2 is different from 2 1:

A	1 2
B	1 3
C	1 4
D	1 5
E	1 6
F	2 3
G	2 4
H	2 5
I	2 6
J	3 4
K	3 5
L	3 6
M	4 5
N	4 6
O	5 6
P	2 1
Q	3 1
R	4 1
S	5 1
T	6 1
U	3 2
V	4 2

W	5 2
X	6 2
Y	4 3
Z	5 3
stop	6 3
space	5 4
comma	6 4
spare	6 5

There is a further complication: in general, more than one output indicator will illuminate. Think about what happens if, for example, lamps are connected to pairs of wires 1 2, 2 3, and 1 3 (as they must be to allow all the combinations to light a lamp). When the 1 2 combination is sent, the 1 2 lamp will obediently light up, and the same will happen for the 1 3 combination. But when wire 1 and wire 3 are connected by the sender to current, all three lamps will illuminate, although 1 3 will shine more brightly than 1 2 and 2 3.

Curiously, with the LED lamps in this configuration, the system will actually operate: because the unselected LEDs won't receive sufficient voltage, only the selected pair will light up. With an operating voltage of approximately 1.5 V for the LED and 3 V supplied, only one LED will illuminate when a wire pair is activated. In fact, even with 4.5 V (three cells), still only a single LED will illuminate. In addition, LEDs can distinguish between 1 3 and 3 1, because they operate on only one current polarity, so connecting pairs in parallel across pairs of the 6 telegraph wires doubles the number of transmittable combinations back

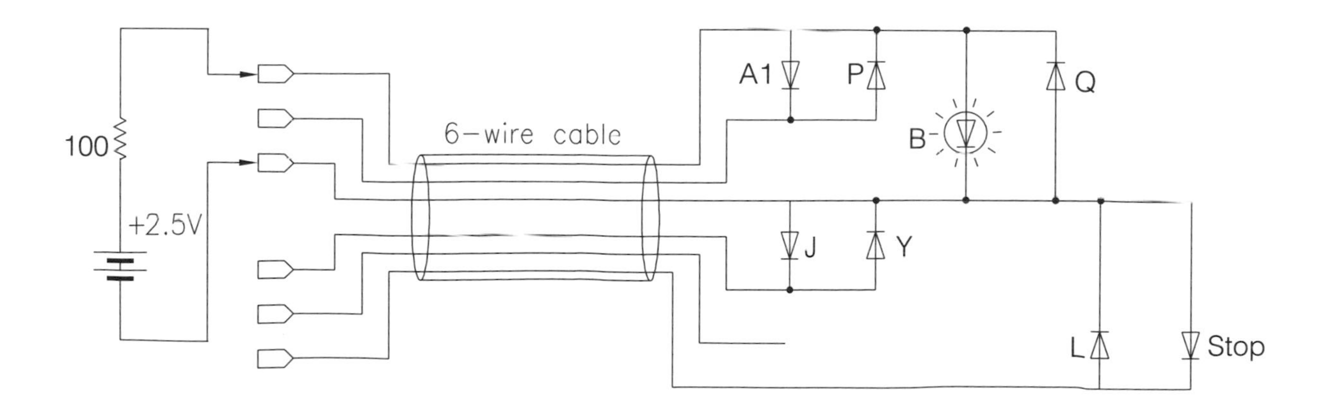

to 30 again, covering the alphabet characters stop and space, as required for a telegraph system.

The 6-wire telegraph I have just described is a bit of a rat's nest, but there are ways to simplify it. At the transmitter end, you could use 2 wands connected to the battery to connect current, rather than the 2-pole switches. However, it is then rather difficult for the transmitting operator, who must memorize or look up all the necessary combinations.

A Simpler Six-Wire Telegraph

Surely there must be simpler approach? In fact there is. You can use the wires as *x-y* grids: an arrangement of *x* and *y* wires with simple selector switches connected via the transmission cable to a matrix of LEDs or lamps is certainly simple. However, with only 7 wires to use, our *x-y* grid has a very limited capacity: just 12 characters (4×3). Our alphabet would end at *L*.

But what if we had bidirectional signals on the wires: switches that gave either zero, a positive signal, or a negative signal, rather than just a positive signal or zero? Bidirectional signals provide a triple-valued logic state on each line. Multivalued logic designs are highly unusual but not unheard of. In *Multiple-Valued Logic Design,* George Epstein describes the math behind multivalued logic and shows how such designs sometimes provide a neater, more reliable, and more economical solution. With two polarities allowed on each wire, we might have 24 characters (2×12). Even allowing for the space character, we still have 23 characters and are missing only 3.

So how about replacing rarely used letters such as *Z* and *Q?* (If you look in books about cryptography, such as Simon Singh's *The Code Book,* you will find statistics on this.) The rarest letters are

J	0.2 percent
Q	0.1 percent
X	0.2 percent
Z	0.1 percent.

We could perhaps use *Kw* for *qu,* so that *quintet* would become *kwintet; ks* for *x,* so that *fox* would become *foks;* and *ts* for *z,* so that *zipper* and *dozen* would become *tsipper* and *dotsen.* Let's look at an actual passage of text:

> The importance of being earnest is a quite key piece of literature. The author, Wilde, once described fox hunting as "the unspeakable in full pursuit of the uneatable," and a cynic as someone who knows the value of dozens of things and the value of none of them.

would become

> The importance of being earnest is a kwite key piece of literature. The author, Wilde, once described foks hunting as "the unspeakable in full pursuit of the uneatable," and a cynic as someone who knows the value of dotsens of things and the value of none of them.

The spelling seems a little eccentric, but the intelligibility of the message is not affected.

With just 6 wires plus ground, 7 wires in all, we could have two 3-by-3 grids, giving 18 characters, 17 plus the space. Could we manage with just 18 characters? We could try, for example, omitting all the vowels:

> English is fairly intelligible, even without the vowels, thinking about it logically.

becomes

> nglsh s frly ntllgbl, vn wtht th vwls, thnkng bt t lgclly.

This also gives us a handy compression. The frequencies of the vowels in English are

A	8.6 percent
E	13.1 percent
I	6.3 percent
O	8.0 percent
U	2.8 percent.

So as an added bonus, we can expect our messages to be about 38.8 percent shorter.

But we're still left with 21 letters and stop and space to transmit. Let's take out the vowels and replace *j, qu, x,* and *z* with *g, kw, ks,* and *ts,* respectively.

> The importance of being earnest is a quite key piece of literature. The author, Wilde, once described fox hunting as "the unspeakable in full

pursuit of the uneatable," and a cynic as someone who knows the value of dozens of things and the value of none of them.

becomes

Th mprtnc f bng rnst s kwt ky pc f ltrtr. Th thr, Wld, nc dscrbd fks hntng s th nspkbl n fll prst f th ntbl, nd cync s smn wh knws th vl f dtsns f thngs nd th vl f nn f thm.

Now maybe we have strayed over that fine line between the cryptic and the gnomic. Perhaps we would be better off replacing more consonants:

B	(1.5 percent) → *P*
J	(0.2 percent) → *G*
K	(0.8 percent) → *C*
QU	(0.1 percent) → *CW*
V	(1 percent) → *W*
X	(0.2 percent) → *CS*
Z	(0.1 percent) → *TS*.

If we also replaced the rarest vowel, *U*, with *O* and replaced *M* with the similar *N*, we would end up with

The inportance of peing earnest is a cwite cey piece of literatore. The aothor, Wilde, once descriped focs honting as "the onspeacaple in foll porsoit of the oneataple," and a cynic as soneone who cnows the waloe of dotsens of things and the waloe of none of then.

Now let's see how we can use these substitutions to make a practical *x-y* matrix 6-wire telegraph system.

What You Need

- ❑ 18 two-pole push switches
- ❑ 6-wire cable
- ❑ 18 LEDs
- ❑ Current-limiting resistors (e.g., 220 ohm)
- ❑ Battery (e.g., 9-V PP3)
- ❑ Battery connector

- ❑ Mounting board
- ❑ 2 circuit boards for display

What You Do

It turns out that you don't in fact need a ground wire in the *x-y* matrix telegraph: each of the receiving LEDs is activated by the 2 (*x* and *y*) signal wires from the

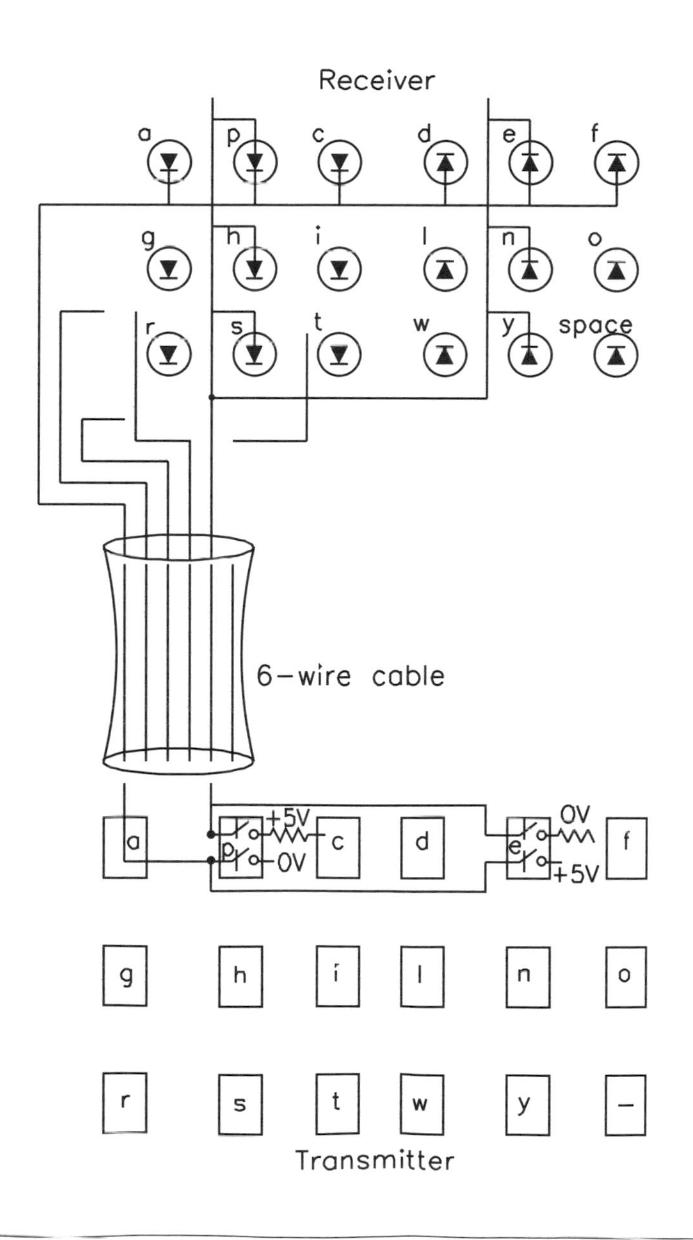

transmitter, so our minimal messaging system uses just 6 wires without ground, just as Wheatstone's did. Construct the display units by using two stripboard circuit boards at right angles for the x and y signals; solder one lead of each LED to the lower x and the other lead to the upper y strips. Make sure you provide some strain relief for the cable so that it is not easily pulled off the circuit boards. The buttons should be assembled into a suitable mounting with a legend on or above each button, indicating the letter that it represents; do the same for the display unit. Now try sending that Shakespeare.

How It Works

Our final design still takes advantage of the fact that LEDs emit light only when current flows in one direction. With the LEDs in pairs facing in opposite directions in the receiver, the positive-going current lights one of the pair, the other acting as an insulator, and the negative-going current lights the other of the pair. The 2-pole switches ensure that we can connect any particular pair of wires to have current in either polarity. The 3-by-3 matrix gives us 9 characters with positive-going current and 9 characters with negative-going current, allowing us our 18-character alphabet.

In addition to single characters, the system can transmit some combinations of the 9 positive current characters and some combinations of the 9 negative characters. (You can also transmit combinations of a single positive character with any negative character that has a different row and column. This means that you can send a positive character with some combinations of up to 4 allowed negative characters: the same applies for a single negative character with up to 4 allowed positive characters.) You can use the combination of characters to transmit other information: space, full stop, shift key, and maybe even the 8 missing letters.

And Finally . . . Doubling the Internet at a Stroke

With the reduction of information content we've developed, we are now in a position to double the capacity of the world's communication system. By replacing ASCII with our very own "NewspeakASC," we can reduce the number of bits per character to as low as 4, instead of the usual 8, so all communication circuits will have twice the capacity! And our minimal messages ought to be readable, if a bit different. We could even think about restoring them to normal English with the aid of a computer spell-checking program. Of course, it is not quite that simple. Communication through the Internet (and probably many other types of communication traffic) uses some form of data compression. Data compression typically involves scanning data for repetitive sequences (such as a long series of space characters): instead of transmitting 640 space characters, you just transmit a single space character, along with the instruction to repeat it 640 times. NewspeakASC is itself a form of data compression, of course. Our more pithy communications won't compress as much as normal data traffic, so our factor of two is optimistic.

REFERENCES

Epstein, George. *Multiple-Valued Logic Design.* Bristol, U.K.: Institute of Physics Publishing, 1993.

Singh, Simon. *The Code Book.* London: Fourth Estate, 1999.

34 Moving Messages

The Moving Finger writes; and having writ,
Moves on: nor all thy Piety nor Wit
Shall lure it back to cancel half a Line.

—Edward Fitzgerald, *The Rubaiyat of Omar Khayyam*

There has always been a need for the temporary display of information. Today, there are massive advantages to keeping information in electronic form. Electronic information takes little space and can be copied, moved, modified, and, perhaps most important of all, searched far more easily than paper-based information. However, information stored in electronic form needs to be displayed. There is a whole science devoted to the display of information from electronic sources, and the industry for making display devices—from projection systems, cathode ray tubes (CRTs; computer monitors), and liquid crystal display (LCD) screens to things as simple as a lamp—is huge, rivaling the computer chip industry itself.

Here we look at some possibilities for simple electromechanical displays. Charles Wheatstone's original 1845 telegraph display used electromagnetically driven compass needles, and although all-electronic systems such as CRTs, LEDs, and LCDs are much commoner, mechanical display devices are still used to some extent today: the large annunciator boards used at some football fields and railroad stations or, easier to get close to, some gasoline pump displays. Many of these displays are based on small devices in which an electromagnet pulls a white or colored patch into view on a dark background, a device something like a relay.

What You Need

- ❑ Electric motors (the small, inexpensive 1–3-V ones)
- ❑ Batteries
- ❑ Switches
- ❑ White disks for mounting on the motor shafts (e.g., 25 mm in diameter)
- ❑ Paper
- ❑ Strong cardstock (or thin plastic sheet)
- ❑ Light gray paper
- ❑ Colored pens, fluorescent colored pens

What You Do

Glue the light gray paper onto the cardstock. Assemble the motors in a row or in an array, push the shafts through holes in the cardstock, and glue them with the shafts projecting in front of the gray paper. Now fix the white disks onto the motor shafts. I used paper disks stuck on top of gearwheels that could be push-fit on the motor shafts. Wire the motors to batteries and the remote switches to operate the display. You don't need to run the motors particularly fast; 1.2 V or 1.5 V should be sufficient. Draw your message images in black on the disks: you can use letters, numbers, short words, or symbols. Each image should be repeated at least three times around its disk. Repeating the images guarantees that one of them will always be the right way up when the disk stops spinning. You can screen off the lower half of the disks with a piece of paper so that you won't be distracted by the incorrectly oriented images. To make a numerical information display, I used ten disks, with the ten numerals.

The display should be wired so that the disks are normally spinning. (When operating, the display sounds like a bee hive!) When the disks are spinning, the areas on which the images are drawn look gray since the images are completely blurred by the motion of the disks. By drawing additional lines on the disks, you can get them to look more uniformly gray. For example, a spinning disk labeled with the number 3 will have three gray lines on it, along with a white border and a white center. By adding lines between the three 3 symbols, along the perimeter and near the center, you can get a more or less uniformly gray result.

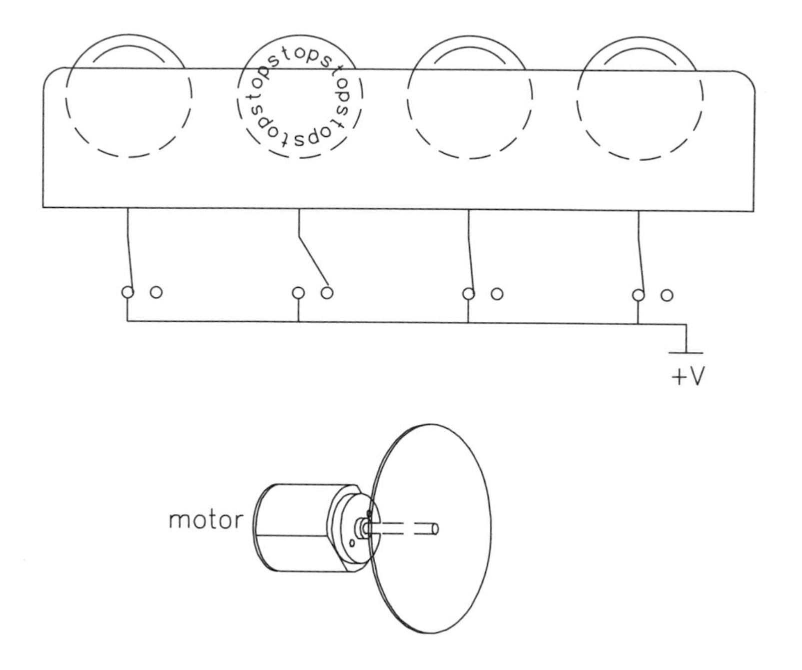

To send an image, stop the appropriate disk, leaving the black image on that disk stationary and visible against the light gray background. By applying different images to the disks, you can send numbers, letters, and even words, although the number of possibilities is limited by the need to have a disk for each character.

The black-and-white display, although readable, is not particularly striking. However, you can achieve bold contrast by using colored disks, like those known as Newton's disks. As before, mount the motors with the light gray paper behind the white disks. Now comes the tricky process of coloring the disks. They must be colored with a pattern of colors—for example, red, green, and blue—that produces a grayish-white impression when the disk is spinning. Pens vary in the vividness of their color and the wavelength of peak light reflection, so I can't prescribe a precise formula. The simplest thing to do is to color them with narrow sectors, like those in a pie chart, of single colors and then spin the disk to check the color. Continue to color, perhaps with a different color, gradually widening the sectors as you go until the disk is nearly full. If the disk looks blue-gray, then add more green and red; if it looks red, then add more green and blue, and so on. Using fluorescent pens should allow you to create color disks that are very light, perhaps almost pure white, when they are spinning. In addition, stripes rather than pie sectors can be used, and these are possibly more striking.

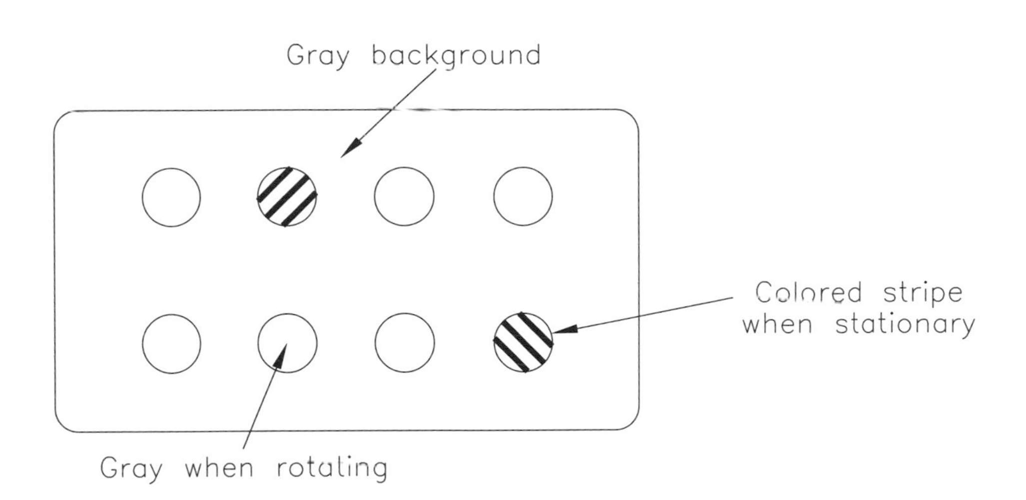

When all the motors are running, numbered disks look almost uniformly gray with what appear to be circular outlines. Only when a motor stops do you see a brightly colored disk on the gray paper, indicating the number communicated. With a little experimentation, you may find that you can draw the symbols in color and still retain the uniform gray color for the rotating disk. By writing the symbol around the disk in colors that are balanced to make the same shade of gray-white within each annular ring on the disk, you can make Newton's disks that display gray-white when spinning and display the desired symbol or message word in color when stopped.

How It Works

Our rotating disk displays rely on a phenomenon called *persistence of vision.* Images that change at a rate faster than about 50–100 ms are averaged and perceived as continuous static or continuously moving images. This phenomenon is also used in modern information displays. Many such displays show moving images; however, even simple displays that appear to be static, such as those on clocks or calculators or watches, are in fact dynamic. It is easy to see this with an LED display by looking at it from a few feet away and loudly humming a low note or blowing a raspberry with your lips. The slight vibration from this action will jiggle your eyes up and down at a minute angle, and you will see the figures wriggle apart weirdly, as if the segments were not fixed down but instead attached to threads that allowed them to wobble to and fro. The CRT screen on a computer or television is the same. Look at a white screen in a dimly lit room and wave a pencil vigorously in front of it. The silhouette of the straight pencil will be bent like a banana!

Larger LED displays and CRTs are scanned. CRTs are scanned in a raster pattern: the illuminating spot of electrons moves rapidly from left to right and then flies back 30,000 times per second, and the line of light produced by this sideways scanning moves down the screen and then flies back to the top 25 or 50 times per second. The "banana" effect on straight objects occurs because the scan illuminates the areas of the back of the object, producing a shadow, at different times: the bottom will be illuminated 20–40 ms after the top. So if the pencil or other object is moving at, say, 1 m/s, the bottom will appear to be displaced by 2–4 cm.

The multicharacter LED display is scanned because each of the seven segments in every numeral in the display is connected in parallel with the same segment in every other numeral. This wiring method minimizes the number of driver circuits (and accompanying wiring). Just one driver circuit per numeral and one circuit per segment are required: thus just ten drivers can be used for a three-numeral display, rather than the twenty-one that would otherwise be needed. A numeral is displayed by selecting segments and numerals using the connecting wires. For example, the sequence of operation might be as follows: select segments and the numeral 1, reselect segments and the numeral 2, and then reselect segments and the numeral 3.

You can track the movement of objects with your eyes to some extent by turning your head and also by means of the surprisingly rapid movement of your eyeballs. In fact, your eyeballs normally make rapid movements, called *saccades,* from side to side over extended objects, to scan the fovea, the part of the eye that distinguishes fine detail. However, once the angular velocity of an object as measured at the eye exceeds a certain value, or when the rapidly moving object is visible—as here—only over a small viewing angle, then the image becomes blurred. This is the case with our rotating disks: we perceive the light they reflect as slightly dimmer than that reflected from a white disk. The fact that the light from any particular part of the disk imaged onto the back of the eye is flashing on and off is not perceived.

The Newton's disk display combines persistence of vision with the science of color vision. Light from the sun, or any other very hot object (like the filament in an electric lamp), is composed of wavelengths spread fairly evenly over the visible spectrum, although with some bias toward the red, long wavelength, end. We sense white light when our eyes receive a correctly balanced mixture of light from at least three colors in the spectrum, such as blue, green, and red. Our color disk is whirling around at such a speed that our eyes simply register the average spectral color of the incoming light and ignore the fact that the light received from any particular part of the disk is constantly changing in color.

Both the persistence of vision effect and the blurring of moving images involve the same time constant, around 50–100 ms, the time constant of the eye receptors. However, as I mentioned, the eyeball is capable of performing fast movements to track objects, so that their image on the retina moves more slowly than this time constant would indicate. (You can easily show the effects of the eye's ability to "freeze" fast movement as follows: look out and down at the road surface beneath you when you are riding in a car. After a while you will find, at least at modest speeds of 30 or 40 mph, that you can see pieces of the roadway surface over a small area, just for a moment as your eye tracks it backwards past you, and then the surface is blurred again as your eye swings forward.) Try sweeping your eyes rapidly over the rotating disk display. You may find that you can "freeze" the motion of the spinning disks enough to see that they are not simple gray disks.

THE SCIENCE AND THE MATH

The perception of color by the human eye relies on just three color receptors, with their peak-sensitivity wavelengths surprisingly close together at 445 (blue-violet), 535 (green), and 570 nm (red-orange). By cross-correlating the sensitivity curves (S_r, S_b, or S_g) with the input spectrum, I, we can obtain the responses, R_r, R_g, or R_b, from each eye sensor and estimate what color will be perceived:

$$R_r = I[S_r(\lambda)I(\lambda)]d\lambda,$$

$$R_g = I[S_g(\lambda)I(\lambda)]d\lambda,$$

and

$$R_b = I[S_b(\lambda)I(\lambda)]d\lambda.$$

With a spectral source, we can judge the color by adding the sensor responses together. With a spectral line at 510 nm, for example, we would expect a response, in arbitrary units, of $R_r = 0.6$, $R_g = 0.8$, and $R_b = 0.2$. This combination would give the impression of green.

Sensor response curves and numerical tables of color impressions are difficult to handle, however. James Clerk Maxwell first defined a "color triangle" for understanding light mixtures in the nineteenth century. Much later, in 1967, an international committee called CIE (Commission Internationale de l'Eclairage) finalized a color triangle that is the basis for color science today. The CIE defined three idealized color sources and adopted a diagram with these pure sources at each corner and different colors represented as points on the triangle. Any color can be defined by the amount of each of the three primary (red, green, blue; RGB) colors, or, equivalently, by the brightness plus the ratios of two colors to the other one. Thus any color

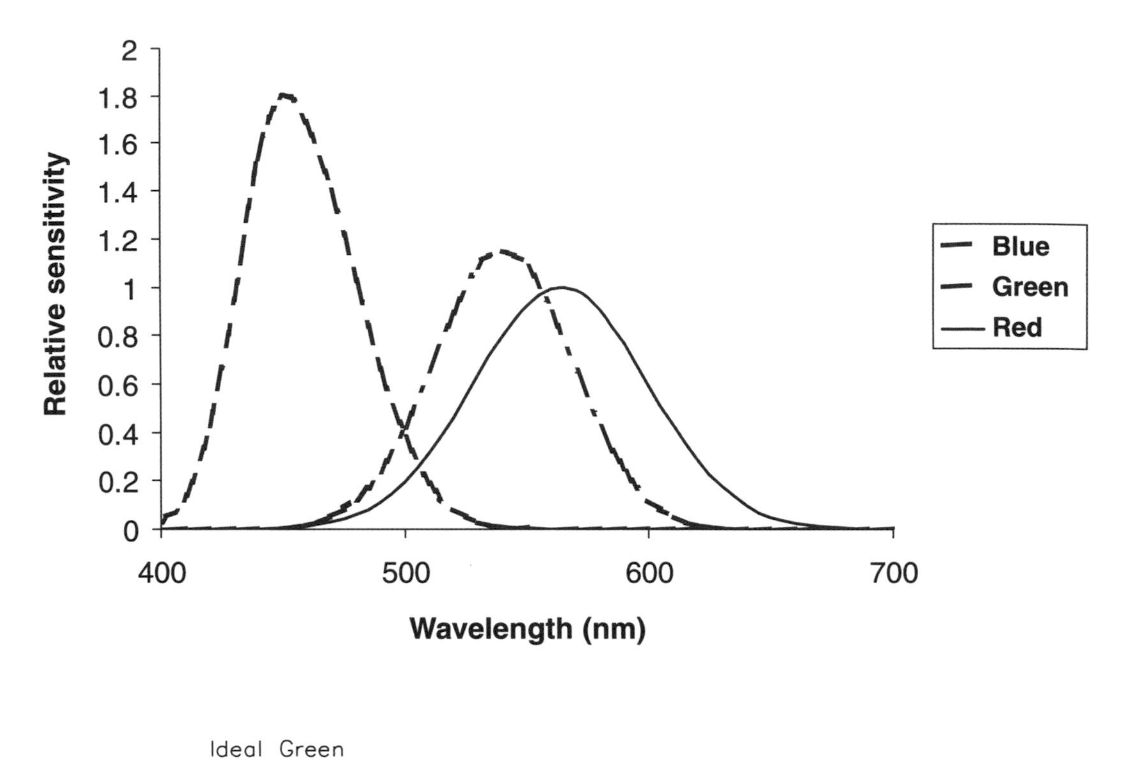

Relative sensitivity
2
1.8
1.6
1.4
1.2
1
0.8
0.6
0.4
0.2
0
400
500
600
700
Wavelength (nm)
Blue
Green
Red

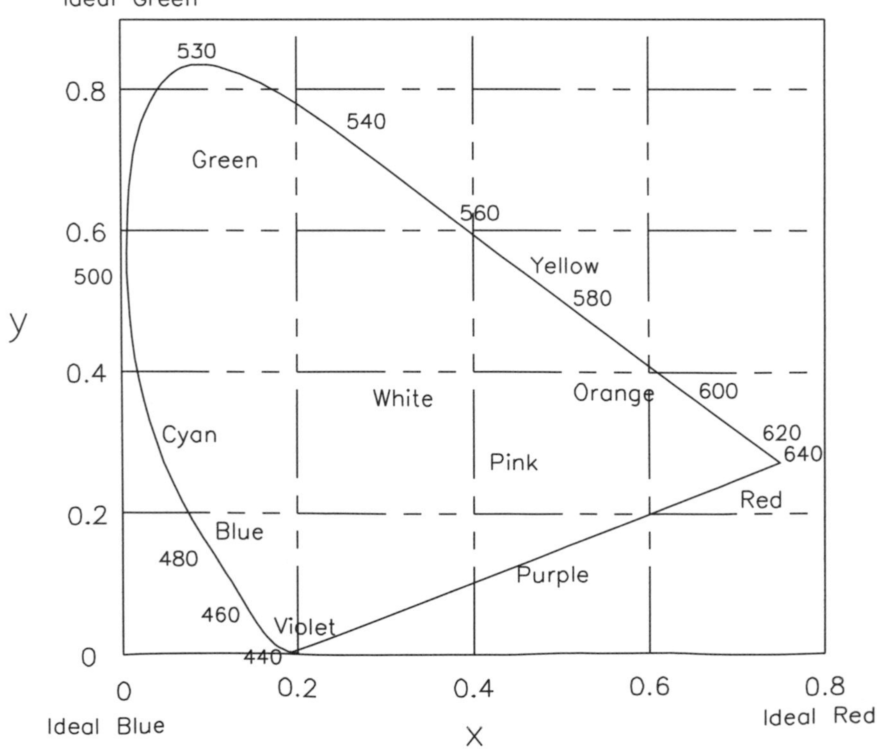

Ideal Green
530
0.8
540
Green
560
0.6
500
Yellow
580
y
0.4
White
Orange
600
Cyan
620
640
Pink
0.2
Red
Blue
480
Purple
460
Violet
0
440
0
0.2
0.4
0.6
0.8
Ideal Blue
Ideal Red
x

can be represented on a two-dimensional plane, discounting the overall brightness. Pure spectral colors lie on the curved line in the diagram, with the conventional names for different colors filled in at various points. The straight line from red to violet represents all the possible shades of a nonspectral color, purple, which can be composed of spectral red and violet mixtures. All physically possible colors are contained in the area of the curve and the purple line. A mixture of two spectral colors will produce a visual impression at a point lying on the line between them. So a mixture of blue and yellow, for example, will give white. In our Newton's disk displays, we are striving to achieve a white impression, the area around the middle of the chromaticity triangle.

The use of fluorescent pens allows a much closer approach to white in the color of the spinning disk. Normal colored inks simply absorb light, converting the photon energy into heat. A green ink, for example, absorbs both blue and red photons, reflecting green photons. Fluorescent colors typically convert high-frequency light (ultraviolet, violet, blue) into lower-frequency light (green, red): a photon is not lost but converted. This is the reason that fluorescent colors appear brighter, and it explains why we can use them to create a much whiter impression from a spinning disk.

And Finally . . . Spinning Seven Segments

Maybe you could make a seven-segment display out of the rotating color disks. However, there may be a better geometry: the disks might be more suited to a "multiblob" representation of the alphabet than to a set of lines, as in the conventional seven-segment display. Could drum-shaped color cylinders be used somehow?

Unusual Actuators

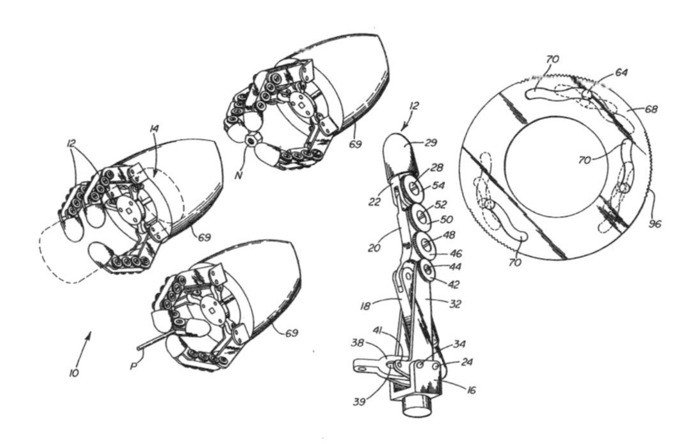

Pangloss disait quelquefois à Candide: "Tous les événements sont enchaînés dans le meilleur des mondes possibles; car enfin, si vous n'aviez pas été chassé d'un beau château à grands coups de pied dans le derrière pour l'amour de Mlle Cunégonde, si vous n'aviez pas été mis à l'Inquisition, si vous n'aviez pas couru l'Amérique à pied, si vous n'aviez pas donné un bon coup d'épée au baron, si vous n'aviez pas perdu tous vos moutons du bon pays d'Eldorado, vous ne mangeriez pas ici des cédrats confits et des pistaches."

"Cela est bien dit," répondit Candide, "mais il faut cultiver notre jardin."

Pangloss said many times to Candide: "All the events are chained together to make the best of possible worlds; for, in short, had you not been kicked out of a fine castle for the love of Miss Cunegund; had you not been tortured by the Inquisition; had you not traveled over America on foot; had you not stabbed the Baron through the body; and had you not lost all your sheep, which you brought from the good country of El Dorado, you would not have been here to eat preserved fruit and pistachio nuts."

"Excellently observed," answered Candide, "but we must work in our garden."

—Voltaire, *Candide*

Actuators are the unsung heroes of control systems in industry. It is all very well to attach hundreds of sensors, install multigigahertz computers, and write gigabytes of software, but if the software doesn't make something happen, then it is futile. You need to do something with information you collect, and to do that you need an actuator. Actuators are often the most troublesome part of a machine: they are complicated, expensive, and sometimes less than perfectly reliable.

The model servo motor in the servo telegraph project is a classic example of an actuator. Before the servo motor came along, model cars and model airplanes were controlled by crude on-off actuator devices that were often unreliable. The model airplanes I had as a kid frequently crashed because the servo motors necessary to make them work properly were too expensive. I made do with systems based on *escapements:* Rube Goldberg gadgets made from alarm clock parts, relays, and wound-up rubber bands. Model servo motors took a long time in coming and were expensive because they are complicated. You need quite a bit of gear to make one—motor, gearbox, feedback potentiometer, amplifier, push-pull output stage, regulated power supply—and all the parts must match up mechanically and electrically.

In *Advances in Actuators,* Tony Dorey and John Moore and their coauthors describe some typical industrial actuators—hydraulic cylinders and the like—as well as the new generation of devices we can now play with: nanoactuators. These use entirely new and unusual modes of operation, some of which may actually be applicable to larger-scale devices. In this part, we look first at an unusual actuator that has rarely been used—although I did discover a machine for delivering anesthetic gas that used a similar mechanism. We then look at an actuator that achieves—with just blobs of ink and a tiny movement, only a millimeter or so—what would normally require a much larger instrument and larger movement.

REFERENCE

Dorey, Anthony P., and John. H. Moore. *Advances in Actuators.* Bristol, U.K.: Institute of Physics Publishing, 1995.

35 Balloon Biceps

> The "silly question" is the first intimation of some totally new development.
>
> —Alfred North Whitehead

Pneumatic devices are a popular way of providing the motive force for fast and powerful systems, and they are often used as the "muscle" in systems in which the "brains" are provided by electronics. Pneumatic devices provide an easy way to amplify an electronic input because the low viscosity of air means that small valves and tubing can be used; a tiny movement of a valve—just a millimeter or two—provides a lot of high-pressure air and a lot of power. In addition, an air tank is a convenient container for storing energy that can be released swiftly when needed. However, pneumatic systems normally require precisely engineered cylinder-piston actuators, as well as expensive valves and tubing.

When someone lifts a heavy barbell, there is a certain swelling of the biceps muscle in the arm, a swelling exaggerated by the cartoonist who drew Popeye eating his can of spinach. Why does the biceps swell? When I asked a biologist I met, I couldn't get a straight answer to this "silly question." Is the swelling the consequence of the muscle contraction, or the cause? Surely the swelling is for the most part simply a consequence. But I surmised that perhaps in some small way it also contributes to the contraction, and this supposition led to the idea that driving the swelling of a balloon might lead to a shortening of a "muscle."

What You Need

- ❑ Net bag (the kind that onions are often sold in)
- ❑ Wooden "arm"
- ❑ Pivot for arm (e.g., hinge)
- ❑ Screws to form attachment points
- ❑ Return spring or rubber band for arm
- ❑ Load for arm (e.g., large weight)
- ❑ Balloon pump
- ❑ Balloon
- ❑ Clip
- ❑ Tubing (12-mm outside diameter)
- ❑ T-piece for tubing

What You Do

The paradox of the balloon biceps actuator is that it pulls when it is "pushed" by the air pressure from the drive system. There are other ways of achieving a tractive force from a positive pressure—in double-acting hydraulic cylinders, for example, the pressure is fed into the connecting-rod end rather than the cylinder end—but the balloon biceps actuator is radically simpler.

I first tried to make the balloon "muscle" by taping strings along longitudinal lines on the surface of a balloon and then allowing it to deflate before tying the strings at each "pole" to the attachment points. However, all this knotting and taping of multiple strings turned out to be quite unnecessary: a net bag does the job admirably well and is far less trouble. Instead of tearing the bag, cleanly cut off one of the knotted or stapled ends and remove the onions. Install a piece of tubing on the open end of the balloon and push the tubing and the balloon through the part of the net that will be next to the stationary end of the assembly. Then glue or staple the ends of the net back together. Anchor the ends to the wood with the small tacks.

The balloon needs to be attached to the balloon pump so that there are no leaks. Make sure your tubing is big enough to allow the balloon to fit snugly around it and small enough to fit tightly on the balloon pump. If necessary, you can improve the seal at the balloon joint by wrapping string or a strong rubber band tightly around it. The T-piece must also fit snugly, and the clip must be capable of crushing the tube sufficiently to form a reasonable block. Otherwise

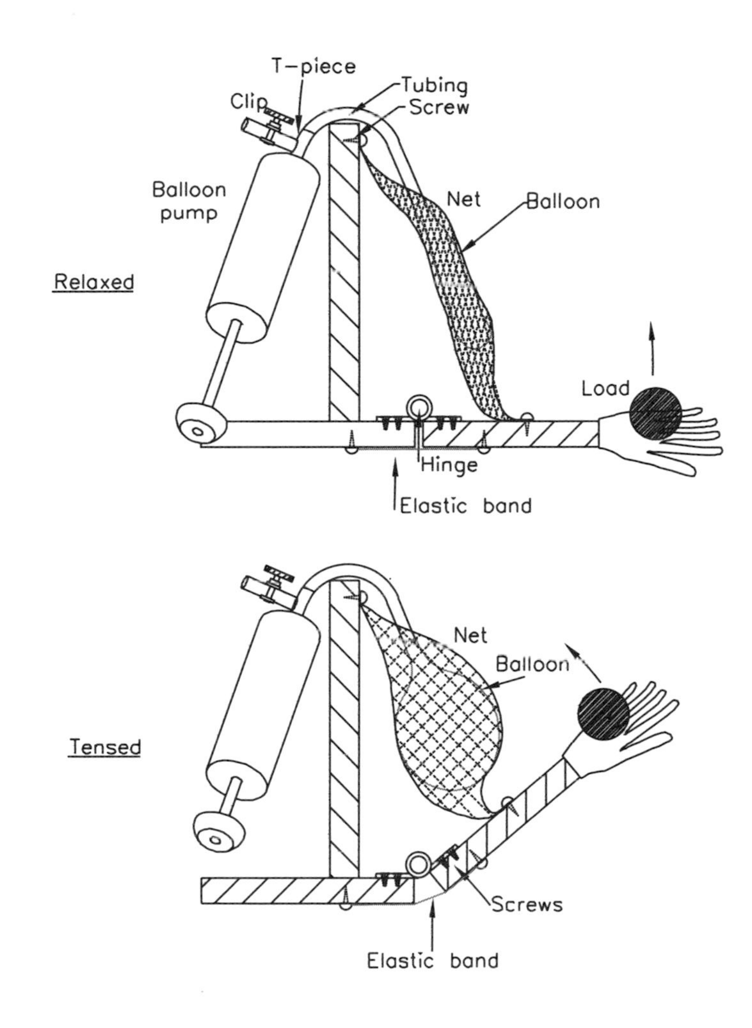

you could use a "cork"—perhaps a plastic ballpoint pen case might seal the end of the tube nicely.

To contract (tense) the muscle, inflate the balloon with the pump. To extend (relax) the muscle, release the clip or pull out the cork to allow the air volume to be forced out of the balloon.

How It Works

When the net is stretched tightly between its attachments, it lies in a straight line that marks the shortest distance between the two attachments. As the balloon muscle inside the net lifts the weight on the arm, the muscle swells enormously, and the swelling forces the straight lines formed by the net's strings into a bent "dog leg" shape. The bending requires that the distance between the attachments become shorter, so the muscle pulls the attachments together. Because of the way the net spreads the force over the surface of the balloon, all the force created by the air pressure on the balloon surface is employed in pushing the strings sideways. The sideways force actually has a "gearing" effect, and a small sideways force can exert a substantial pull along the axis of the balloon biceps (the effect is discussed in my *Vacuum Bazookas* in the "String Nutcracker" chapter).

THE SCIENCE AND THE MATH

The balloon biceps actuator works because the string in the net is essentially fixed in length while it is under tension. If we consider the balloon to be a circle at the middle of the net and we view the net as a set of strings, then we can calculate the change in the distance between the attachments, ΔL. If we assume that the balloon is a sphere with radius R, then its volume, V, equals $4/3\pi R^3$. The geometry of the device is described by Pythagoras' equation,

$$(L - \Delta L)^2 + R^2 = L^2,$$

and solving for ΔL gives

$$\Delta L = L[1 - \sqrt{(1 - R^2/L^2)}].$$

When $R << L$, then

$$\Delta L \sim LR^2/(2L^2) = R^2/(2L),$$

$$\Delta L \sim (3V/4\pi)^{2/3}/2L,$$

or

$$\Delta L \sim 0.192V^{2/3}/L.$$

This equation is not strictly accurate: for example, as the balloon swells to fill the net, its volume changes from that of a sphere to that of a sphere with truncated cones attached. Also, depending upon the type of net you use, and how tight it becomes, the fact that the elongated diamond-shaped net apertures become nearly square may also have an effect. Nevertheless, the dimensions and general form of the equation for ΔL still seem about right.

It now becomes clear why the action of the balloon muscle is relatively smooth and linear. An actuator that simply decreased R (such as a piston-cylinder hydraulic or pneumatic device) would have an action that started slowly and then sped up (R^2

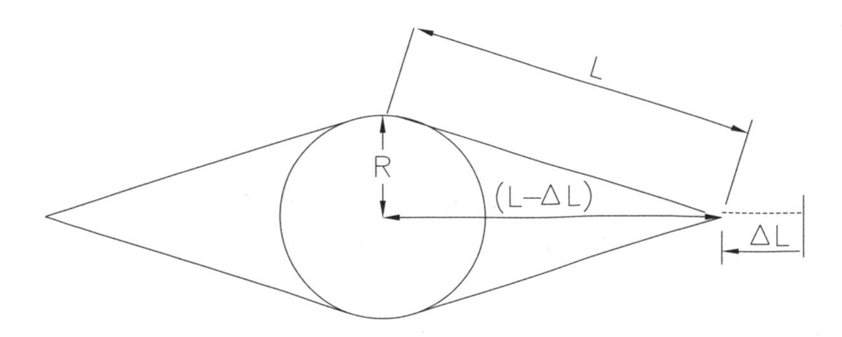

law). However, because the volume of the balloon varies as R^3, the length change varies as $V^{2/3}$, which is much closer to linear with respect to the volume of air supplied.

I mentioned the convenience of storing energy in the form of compressed air. How much energy can be stored, and how swiftly can it be released? Consider a volume V at pressure P that is allowed to expand isothermally—following Boyle's law, $PV = k$, where k is a constant—until its pressure is P_f. The stored energy, E, in this system is given by

$$E = \int P dV = \int (k/V) dV$$

or

$$E = PV \ln (P/P_f).$$

This is a substantial but not spectacular amount of energy. A soda bottle (which weighs 50 g) containing 2 L of gas at 4 bar has a stored energy of $4 \times 10^5 \times 2/1{,}000 \times \ln 4 = 1{,}000$ J.* But a sub-C-sized NiCd battery cell weighing 55 g and contains 1.2 Ah at 1.2 V, which corresponds to 5,000 J. The real advantage to storing energy as compressed air is in the release rate. The peak output of the sub-C cell will depend upon its design, but the ultra-high-current versions used, for example, in electric model airplanes, can produce about 20 A, or 24 W. A more typical peak current would be a mere 2 A, or around 2.4 W. Our soda bottle, however, could easily be designed to release its energy in just half a second. Hence our soda bottle is capable of a 2 kW power output! This power output is exactly what the air-powered water rockets that have been popular with kids for the past thirty years take advantage of. The rockets use the air energy to force water out the back at high speed, propelling the rocket body upward.

*The humble soda bottle is surprisingly efficient at storing energy— weight for weight about 2.5 times as efficient as high-strength steel. If we scaled the soda bottle up in pressure and volume, it would be comparable to an industrial compressed-gas cylinder. A gas cylinder of volume 10 m^3 at 300 barg contains about 5 MJ, which is equivalent to a mere 500 g of trinitrotoluene (TNT) and weighs about 100 kg.

And Finally . . . Linear Balloon Biceps

You could try cylindrical balloons inside the net, of course.* After an initial tautening of the elastic surface, a cylindrical balloon fills the net by forming a sphere at one point along the axis, and this sphere then elongates to form a cylindrical shape with two spherical end caps. But how would the length change, ΔL, of this balloon biceps vary with volume, V? Is the behavior of the cylindrical version more complex? Would a cylindrical assembly be as effective in practice as the version based on a round balloon?

*I recently found that a firm in the United Kingdom, Learning Resources Ltd. in London, is currently offering an "air muscle" assembly based on a linear tube that swells when inflated to shorten a length of braided-hose armor. This miniature device works at much higher pressure (100 psi, or 7 bar) but is clearly related to our balloon biceps actuator.

REFERENCE

Downie, Neil A. *Vacuum Bazookas, Electric Rainbow Jelly, and 27 Other Saturday Science Projects*. Princeton, N.J.: Princeton University Press, 2001.

36 Ink Sandwiches

I must go seek dew-drops there
And hang a pearl in every cowslip's ear

—Shakespeare, *A Midsummer Night's Dream*

When rain lands on a window—or a cowslip petal—it does not flatten itself into a thin film, but instead forms droplets, often in roughly round shapes, like distorted M&M's candies. You might be thinking that the rain forms droplets on the window because it fell as droplets. But spray the window with a hose and you get the same droplets in a suspiciously similar pattern. If you write a message on the window with dilute soap solution and allow it to dry, the message will disappear. However, you will find that the minute amounts—probably only a few milligrams—of soap left on the window will affect the way the water droplets form. If you spray the window again, the message that you wrote with the soap will reappear.

Phenomena like soap-writing are caused by amplifying effects. Tiny variations in the surface of the window cause droplets to form preferentially at certain places and not at others. Just a fraction of a microgram of soap will increase the attraction of the glass for the water, and a droplet sliding down the window will stop when it reaches the soaped area. Similarly, when water vapor condenses on the window, the condensation droplets will appear on the soaped area first. In the absence of any traces of soap, the amplification effects will lead to chaos. The situation a bit like that when a microphone connected via an amplifier to loudspeakers is too close to the loudspeakers. If you talk into the microphone, then the system will amplify your speech, but, if you don't speak, then the system will create its own sound—usually the howling or whistling we call "feedback."

Similar effects happen in the ink sandwich that we will construct in this project. If you squeeze a layer of ink hard enough between two sheets of rigid plastic, you'll get an almost perfectly even layer of ink. The ink is a powerful dye, and even a thin layer gives a highly colored window in this way. But when you pull the sheets apart, the ink layer breaks up into chaotic islands and then into droplets, giving us an actuator for changing the intensity and color of transmitted light.

What You Need

- ❏ Two pieces of rigid, clear plastic sheet (e.g., Plexiglas, Perspex, Lucite, or maybe polycarbonate glazing), 100 mm square, 3 mm thick
- ❏ Glue (I used a powerful solvent glue)
- ❏ Metal strips for lever arms (e.g., from a Meccano or Erector set)
- ❏ Nuts and bolts
- ❏ Soluble ink (e.g., Parker Quink Royal Blue or Pelikan Royal Blue artist's ink)
- ❏ Tape
- ❏ Bulldog paper clips
- ❏ Rubber bands

What You Do

Before you start, put some ink between the two sheets of rigid plastic and watch how the ink behaves as you squeeze the sheets together and pull them apart. This experiment will give you insight into the more complex assembly with lever arms and the ink sealed in.

Glue the metal lever arms to the plastic sheets. I found it useful to treat the plastic with some solvent and then attach the levers with hot-melt adhesive to the middle of the sheets. Seal most of the perimeter with tape, leaving room to inject some ink. Inject the ink and finish the taped seal. Attach several bulldog clips, one at each corner, for example. Now wrap some powerful elastic bands around the levers to squeeze the sheets together. Squeeze the sheets together with your hands too and then remove the pressure for a moment, repeating a few times until the sheets are evenly covered with ink.

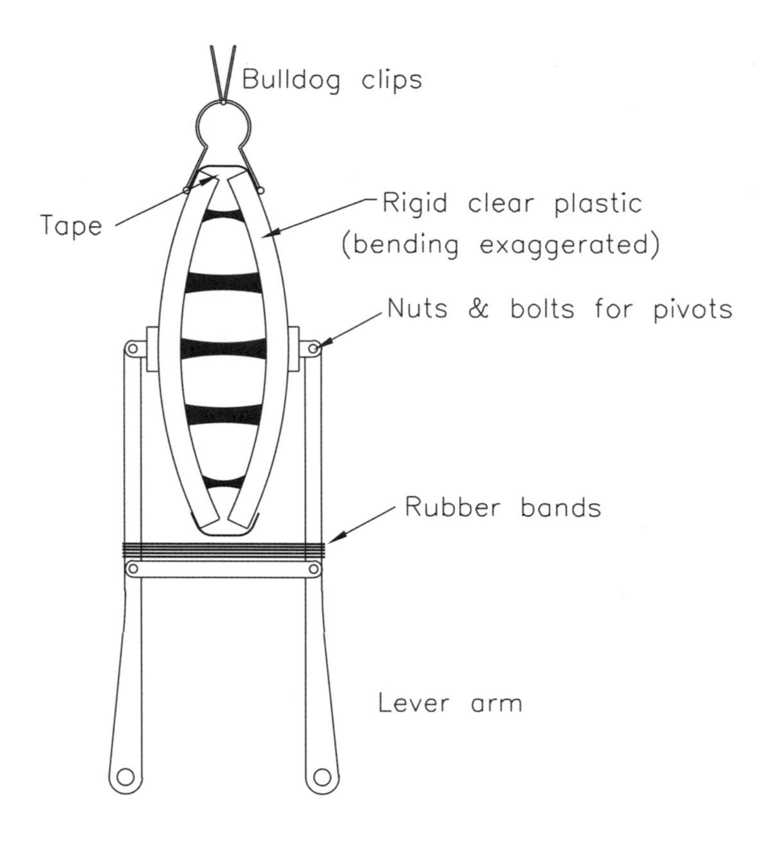

Now when you squeeze the levers, you should find that bubbles and fingers of gas expand into the space between the sheets and that the ink then runs into dark, almost black droplets. When you relax the pressure, an even blue layer of ink, with just a few bubbles, forms again. The device is undoubtedly best operated horizontally, although it also works surprisingly well vertically. Don't squeeze the levers too hard, or they will pull the plastic sheets apart and suck in a lot of air. Conversely, don't press the sheets together so hard that you squeeze ink out. Be prepared to get ink all over your hands!

Shine a lamp or flashlight through the ink sandwich and look at the transmitted light intensity and color. How well do the droplet shapes re-form each time the device is operated? How do they vary with the pressure on the levers?

How It Works

Inks absorb a constant proportion of incident light for each millimeter of thickness. If 1 mm of ink absorbs 50 percent of incident light, leaving 50 percent

Squeezing levers (plates apart)

Let go levers (plates together)

transmitted, then 2 mm will transmit 50 percent of 50 percent (that is, 25 percent) of incident light, and 3 mm will transmit 12.5 percent, and so on.

When the sheets are close together, the ink occupies an even but thin layer. With blue ink, the ink sandwich looks clearly blue, and a flashlight beam shone through the sheets becomes blue. When the sheets are pulled apart, the ink now occupies only a small proportion of the window area but in a much thicker layer, so that the colors turn to black and white. The color of a flashlight beam remains unaltered by the ink sandwich, but the beam is slightly attenuated. Between these two extremes, you will find the ink acting as an absorber—in the inked areas, it colors but also absorbs much of the light; in the other areas, it allows white light through, giving a net effect of white with a blue hue.

The precise patterns formed by the ink as it moves away from even coverage depend upon where you have clamped the edges of the sheets with the bulldog clips, and on the flexibility of the plastic sheets you are using. In general, though, the ink will disappear first from the center, since that is where the levers act. The flexibility of the sheets is such that they pull apart here, and many bubbles and fissures form in the middle and spread out as you squeeze the levers together more.

THE SCIENCE AND THE MATH

Let's think about a simplified version of the ink sandwich, in which the plastic sheets remain parallel to each other and simply move in and out. When the ink is thick enough, it absorbs almost all the incident light: the intensity of light transmitted, *I*, is exponentially proportional to the ink thickness. The Beer-Lambert law describes how light is transmitted when it is continuous:

$I = I_0 \exp(-kx)$,

where I_0 is the incident light intensity, k is a constant, and x is the thickness of ink. However, here we have a constant volume of ink, V. That means that we have reduced light transmission in the areas of thicker blobs of ink but 100 percent transmission in the areas where the ink has been lost. If, for simplicity, we avoid the problem of color and assume black-gray ink, what is the transmission intensity? We can use the following equations to find out:

$I_{ink} = (A_i/A_{tot})I_0 \exp(-kx)$

and

$I_{noink} = A_n I_0/A_{tot}$,

where I_{ink} is the intensity of light transmitted through the inked regions, A_i is the area of those regions, A_{tot} is the total area, I_0 is the incident light intensity, k is a constant for the ink, x is the spacing between the plastic sheets, A_n is the area of the regions with no ink, and I_{noink} is the intensity of light transmitted through the regions with no ink. Putting these two equations together gives us

$I_{tot} = I_{ink} + I_{noink} = (A_i I_0/A_{tot}) \exp(-kx) + A_n I_0/A_{tot}$.

But $A_i = V/x$ and $A_n = A_{tot} - A_i$, so

$I_{tot} = [VI_0/xA_{tot}] \exp(-kx) + (A_{tot} - A_i)I_0/A_{tot}$.

If you plot this function, you will see that you should be able to get from 0 to 85 percent transmission with completely black substances. (Mercury might be such a substance, although because of its toxicity, I would not suggest using it in practice.) With an intensely colored ink offering 50 percent transmission through a 1-mm layer, you should be able to get 50 to 85 percent transmission as you scan across the device from gaps close to zero to gaps a couple of millimeters thick. With a blue ink, for example, you should be able to change the color from virtually pure blue to virtually pure white. With some further analysis, you could show how the color of the transmitted light ought to vary with gap size.

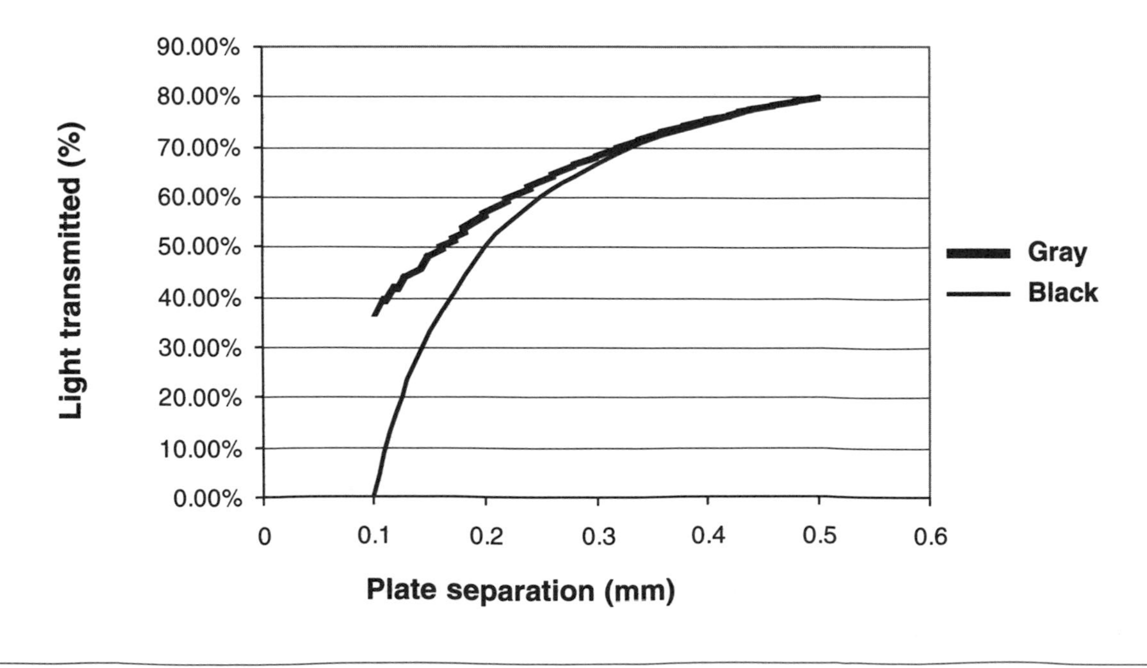

Droplets form between the two plastic sheets when they are separated because when you pull them apart, bubbles that were still present in the ink grow; dissolved gas and vapor are released; and air flows in at the edges to separate the sheets. At first the gases will form long fingers, but as soon as the plastic sheets are a couple of millimeters apart, the sheet of ink will break up into droplets. The details of the process—where fingers of gas form, where droplets form—are chaotic. Small differences between points—places where the surface is slightly different from others or where there is a minute bubble—will be amplified.

And Finally . . . Double-Decker Ink Sandwiches

Why not try two ink sandwiches in different colors in series? With a yellow sandwich and a blue sandwich on top of each other, you could produce green, white, yellow, and blue and all shades in between. Manufacturers of dyes and color filters for photographic and other optical purposes publish the absorption spectra of their products, which can be helpful in choosing materials to test. Could the theater lighting of the future use these handy retractable filters instead of old-fashioned gels? How about a pneumatically operated ink sandwich? Such a device would avoid the problem whereby part of the light-transmission window is occupied by the lever mechanism.

Electrochemical Magic

It is the opinion of several philosophers that the presence of water is essential in electrochemical decomposition, and also for the evolution of electricity in the voltaic battery itself. This opinion has, I think, been shown by other philosophers not to be accurate.

—Michael Faraday, "Electricity"

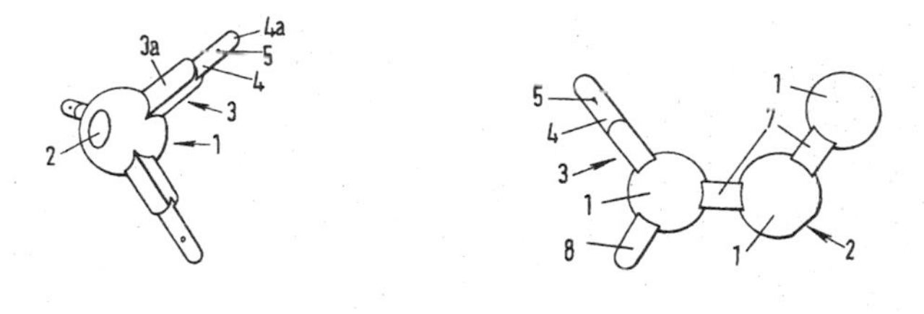

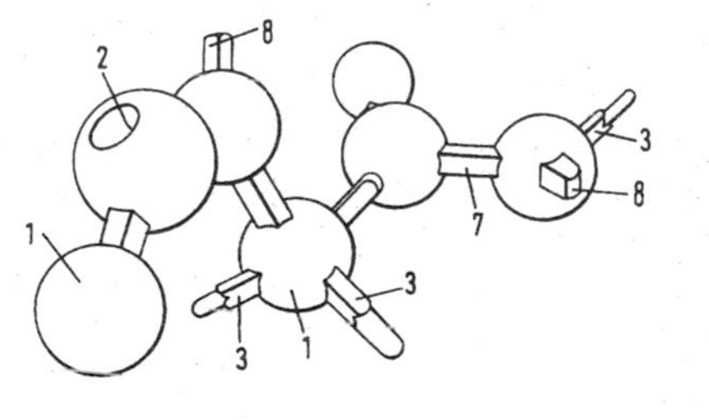

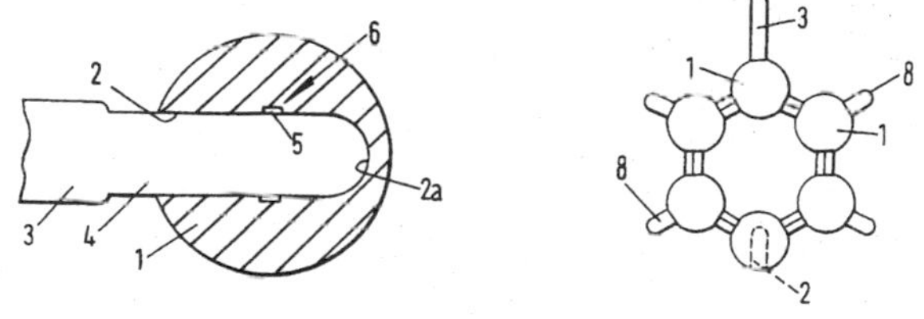

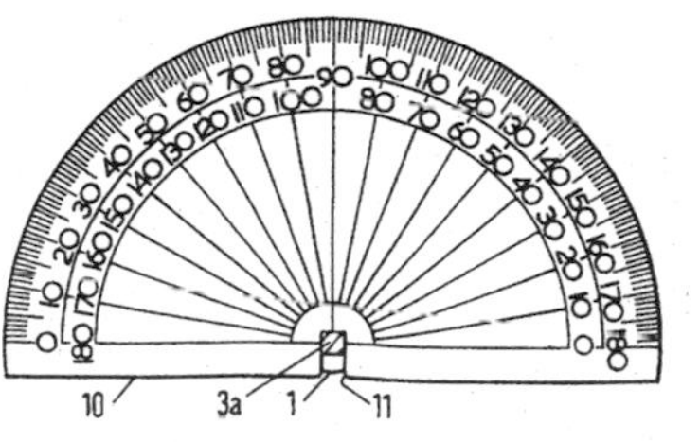

As Michael Faraday points out in his *Experimental Researches in Electricity,* many nineteenth-century scientists wrongly believed that water was essential to electrochemical processes. Faraday goes on to discuss how this idea was not only inaccurate but in fact quite wrong! He even reminds readers that his old boss, English chemist Sir Humphrey Davy, was among those who erroneously thought water to be essential, despite the fact that Davy himself had discovered exceptions to that notion. We follow in Davy's footsteps in "Red-Hot Batteries" and demonstrate a battery that operates completely without water.

During Faraday's time, scientists such as French physicist Joseph Niepce and Louis Daguerre (who developed the earliest photographic processes), Edmond Becquerel, and English physicist William Henry Fox Talbot (who devised the first efficient photographic process in 1839) were studying light-induced effects. In "Wet Solar Cell," we undertake our own study of the effect of light on jars of chemicals and electrodes. In the penultimate chapter, "Unusually Cool Sunglasses," we look at a little scientific surprise from the curious chemistry of corrosion.

REFERENCE

Faraday, Michael. *Experimental Researches in Electricity.* 1 vol. Chicago: Benton/Encyclopedia Britannica, 1952. Originally published in *The Philosophical Transactions of the Royal Society, The Journal of the Royal Institution,* and *The Philosophical Magazine.* A new edition was published in 2000 by Green Lion Press of Sante Fe, N.M.

37 Red-Hot Batteries

> William Gladstone: "This electricity . . . what use is it?"
> Michael Faraday: "I do not know. You may as well ask 'Of what use is a new-born baby?' Why sir, there is every possibility that you will soon be able to tax it!"
>
> —Michael Faraday to William Gladstone, possibly apocryphal

In the nineteenth century, no one could have known that the work of Michael Faraday would be responsible for much of the science behind today's electrical industry. British prime minister William Gladstone may have been suffering from an overdeveloped sense of skepticism when he questioned Faraday about the utility of electricity, but it was not at all obvious at the time that anything substantial would come of the huge numbers of abstruse experiments that Faraday was carrying out.

At first, electrochemical batteries were the only available source of electricity at substantial current, and early batteries were unimpressive in their capability, often managing only a few milliamps on a continuous basis. Until recently, practically all batteries used a solution of salts, acids or alkalis, in water as the electrolyte, the liquid conducting medium essential for their operation. Often the aqueous salt solutions caused considerable problems. Early batteries lost power as the electrolyte got cold, or even froze (which is still a problem for today's standard batteries). They often suffered severe corrosion, or "local action," in which the active materials, such as the zinc anodes, disintegrated even when the battery was not being used. Early battery designs frequently avoided this problem by means of arrangements for winching the battery electrodes out of the electrolyte in lieu of an on-off switch (see, for example, Lieutenant L. Fitzgerald's letter in *Model Engineer & Electrician* magazine).

In recent years, batteries that use exotic electrolytes—liquid solvents other than water, organic materials, even solids—have been steadily creeping into use, notably, the now relatively common lithium batteries, whose high capacity and light weight stem from their use of the highly reactive and ultra-lightweight metal lithium. Because lithium reacts spontaneously and violently with water to form hydrogen gas (which could cause further excitement), solvents such as thionyl chloride have replaced water in these batteries. The polymer batteries that have also become available use as their electrolytes substances that are not only not water but also not even liquid. They use solids—conducting polymers.

Even more exotic than these are sodium-sulfur batteries, which operate only at several hundred degrees centigrade. The high temperature requirement and the fact that sodium explodes on contact with water may appear to rule out their practical use. However, electric vehicles need such huge amounts of electrical energy that exotic batteries may be the only batteries able to meet the demands. The sodium-sulfur battery has not yet been commercialized. However, it has been tamed sufficiently for use in full-size demonstration electric vehicles.

For several decades now, unbeknown to most of us, another type of battery that runs at elevated temperatures has frequently been used in small-scale practical devices: the thermal battery. Thermal batteries use a lithium anode and an iron sulfide cathode with a caustic lithium-based electrolyte. They are highly reliable and can be stored for many years without losing even the tiniest part of their capacity, because the electrolyte, when it is solid and is stored at temperatures below 100 °C or so, is completely nonconducting. When the batteries are needed, however, a layer of "firework" material is ignited to heat them in seconds to the 500 °C or so needed for their operation. They can then supply large currents, offering a good capacity on account of the lithium in the anode. The firework material is a special type containing lots of iron powder, which absorbs the gas normally emitted by a firework and which itself has to be a highly reliable subsystem. Lithium thermal batteries are used principally in military and emergency applications.

Our red-hot battery is a simple and safe small thermal battery with surprisingly large capacity and capability for its size. Owing to polarization and internal resistance effects, not many small homemade batteries of simple construction are capable of running even the smallest electric motor. However, even a small, crudely made version of this battery will run a motor.

What You Need

- ❑ Sodium hydroxide, NaOH (caustic soda)
- ❑ D-size battery (for zinc cup)
- ❑ Copper wire (multistrand connecting wire)
- ❑ Iron cup or plate to contain zinc cup (e.g., tinplate cap with plastic parts and varnish scraped off with a wire brush)
- ❑ Alligator clip connecting wires
- ❑ Electric motor ("solar" low-current type)
- ❑ Multimeter and resistor (optional)
- ❑ Electric hot plate
- ❑ A few tiny pieces of gravel

What You Do

Cut the D-size battery up, retaining the bottom of the zinc outer casing to form a cup, perhaps with a strip of zinc to use for making connections; this is the negative electrode, or anode. Now make the cathode by stripping off 15–30 cm of the multistrand wire and crumpling it into an untidy ball of about 2 cm in diameter at the end of the wire. Now squash the ball of wire down into a thin "coin," or "pill," small enough to fit in the zinc cup without touching its sides. Heat up the wire pill in a gas flame until it is red-hot and leave it there for a minute or two to oxidize. A brownish-black oxide tarnish layer, composed mostly of copper(I) oxide (Cu_2O) and copper(II) oxide (CuO), will form on its surface. Your copper–copper oxide cathode is complete.

Now the cell can be assembled on top of the electric hot plate. Place the iron cup on the hot plate, and then place the zinc cup inside the iron cup. Next put a few tiny pieces of clean gravel in the zinc cup to prevent the cathode from drooping down onto the bottom of the zinc. Add a half teaspoon of sodium hydroxide, and follow that with the cathode, clipping the wire of the cathode to the iron cup or the zinc cup so that the wire is positioned correctly. Use the alligator clips to

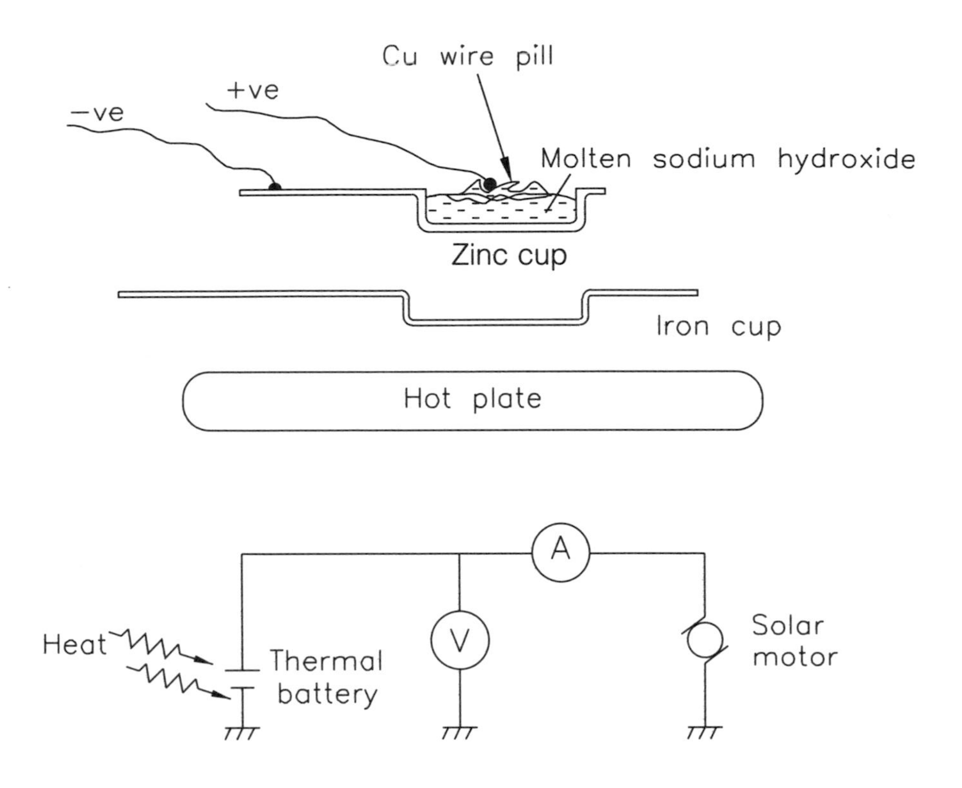

make the electric connections to the meter (with a resistor across its leads) or to the solar motor.

Now warm up the hot plate. I found that the setting needed was somewhat below the maximum. If you overheat the apparatus, you will melt the zinc. (When the hot plate elements are red-hot, they are hot enough to melt the zinc.) But you've still got to get the elements hot enough to melt the sodium hydroxide, which is a lot hotter than the "simmer" setting. Once the hydroxide is melted, you should find that the motor rotates rapidly. After a few minutes, it slows down and then carries on at the lower speed.

Wear goggles when leaning over the hot plate, just in case the cell emits a tiny droplet of hydroxide.

How It Works

In the cell, copper oxide is being reduced to copper while zinc is being converted to compounds equivalent to zinc oxide. In the hydroxide electrolyte, zinc oxide forms sodium zincate:

$Cu_2O + Zn \rightarrow 2\ Cu + ZnO$ + electricity.

But this process works only if the cell is hot enough that the sodium hydroxide forms a liquid, in which electrical current is carried by the electrically charged sodium and hydroxide ions:

$NaOH \rightarrow Na^+ + OH^-$.

THE SCIENCE AND THE MATH

The sodium hydroxide in our red-hot battery is an insulator at room temperature, but it conducts very well once melted, at only 300 °C or so. At the melting point, the positive sodium ions and negative hydroxide ions, normally frozen into regular arrays in the solid state, are free to move.

The red-hot battery should be capable of at least 500 mV at currents up to 50 mA, or perhaps more. The motor will need about 300 mV to run, if my example is typical, and will draw a current proportionate to its 15-ohm resistance, at least at low speed (e.g., 30 mA at 500 mV). The cell will run for an hour or two. Several effects compete to stop it from working. The zinc metal can be exhausted, but this probably won't happen until after the electrolyte has stopped functioning. The formation of zincate reduces the electrolyte efficiency, but the electrolyte also suffers from an important parasitic reaction from the atmosphere. The sodium hydroxide electrolyte absorbs CO_2 to form sodium carbonate, which is a nonconducting solid at the temperature of operation. The sodium zincate formed may also add problems. (However, the fact that a mixture of salts will usually melt at a lower temperature than the highest-melting constituent may help a little.)

The atmosphere also provides a beneficial effect: although the copper oxide from the gas flame treatment is lost from the wire surface in a few minutes, the cell will continue to operate at reduced voltage and current, presumably because oxygen from the air takes part in the cell reaction.

Our cell actually uses nearly the same reactants as the nineteenth-century Lalande cell, which used copper oxide and zinc but had an aqueous hydroxide electrolyte. In our cell, sodium and hydroxide ions enable the cell to conduct electricity by supplying the electrons that allow the following reactions to take place:

anode: $Zn \rightarrow Zn^{2+} + 2e^-$

cathode: $2Cu + \frac{1}{2}O_2$ (from air) $\rightarrow Cu_2O$

$Cu_2O + e^- \rightarrow 2Cu + O^{2-}$

overall: $Zn + \frac{1}{2}O_2 \rightarrow ZnO$.

(This is a simplified scheme: the zinc oxide actually reacts to form sodium zincate in the molten hydroxide, and the O^{2-} ions exist as OH^- ions in the molten hydroxide: $H_2O + O^{2-} \rightarrow 2OH^-$.)

And Finally . . . More Red-Hot Batteries

There does not seem to be any real obstacle to the use of our cell in a practical device. Perhaps a distress rocket or parachute flare, for example, could be made with a radio beacon built in; the firework rocket exhaust would provide the heat to melt the hydroxide.

The reaction of copper with oxygen to form copper oxide is the only way that the battery can provide power for more than a few minutes, so a supply of air (oxygen) is essential. But air contains carbon dioxide, which may be responsible for the electrolyte's tendency to solidify. Perhaps we could eliminate the CO_2 from the air by adsorbing it in soda lime, a highly porous mixture of at least 90 weight percent dry calcium hydroxide ("slaked lime") and just a little sodium hydroxide.* Although soda lime is relatively safe, it is highly alkaline, and contact with the skin should be avoided. It reacts as follows:

$$Ca(OH)_2 + CO_2 \rightarrow CaCO_3 + H_2O.$$

The tiny volume of gas consumed by our hot battery and the relatively low concentration of CO_2 in the atmosphere—just a few hundred parts per million—mean that a very small quantity of soda lime granules, just a few grams in the air-supply tubing, should suffice to eliminate CO_2 from the air supply to the battery. The CO_2-free air supply would have to be connected to the battery top so that the flow swept away any ambient air. How much longer would the battery run with this modification?

*Soda lime is specially prepared for use in the closed-circuit breathing systems used in firefighters' breathing sets and, more rarely, in underwater scuba sets. The adsorption of the CO_2 breathed out by the user (3–5 percent), and the addition of a relatively small flow of oxygen, allow the user to rebreathe the air just breathed out. This recycling process enormously reduces the amount of gas that has to be carried in a compressed gas cylinder, since only the oxygen gas need be supplied. Small disposable canisters with 200–500 g of soda lime are also available for use in closed-circuit anesthetic machines in hospital operating theaters.

REFERENCES

Periodicals such as *Technics* and *Model Engineer & Electrician* often describe late-nineteenth-century batteries and some of their drawbacks.

Fitzgerald, Lieutenant L. Letter. *Model Engineer & Electrician,* December 22, 1904, p. 595.

Richie, Andrew. "Military Applications of Reserve Batteries." *Philosophical Transactions of the Royal Society of London A* 354 (1996): 1643–1652. Describes specialized military applications of thermal batteries and other devices.

38 Unusually Cool Sunglasses

If scientific progress were predictable, it would become a sort of engineering, useful perhaps, but much less fun.

—Donald R. Griffin, *Bird Migration*

Engineering often gets a bad rap, but actually it has a generous share of its own surprises. Consider the problem of corrosion, for example. Corrosion is a constant concern of engineers, and with dozens of different modes of corrosion and hundreds of metals and metal alloys in everyday use, engineers often encounter unexpected phenomena. I was once presented with failures in brass components to investigate. (A brass is basically an alloy of copper and zinc, often with smaller amounts of other elements.) I was surprised to discover that some strong brasses—high tensile (HT) brasses—corrode easily when stressed, even in the presence of something as ordinary as moisture. This process leads to what is known as "stress corrosion cracking," which is a frequent source of failure. Ordinary brass, although weaker to start with, is resistant to moisture. Other brasses are largely resistant to corrosion, even by seawater, and are used in all sorts of marine applications. But these alloys were developed only after long and bitter experience. The condensers of steamships were stressed by steam pressure, and with corrosive seawater on one side and hot steam on the other, the condensers suffered frequent failures. The great battleships of the British naval fleet of World War I, for example, were often crippled by "condenseritis." Corrosion left these leviathans wallowing in the waves, their thick armor plate and powerful guns rendered impotent by their inability to move. In his *Engineering Materials 3,* David Jones describes a number of famous corrosion failures, some due to the pitting type of corrosion we make use of in this project.

Here we use the pitting effect of a particular type of electrolytic corrosion to create shades that are really cool: not only do they keep the rays of the sun from beating down mercilessly on your eyes, they also allow cooling breezes to pass right through them.

What You Need

- ❑ Salt water
- ❑ Heavy-gauge aluminum foil dish
- ❑ Plastic bowl
- ❑ 12-V battery charger
- ❑ Electrode (e.g., thick solid copper wire)

What You Do

Make a concentrated solution of salt in water, and put the solution in the foil dish along with the copper electrode. Don't put the electrode too close to the aluminum surface—try about 25 mm away to start with. Place the dish assembly in a plastic bowl, and connect the battery, negative to the copper and positive to the aluminum. Periodically remove the foil from the solution and hold it up to a

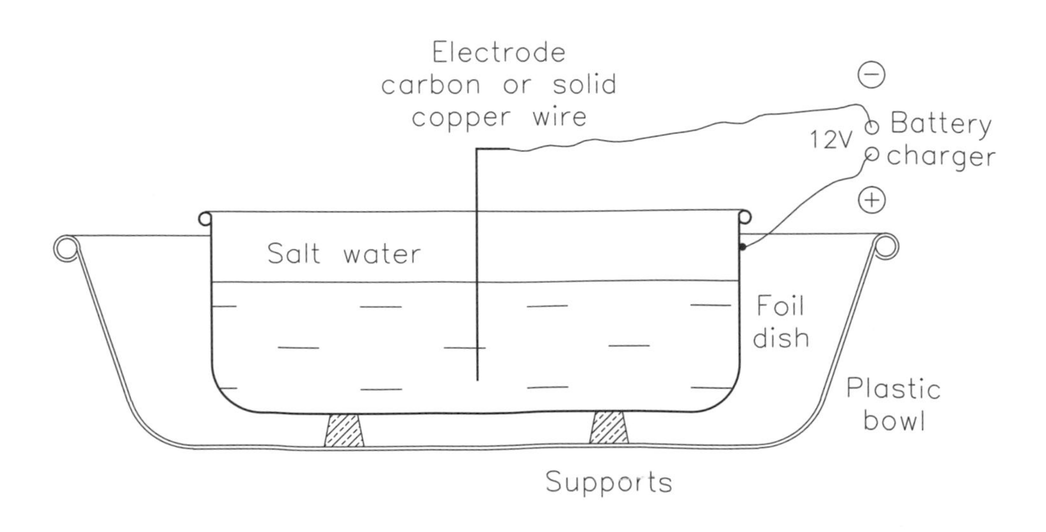

strong light to check for progress. As time passes, you should find that the surface of the foil turns duller and then begins to transmit light. When you think it is ready, wash the foil carefully and then look through it. When you hold it close to your eye, you'll be surprised to find that you can see right through it: it will transmit not just light but also an excellent image.

How It Works

If you look at the foil under a powerful magnifying glass or a low-power microscope, you will begin to understand how the foil is able to transmit light. There is relatively little overall thinning of the metal, but pitting corrosion occurs to such an exaggerated extent that the metal actually becomes perforated. Light can pass through the foil because the dish is now a colander, punctured by millions of tiny holes.

You might not expect to see a good image through an array of holes, but the optical quality of our microcolander is in fact quite good. Essentially, the holes are so close to your eye that they are out of focus, and their only effect is to reduce the amount of light, which is exactly what we require from sunglasses.

THE SCIENCE AND THE MATH

Pitting corrosion is what makes our unusually cool sunglasses work. If corrosion were evenly spread over a metal surface, the corrosion would in many cases be almost imperceptible and would certainly be unimportant with regard to the structural failure of machine parts. However, Murphy's law ensures that metals that are relatively resistant to corrosion—stainless steel, for example—are subject to pitting.* Many grades of stainless steel corrode by pitting in the presence of chloride ions, such as those in seawater.

Pitting corrosion occurs when tiny electric batteries are set up on the surface of a metal. A typical metal surface has slight variations due to differences in alloy content or to contamination, for example, and corrosive media also vary in the amount of dissolved oxygen or the salt concentration. Each part of the surface can be considered as a battery pole; some parts will behave as anodes, some as cathodes. Surface differences will be amplified once the area has been attacked. Areas that have become more cathodic or anodic or conductive will become more cathodic or anodic or conductive, and then these areas will be subject to more attacks, whereas neighboring areas will be less affected.

Aluminum is one of the few reactive metals that react only slowly with water or air. The reason for the slow reaction is the formation of a thin but tenacious and impervious layer of aluminum oxide

*Murphy's first law: If it can go wrong, it will. This law seems to have been named for Colonel Edward A. Murphy Jr. of the U.S. Army. Murphy was an expert in testing rocket sleds, which were used in the 1940s to evaluate human performance under high acceleration. They often went wrong. Murphy's first law now comes with an anonymous codicil, Murphy's second law: Murphy was an optimist.

on its surface. A layer just a nanometer or so thick is sufficient to stop aluminum from reacting. When this layer is absent or inactivated, aluminum reacts swiftly. The addition of a trace of mercury to the surface of aluminum reveals its true chemical nature, and renders it more reactive than iron. In damp air, aluminum treated with mercury swiftly turns dull, and a white crust of corrosion products forms in a few days, severely damaging the metal. This is the reason why you should not take a mercury thermometer on an airplane. If the thermometer fractured and released mercury, then corrosion of the aircraft's aluminum alloy structure could occur, with expensive and possibly even disastrous consequences. The adherent oxide layer of aluminum amplifies the pitting effect and leads to the spectacular pitting that we see in our cool sunglasses. The holes look surprisingly neat, as though they could

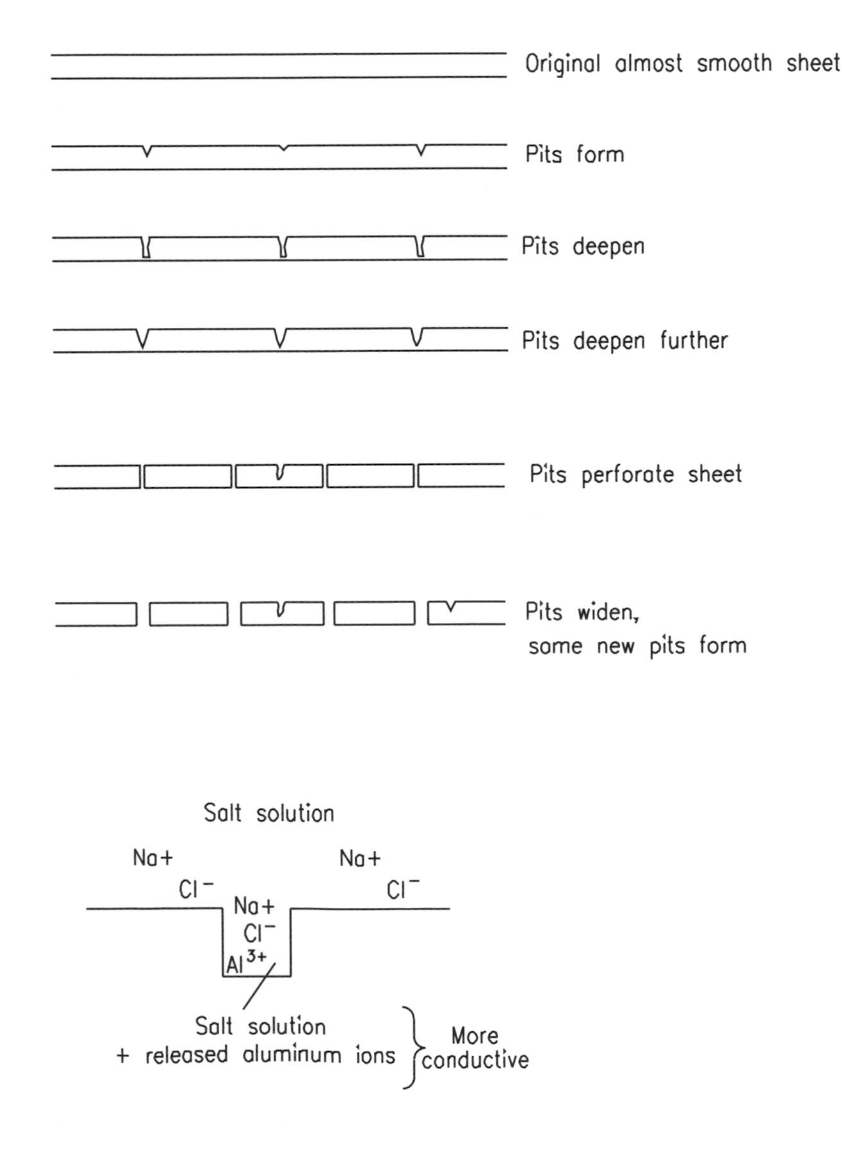

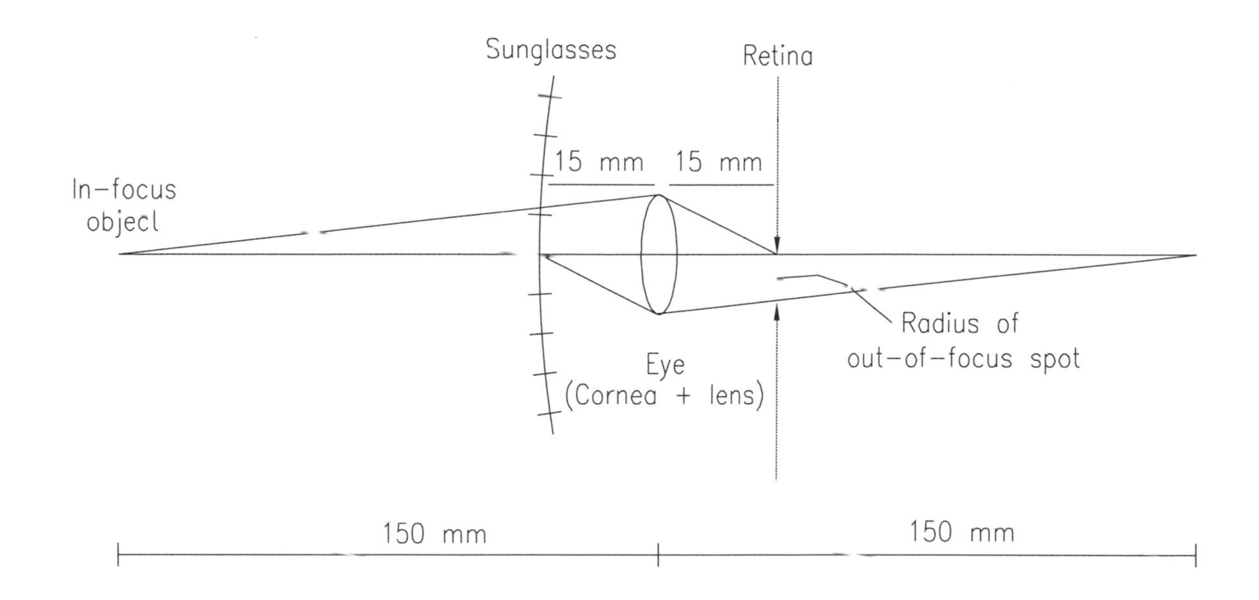

have been created by an army of elves busy with microscopic drills.

The quality of the image we see through our microcolander sunglasses is surprisingly good. This is because the holes are close enough to the eyes to be out of focus. In addition, the holes are small enough so that the fuzzy image of the out-of-focus hole appears much larger than it actually is, and all these fuzzy images overlap completely, giving us a complete image through the screen. You can understand this by thinking about the limits on the depth of focus of your eyes. The lenses in your eyes can focus from approximately 150 mm to infinity, so very small spots nearer than 150 mm will simply not be in focus. If you assume that the sunglasses are 15 mm away from the lens and that the focal length inside your eye is just 15 mm, you can calculate the size of the out-of-focus spot. The geometry makes it clear that the spot will be huge and not easily perceived. Perhaps we shouldn't be surprised at this. If you wear glasses, you don't notice the dust particles covering your lenses, except in certain circumstances, such as in bright sunlight. (In bright sunlight, the pupil contracts hugely, increasing the depth of field of the eye lens and making the dust particles much more visible, even if still slightly out of focus.)

Why don't the holes in our sunglasses diffract the light and thus spoil the sharpness of the image? The answer is that the holes are much larger than the wavelength of light, so diffraction will affect only a tiny proportion of the light. Diffraction will occur at the edges of the holes, however, scattering the light coming from the scene viewed through our microcolander shades, giving a kind of "soft focus" effect. The angle through which light is diffracted depends upon the size of the hole: in our case, with holes of about 100 μm in diameter and light wavelength of about 0.5 μm, the angle, α, of the diffraction is only $\tan^{-1}(0.25/100)$, or approximately 0.28 degrees. (But see an optics book, such as Eugene Hecht and Alfred Zajac's *Optics,* for details about diffraction through a circular hole.)

And Finally . . . Microscopic Sieves

How else might you use this curiosity of the electrochemistry lab? Well, certain types of plastics, such as polycarbonates, can be etched, and they show tendencies to deep pitting and holes just like our electrolyzed aluminum. However, with the right etchants, the plastic will be pitted only where it has been ionized by the passage of a high-speed particle. The tiny number of charged terminal groups and dangling chemical bonds left in a polymer after a high-energy proton, electron, or other particle has passed through are sufficient to cause that area of polymer to be preferentially etched. The etch pits can be as small as 100 nm across. Etched polymers created in this way have been used (as an ultra-high-resolution alternative to photographic emulsions) for recording cosmic rays (see Michael Friedlander's *A Thin Cosmic Rain*).

However, the etched polymer membranes have also found a more practical use as precise filters that, at the microscopic level, act as sieves.* Nuclepore membrane filters are made by irradiating plastic film in a nuclear reactor or some similar device and then etching the film to create billions of holes with identical diameter; the etching conditions determine the hole size. These membranes are used where a very precise filter is needed—for example, in the separation of biological cells or cell microcomponents, or in the collection of air samples. Can we use our etched aluminum sheet as a sieve filter? Perhaps our aluminum sieve would be useful for filtering bugs—pathogenic organisms—out of impure water?

*Most filters don't work as sieves; their operation is in fact more complicated than that (see the discussion in my *Industrial Gases,* 213–217, or in Brauer and Yalamanchili's *Air Pollution Control Equipment,* 162–173).

REFERENCES

Brauer, Heinz, and B. G. Varma Yalamanchili. *Air Pollution Control Equipment.* Berlin: Springer-Verlag, 1981.

Downie, Neil A. *Industrial Gases.* London: Chapman and Hall, 1997.

Friedlander, Michael W. *A Thin Cosmic Rain: Particles from Outer Space,* 92–95. Cambridge, Mass.: Harvard University Press, 2000.

Hecht, Eugene, and Alfred Zajac. *Optics.* Reading, Mass.: Addison-Wesley, 1974.

Jones, D.R.H. *Engineering Materials 3: Materials Failure Analysis.* Oxford, U.K.: Pergamon, 1993.

39 Wet Solar Cell

> Doublethink means the power of holding two contradictory beliefs in one's mind simultaneously, and accepting both of them.
>
> —George Orwell, *1984*

When photons of light strike certain objects, electricity can be caused to flow. This photoelectric effect is at the heart of the generation of electricity directly from sunshine, a technology that could be key to the energy industry in the future. However, even if the photoelectric effect had no economic significance, it would still be studied intensively. To understand it, modern physicists have to employ a kind of "doublethink." The photoelectric effect (and, ultimately, the whole of quantum mechanics) requires that photons be both particles and waves simultaneously. If light is not composed of waves, then how can we explain optical interference? But if light is made of waves, then how can we account for the photoelectric effect? Albert Einstein's 1905 theory about the effect, combined with Max Planck's earlier work on black body radiation, laid the basis for what we now call quantum mechanics, one of the foundation stones of modern physics. (John Stachel tells part of the story in *Einstein's Miraculous Year.*) It was for his work on the photoelectric effect, incidentally, that Einstein was presented with the Nobel Prize for Physics. (His more famous work on relativity and the mass-energy equivalence equation, $E = mc^2$, did not receive the prize.)

Early-nineteenth-century scientists like Edmond Becquerel in France, playing with platinum electrodes in different solutions and later with silver iodide–coated silver electrodes in acidified water, were probably the first to stumble on the phenomenon of photoelectrochemistry: electricity generated by the effects of light on electrochemical cells. In a paper titled "Origins of Photoelectrochemistry," Mary Archer tells how young Becquerel's discovery in 1839 was greeted with disbelief by other scientists of the time; the problem was exacerbated by the fact that no one could explain his results, quantum mechanics being very much in the future. This class of discoveries in photochemistry was first used by inventors to enable the capture of images—photography—perhaps a more glamorous activity than the more everyday business of power conversion.

Liquid solar cells—such as those using ferrous salts and iodine, thallous sulfide, or selenium, and the copper/copper(I) oxide system we use in this project—had been discovered by the early 1900s; the first recorded scientific paper on electricity from copper in sunshine was published by Wilhelm Hallwachs in *Physikalische Zeitschrift* in 1904. Although there have been solid-state solar cells since the early days of the space race (almost all the U.S. spacecraft have used them), and they are quite common now even in consumer items like calculators, nobody has ever marketed "wet" solar cells. Why this should be is a bit of a mystery to me, since they are so simple to make. Perhaps the answer lies in mundane but important issues like corrosion, sensor lifetime, leak-proofing, and so forth.

At the time that the photoelectric effect was discovered, demonstrating it required the leading technology of the day, exotic metals, and high-vacuum pumps. Today, photoelectric devices such as photovoltaic solar panels can be made only in well-equipped semiconductor factories. Nevertheless, you can demonstrate quantum physics and the possible future of power generation on planet Earth using only a multimeter and a few things you might find in your kitchen. The solar cell in this project must be the world's simplest. Ancient Greeks could have constructed this masterpiece of technology.

What You Need

Solar cell

- ❑ Multimeter (current meter 0–100 μA or 0–200 μA)
- ❑ Solid copper wire (1–2-mm diameter)
- ❑ Tape

- ❑ Reading lamp
- ❑ Clear container (e.g., small wine, sherry, or liquor glass)
- ❑ Sodium bicarbonate (baking soda), or table salt
- ❑ Black cloth or paper (just a few square inches)
- ❑ Water

Photoelectric alarm

- ❑ Circuit board
- ❑ Operational amplifier (Many will work, but I suggest CA3140.)
- ❑ Resistors, 330 kohm, 1 kohm
- ❑ Capacitor, 1 μF (optional)
- ❑ 9-V battery
- ❑ Battery connector
- ❑ 9-V DC piezoelectric beeper
- ❑ 2 diodes (e.g., 1N4004 type)
- ❑ Transistor, BC107

What You Do

Hold two pieces of copper wire in pliers so you don't burn your fingers, and heat the wires simultaneously in a flame until they glow red; continue heating them for a few seconds longer, and then remove them from the flame.

Dissolve approximately 5 g of bicarbonate (sodium bicarbonate works best, but ordinary table salt works well too) in a cupful of warm water, and then put the solution in the sherry glass. Now dangle the wires in the glass. Position one wire so that the piece of black cloth will largely prevent light from falling on it, perhaps forming the black cloth into a tube to partially enclose the electrode. Connect up the multimeter. That's it. Surely there must be more to it than this? Nope, its that simple!

In the dark, the cell should give an output of less than 10 μA, but with a strong light (sunshine or a reading lamp at point-blank range, for example), you will get at least 30 μA and maybe as much as 100 μA of electric current. Although this is not much current, it's plenty to operate an electronic circuit like the one I'm going to suggest in a moment. With larger electrodes you will get much more current.

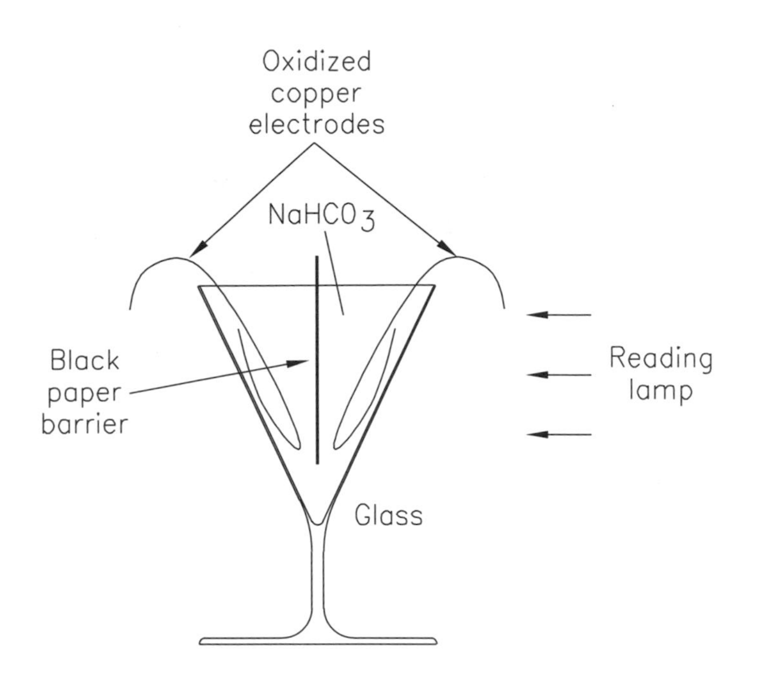

How It Works

What's going on here? A photon—a quantum, or "particle," of light—collides with the electrode surface and is absorbed; the photon's energy pushes a negatively charged electron out of its normal orbit around the positively charged nucleus and allows the electron to drift freely in the conductive metal of the wires. As a result, the electrode acquires a small negative voltage. Since the charged electron is loose in the wire, that voltage causes the electron to move, and we observe an electric current. If the electrode were isolated, this process would stop after a very small number of electrons had been freed because the electrode would soon be charged up to a relatively high negative voltage, repelling any additional electrons. However, the electrons created by the light are conducted through the wire to the dark electrode via the electric circuit, and then the current passes through the conducting solution with its charged bicarbonate (HCO_3^-) and sodium (Na^+) ions, forming a complete circuit and allowing current to flow continuously. A more intense beam of light contains more photons than a less intense beam and will eject more electrons and produce a larger current.

This is not an obscure phenomenon, destined only for dusty academic bookshelves. Solar cells are a hot research topic. Even in the chilly north of Europe, sunlight still offers a great abundance of power that just requires conversion into useful form. Recently, some important forward steps have been taken in research on liquid solar cells. Photoelectrochemical solar cells that just might one day rival their silicon cousins are being tested. The Swiss researchers Michael Graetzel and Anders Hagfeldt have made cells using titanium dioxide electrodes stained with organic dyes that generate power almost as efficiently as silicon solar cells and that could be much cheaper to manufacture.

The Wet Photoelectric Alarm

Just to show that this phenomenon can actually be useful for something, why not try a photoelectric alarm? The current from the cell will be too small in general to operate even the smallest low-power "solar" electric motors. If you made a big version of the cell (say, 30 cm square), "tuned" the state of the conductors and the salt concentration, and waited for strong sunlight, then you might be able to operate a motor; but this would be a difficult project. However, a device that uses our wet solar cell as a sensor is quite straightforward and will work with a liquor glass–size cell.

Although the circuit I've drawn could be made more sensitive, it is intended to trigger at 10 μA or so, to avoid false triggering by stray currents that flow even in the dark. The circuit triggers when it receives light. If you wanted to make a

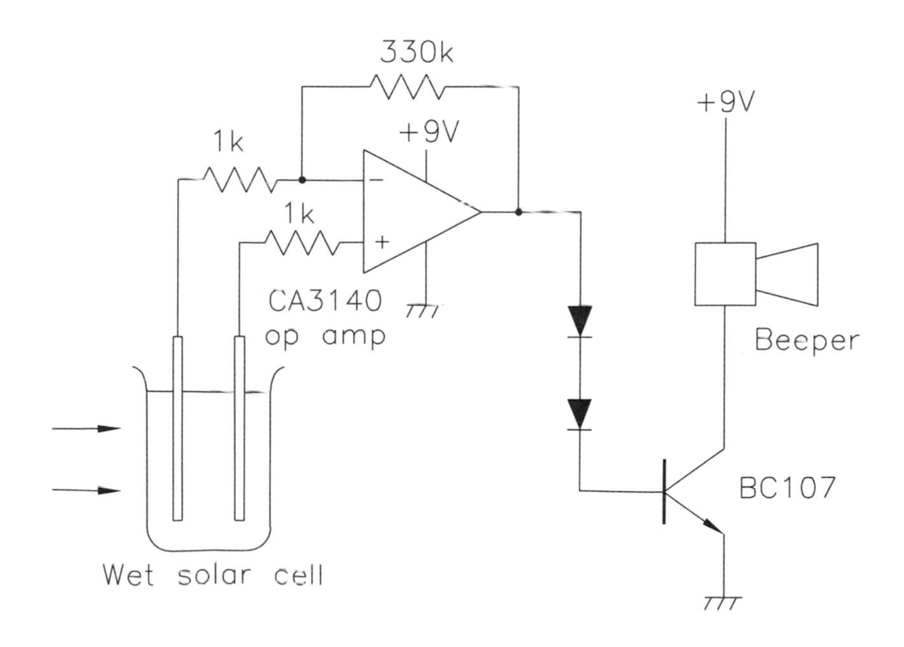

"break-the-beam" burglar alarm, you could change the circuit so that the beeper turned off at 10-μA input current and switched on when the current fell below this.

THE SCIENCE AND THE MATH

Light is electromagnetic waves. To explain the photoelectric effect, Max Planck and Albert Einstein figured out that light must come in little packets, or quanta. The light quanta, photons, each have an energy given simply by Planck's constant, h, and their vibration frequency, f:

$$E = hf.$$

For example, blue light, $f = 6.2 \times 10^{14}$, has a quantum energy of 4.1×10^{-19} J, or 2.6 eV; and red light, $f = 4.3 \times 10^{14}$ J, has a quantum energy of 2.8×10^{-19}, or 1.7 eV.

What convinced Einstein that light must come in quanta? Many experiments showed that a less intense beam of light contained fewer photons, ejected fewer electrons, and thus produced a smaller current than a more intense beam; whereas a beam of light with less energy in each photon (red light, rather than blue light) would often, depending on the electrode material, produce no current at all. Experiments showed that there was a minimum energy needed to propel the electron out of electrodes such as copper oxide. If the quantum of light had too low a frequency, no electron would be emitted and no current would be measured, no matter how intense the beam of light. It was this particular feature that led Einstein to his theory of photoelectricity and, ultimately, to quantum mechanics.

What does all this have to do with our wet solar cell, you ask? Well, the key to our cell is the bicarbonate solution, which essentially acts as a contact to the CuO layer on the electrode. The brown copper(I) oxide is the photosensitive element that emits electrons when it receives photons.

You could modify your cell by using a different electrode for the dark electrode, which, after all, is there simply to collect the current. However, using two identical electrodes works well because the cell generates no current when both electrodes are in the dark. With two different electrodes, the system tends to behave as a regular chemical battery and produce current even in the absence of light. This probably explains why the effect of light on the electrodes and chemical solutions of a battery was not noticed by the pioneers of electricity. Even the great Michael Faraday fails to mention these effects in his *Experimental Researches in Electricity,* the world's first real textbook on electricity. Given how simply the effect can be demonstrated, its absence from this early work is curious. Today there still seems to be relatively little work done on wet solar cells compared with the well-known silicon and photovoltaic solar cells.

How does all this apply to our alarm circuit? The circuit simply amplifies the current from the electrodes and feeds it to the beeper circuit. An idealized operational amplifier (op amp) can be considered to be a black box that tries to zero the current into its inputs and keep their voltages the same by pushing current out of its output. In our circuit, the current in the input leads from the cell is drained away by the 330-kohm resistor, such that the idealized op amp has zero current in and the same voltage at each input terminal. To achieve this, the op amp moves its output voltage until the negative input terminal is once again at the same potential as the positive input, with zero current flowing into either terminal. Suppose the photoelectric cell is giving a 3.3-μA short-circuit current. When the op amp output gives 1 V on the output (plenty to operate the beeper via the transistor), it will push 3.3 μA down the 330-kohm resistor, a current exactly equal to the photoelectric current, assuming that the input terminals can be considered to be near ground potential. In effect, this circuit converts an input current, I_{in}, into a voltage, V_{out}, as described by the following equation:

$$V_{out} - I_{in}R,$$

where R is resistance (330-kohm, in this case). So

$$V_{out} = 330{,}000 \times I_{in}.$$

You might be mistakenly thinking that this is Ohm's law and that you could achieve exactly what this circuit does with just a resistor! If so, you are missing the fact that the output voltage of the op amp here is provided with a considerable current capability, 20 mA or so, whereas a 330-kohm resistor on its own could provide only a few microamps before the equation would cease to apply. The transistor boosts the op amp current to the 200 mA or so that the beeper needs (although some piezoelectric beepers take less current than this). The capacitor across the beeper helps to stop it from interfering with the circuit that switches it on—in this case, our photoelectric cell.

And Finally . . . Other Electrodes

You could try other electrodes. Crystals of fool's gold, iron pyrite, are conductive, and, even more important, they emit electrons when illuminated. Try to find two elongated lumps of pyrite. Check for electrical conduction, and then clamp alligator clips to the lumps and immerse them in a sodium bicarbonate or salt solution, as you did with the copper wire. The photoelectric effect will probably be weaker than for the copper, so try hiding the cell underneath a blanket with a reading lamp inside and switching the reading lamp on and off. A few microamps of photocurrent should be possible with lumps of pyrites a few centimeters long.

Naturally enough, pieces of silicon will also function well as photosensitive electrodes—although perhaps not as well as they do in a solar cell. A solar cell has a structure in which the electrodes gather electrons and "holes" (missing electrons in the band of bound electrons; in a semiconductor these holes behave like positive-charge carriers) generated by incoming light. Dunking pieces of solar cell in a conducting solution effectively shorts out this structure and leaves only a limited functionality. Two 5-mm squares of silicon cracked from the edge of a solar cell should be capable of generating 20–30 μA under a reading lamp.

If you know someone who works for a semiconductor firm, you may be able to get hold of broken pieces of the wafers of silicon crystal used to make silicon chip integrated circuits. However, pieces of commercial raw unprocessed silicon are, like the solar cell pieces, rather low in performance and will typically manage only a fraction of a microamp. The problem is that ultrapure silicon—and the manufacturers pride themselves on purity—has a high resistivity, giving a resistance of megohms for a small piece and restricting the current available. The makers of semiconductor products add special impurities—dopants—to modify the pure silicon and also render it more conductive, but they like to start with as pure a crystal as possible.

With commercial silicon, you may see an initial pulse followed by zero current. This behavior indicates a photocurrent flowing into a electrical capacitance; however, once that capacitance has charged, then the high resistivity of the silicon prevents further current flow. Solar cell silicon is much more conductive than pure silicon because it has already been treated with dopants that donate extra electrons to act as charge carriers.

Silver compounds have been used in photography since soon after its inception because of their sensitive reaction to light, and therefore silver-based electrodes might be expected to give photoelectrochemical effects. If you can commandeer some silver, some broken jewelry perhaps, try silver electrodes in salt solution. Although perhaps lower in performance than, say, the iodide salts first tested by Edmond Becquerel and other nineteenth-century scientists, any silver with a salt should also work and may give a substantial current. If you can get a really good performance from one of the systems, can you make a useful solar battery out of it? How many volts can you get?

REFERENCES

Archer, Mary D. "Origins of Photoelectrochemistry." Paper presented at the Fifth BOC Priestley Conference, Birmingham, U.K., 1989.

Einstein, Albert. "On a Heuristic Point of View concerning the Production and Transformation of Light." In *Einstein's Miraculous Year,* ed. and with an introduction by John Stachel with the assistance of Alice Calaprice, Sam Elworthy, and Trevor Lipscombe; foreword by Roger Penrose. Princeton, N.J.: Princeton University Press, 1998. Originally published in German in *Annalen der Physik* 17 (1905): 132–148.

Faraday, Michael. *Experimental Researches in Electricity.* 1 vol. Chicago: Benton/Encyclopedia Britannica, 1952.

Graetzel, Michael, and Anders Hagfeldt. "Light-Induced Redox Reactions in Nanocrystalline Systems." *Chemical Reviews* 95 (1995): 49–68.

Hallwachs, W. *Physikalische Zeitschrift* 5 (1904): 489–499.

Hints and Tips

Suppliers

Electronic component suppliers these days often stock not just electronics but also lots of other useful small parts, and they usually deliver very quickly. Premier Components and Tandy/Radio Shack in the United States and Farnell Electronics and Rapid Electronics in the United Kingdom are examples. Rapid Electronics is a smaller firm, with a smaller, lower-cost range of supplies. Edmund Scientific Co. supplies an especially good range of optical odds and ends, and Polaroid Corporation will supply polarizing filters at a good price. However, most of my parts and materials come from a local hardware store.

Local pharmacies will often provide chemical supplies if you place an order with them. Farm suppliers and animal-feed suppliers keep useful chemicals, though often under nonscientific names; you have to read the small print to know what you are getting. You can also try mail-order chemical companies like Flinn or, in the United Kingdom, Merck: these firms often deal only through pharmacies or other agents.

Toy stores sell all sorts of goodies, such small electric cars, that can be converted to other things. The really good toy stores will also sell construction kits. I particularly like Meccano and Erector because you can interface the parts to other materials so easily, but I find that K'nex, Lego, and even Tinkertoy have their uses too.

Finally, there are conventional school suppliers. They don't generally provide that wide or imaginative a range of products, but their prices are usually competitive, at least if you want a whole bunch of something—a load of small DC motors with propellers, for example, as used in railroad yachts.

DC Electric Power and Other Electrics

I use nickel-cadmium (NiCd) or nickel metal hydride (NiMH) rechargeable batteries for most things. They cost, for a AA cell, just a dollar, and they last five hundred times as long as a regular cell; you just have to put them in a charger unit once in a while. They also avoid the annoyance you get with regular cells when you mistakenly leave a device switched on overnight and discover the next morning, when you pick up the device again, that the batteries have discharged and are junk. With a NiCd battery, you can just recharge it.

For larger currents, I use a car battery charger, one of the very simple inexpensive types that provide a constant voltage output (around 14 V) for charging standard auto batteries, not one of the electronic "intelligent" chargers. Simple battery chargers just provide an output that is AC mains transformed down to 12 V and then full-wave rectified to a 100-Hz modulated DC. Be aware that you can get weird readings from a multimeter connected to a circuit supplied by a battery charger if you accidentally switch the multimeter to AC. It still reads a voltage, but the figures typically don't make any sense.

Finally, don't forget to purchase a digital multimeter if you don't already have one: these really are versatile and accurate pieces of equipment at a bargain basement price. A simple one costing less than $20 will provide measurements of voltage from millivolts to 1,000 volts and current from microamps to 10 amps, as well as providing resistance measurement and a transistor tester. Pay slightly more and you can have a meter with a thermocouple that will measure up to near molten iron temperature, and other meter settings that give frequency, capacitance, and a barrel of other goodies.

Using an Analog Input Board or Datalogger for Your PC

An analog input board, although rather clumsy to use, is invaluable: it will provide oscilloscope, chart recorder, and "data massaging" services.

Science from Your Spreadsheet

I used to write short computer programs in Basic or QBasic to plot graphs or prove particular principles, given some kind of understanding of the math associated with the phenomena. You can still get QBasic from the MS-DOS part of older PCs and install it on a modern Windows PC. A version of Basic—Visual Basic or VBA—is included in Microsoft's Excel spreadsheet package, hidden in the macro editor. However, these days I find that I rarely need Basic, and I do most things using a spreadsheet like Lotus 123 or Microsoft Excel. To plot a graph of a simple function in a spreadsheet, just fill a column with successively increasing x values, put the formula for the function $y = f(x)$ in the next column, and then click on the x-y scatter graph facility. Plotting a two-dimensional function is slightly less obvious. Probably the easiest way is to fill a row and a column at the edge of an array of empty cells with successive x and y values. Then put the function $z = f(x,y)$ into each of the array cells, being careful about where you put the $ signs for the x and y values. Finally create a plot using the isometric surface plotting facility. Some spreadsheets will also plot contour plots.

You can adjust the look of your graphs by substituting constant values or different formulas in different plot areas, if needed, and you can switch between formulas using the @if function.

Monte Carlo Programming

The @rand() function is a powerful tool for computer simulation with a spreadsheet, or what is sometimes dubbed *Monte Carlo programming*. Suppose you want to know the area of a two-dimensional shape: simply generate random paired values of x and y in ranges that enclose the unknown shape, and note whether the values fall within or outside the shape. When you have finished, simply count the number of pairs that fell inside the shape, divide by the total number of pairs, and multiply by the area of the rectangle. In this case, there are more efficient procedures, and the Monte Carlo technique will give an answer of limited accuracy. But for more complex phenomena, more efficient procedures may be difficult to identify. The Monte Carlo procedure will give an answer without any deep analysis.

Analog Quartz Clocks

If you take apart a quartz clock with an analog display, you will uncover quite a little treasure trove of useful parts (considering that a clock sells for just a few dollars). It is well worth thinking about reusing some of that gadgetry in your projects. Inside the clock you will discover a small circuit board, containing a single electronic chip (hidden underneath a drop of black epoxy resin), and a tiny silver cylinder, just 2 mm in diameter and 5–6 mm long; this is the quartz crystal. The crystal is maintained in oscillation at 32 kHz or so by the chip, which also divides its output down to a 1-Hz signal. The rest of the clock is an electromechanical counter that moves a solenoid on a pawl, which pulls a gearwheel around one tooth, rotating the second hand 6 degrees and the hour hand 0.1 degree each time it receives an impulse from the chip.

By removing the circuit board and substituting your own current pulses, you can operate the clock display from another drive mechanism. (You could also use the 1-Hz pulses from the circuit board to activate an experiment that needs such pulses. The frequency of the pulses will of course be as accurate as the crystal—a few tens of parts per million or better.)

English and Metric Units

Mistakes with units have caused some famous bloopers. There were spectacular mistakes even in the hallowed ranks of NASA rocket scientists with a couple of Mars space missions. I still find the old English units of measurement useful to get a feel for things, but of course the consistent metric system is much better for calculation. Here are some useful conversion factors:

1 inch	25.4 mm	1 hp	745 watts
1 ft	300 mm	Temp (F)	(1.8 × Temp (C) + 32)
1 mile	1.609 km	–40 °F	–40 °C
1 in^3	16.4 cm^3	32 °F	0 °C
1 ft^3	28.3 liters	68 °F	20 °C
1 oz	28.35 g	212 °F	100 °C
1 mph	1.609 km/h		

Suggested Further Reading

Books

Alder, Henry L., and Edward B. Roessler. *Probability and Statistics.* 3d ed. New York: W. H. Freeman, 1964.

Bader, Paul, and Adam Hart-Davis. *Local Heroes: The Book of British Ingenuity.* Stroud, U.K.: Sutton, 1997.

Balakrishnan, A. V., ed. *Advances in Communication Systems.* Vol. 1. New York: Academic Press, 1965.

Banks, Robert B. *Slicing Pizza, Racing Turtles, and Further Adventures in Applied Mathematics.* Princeton, N.J.: Princeton University Press, 1999.

Bleaney, B. I., and B. Bleaney. *Electricity and Magnetism.* 3d ed. Oxford, U.K.: Oxford University Press, 1976.

Braddick, H.J.J. *Vibrations, Waves, and Diffraction.* London: McGraw-Hill, 1965.

Bunch, Bryan, and Alexander Hellemans. *The Timetables of Science.* London: Sidgwick and Jackson, 1989.

Burke, James. *Connections.* London: Macmillan, 1978. "Connections" is also the title of Burke's monthly column in *Scientific American.*

Clark, Ronald W. *Edison: The Man Who Made the Future.* London: Macdonald & Janes, 1977.

CRC Handbook of Chemistry & Physics. 76th ed. Boca Raton, Fla.: CRC Press, 1995.

De Bono, Edward. *The Use of Lateral Thinking.* New York: Harper and Row, 1970.

De Vries, Leonard, comp. *Victorian Inventions.* Trans. Barthold Suermondt. 1971. Reprint; London: John Murray, 1991.

Doherty, Paul, and Don Rathjen. *The Exploratorium Science Snackbook Series.* New York: Wiley, 1991–1996.

Dummer, Geoffrey W. A. *Electronic Inventions and Discoveries.* 4th ed. Bristol, U.K.: Institute of Physics Publishing, 1997.

Eastaway, Rob, and Jeremy Wyndham. *Why Do Buses Come in Threes? The Hidden Mathematics of Everyday Life.* London: Robson Books, 1998. A fun math book.

Ehrlich, Robert. *Turning the World Inside Out.* Princeton, N.J.: Princeton University Press, 1990.

———. *Why Toast Lands Jelly Side Down.* Princeton, N.J.: Princeton University Press, 1997.

Elmore, William C., and Mark A. Heald. *Physics of Waves.* New York: McGraw-Hill, 1969.

Fauvel, John, Raymond Flood, Michael Shortland, and Robin Wilson, eds. *Let Newton Be! A New Perspective on His Life and Works.* Oxford, U.K.: Oxford University Press, 1988.

Feynman, Richard P. *Feynman Lectures on Computation.* London: Penguin Books, 1999.

Fleming, J. A. *Electric Wave Telegraphy and Telephony.* London: Longmans, 1916.

Gregory, Richard L. *Eye and Brain: The Psychology of Seeing.* 5th ed. Princeton, N.J.: Princeton University Press, 1997.

Grosvenor, E. S., and Morgan Wesson. *Alexander Graham Bell.* New York: Harry N. Abrams, 1997.

Hare, Mick, ed. *The Last Word: Questions and Answers from the Popular Column on Everyday Science.* Oxford, U.K.: Oxford University Press, 1998.

———. *The Last Word II: More Questions and Answers on Everyday Science.* Oxford, U.K.: Oxford University Press, 2000.

Highfield, Roger, and Peter Coveney. *The Arrow of Time.* London: Flamingo, 1991.

Hill, Winfield, and Paul Horowitz. *The Art of Electronics.* Cambridge: Cambridge University Press, 1989.

James, Peter, and Nick Thorpe. *Ancient Inventions.* London: Michael O'Mara Books, 1994.

Jargodski, Christopher P., and Franklin Potter. *Mad about Physics: Braintwisters, Paradoxes, and Curiosities.* New York: Wiley, 2001.

Jewkes, John, David Sawers, and Richard Stillerman. *The Sources of Invention.* London: Macmillan, 1958.

Jolly, W. P. *Marconi.* London: Constable, 1972.

Jones, David E. H. *The Inventions of Daedalus: A Compendium of Plausible Schemes.* Oxford, U.K.: Freeman, 1982. A collection of witty imaginary inventions from a long-running but now extinct column in *New Scientist* magazine.

Kirk-Othmer Concise Encyclopedia of Chemical Technology. 4th ed. New York: Wiley, 1999.

Koerner, Thomas W. *The Pleasures of Counting.* Cambridge: Cambridge University Press, 1997.

Laithwaite, E. R. *Propulsion without Wheels.* 2d ed. London: English Universities Press, 1971.

Landes, David S. *Revolution in Time: Clocks and the Making of the Modern World.* Cambridge, Mass.: Harvard University Press. London: Viking, 2000. This superb book covers clocks and their impact on society in history.

Lavoisier, A. L. *Elements of Chemistry.* France, 1789. For a more recent edition, see Hutchins, Robert M. *Great Books of the Western World.* Vol. 45, *Lavoisier, Fourier, Faraday.* Trans. Robert Kerr. Chicago: William Benton/Encyclopaedia Britannica, 1952.

Lovelock, James. *Gaia: A New Look at Life on Earth.* 1979. Reprint with corrections and a new preface, Oxford, U.K.: Oxford University Press, 1995.

Michels, W. J., C. E. Wilson, and J. P. Sadler. *Kinematics and Dynamics of Machinery.* New York: HarperCollins, 1983.

Moroney, M. J. *Facts from Figures.* London: Pelican/Penguin Books, 1951.

Sekuler, Robert, and Randolph Blake. *Perception.* New York: McGraw-Hill, 1990.

Sharpe, Carill, ed. *Kempe's Engineers' Yearbook.* Tonbridge, U.K.: Miller-Freeman, 1996.

Sokolnikoff, I. S., and E. S. Sokolnikoff. *Higher Mathematics for Engineers and Physicists,* chap. 11. London: McGraw-Hill, 1941.

Soucek, Ludvik. *The Story of Communications.* London: Mills & Boon, 1969. A curious but amusing illustrated book translated from the Czech.

Standage, Tom. *The Victorian Internet: The Remarkable Story of the Telegraph.* New York: Walker Publishing, 1998.

Stephenson, Geoffrey. *Mathematical Methods for Science Students.* 2d ed. Harlow, U.K.: Longman, 1973.

Stewart, Ian. *Game, Set, and Math.* Oxford, U.K.: Basil Blackwell, 1989.

Stremler, Ferrel G. *Introduction to Communication Systems.* 2d ed. Reading, Mass.: Addison-Wesley, 1982.

Tables of Physical and Chemical Constants. Originally compiled by G.W.C. Kaye and T. H. Laby, now prepared under the direction of an editorial committee. 16th ed. Harlow, U.K.: Longman, 1995.

Thomson, J. J. *Conduction of Electricity through Gases.* Cambridge: Cambridge University Press, 1906.

Thomson, N., ed. *Thinking Like a Physicist.* Bristol, U.K.: Adam Hilger, 1987.

Usher, Abbott Payson. *A History of Mechanical Inventions.* Cambridge, Mass.: Harvard University Press, 1934. New edition, New York: Dover, 1988.

Vogel, Steven. *Life in Moving Fluids.* 2d ed. Princeton, N.J.: Princeton University Press, 1994.

Web Sites and Periodicals

The U.S. Patent Office, the European Patent Office, and to a lesser extent, the British Patent Office are treasure troves of more than 30 million inventions and discoveries, mostly in English or with English abstracts, from around 1600 to the present day. All three have searchable Web sites: www.uspto.gov/; http://ep.espacenet.com/; and www.patent.gov.uk/.

Scientific American has been around since 1850, and I find the issues before 1930 or so to be particularly interesting today. *New Scientist* is a highly readable United Kingdom–based science weekly. The *Proceedings of the Royal Institution of Great Britain* (London, 1799 on) publishes tidied-up transcripts of lectures aimed at people with a scientific background but no specialist expertise in a field; this journal often provides masterful overviews of a subject. The *Proceedings of the Royal Society* (London, 1661 on) often includes more maverick contributions than most scientific journals. *Eureka Innovative Engineering Design,* a United Kingdom–based magazine, is basically an advertising vehicle for engineering suppliers, but it has short articles on genuinely new technologies.

Late-Victorian technical and scientific journals offer a fascinating glimpse into a world that may have looked rather modern in photographs but was, underneath, vastly different from today's world. One such periodical is *Technics* (1904–1905; it may have continued on after 1905, but these are the only issues I have found).

Index

acoustic emission, from contacts (Singing Contacts), 65
actuators, 286; analog actuator bus, 231
Adams, Douglas, 185
addressing system, in data bus. *See* analog bus system (AMIPLEX)
aerodynamics, development of, 32
air-lift pump, 246
air muscle, 287, 291
Air Products & Chemicals, Inc., 113
aluminum, corrosion of, 309
American Standard Code for Information Interchange (ASCII), 265
Amontons, Guillaume, 94
Ampère's law, 67, 155
amplifier, 148; class-A, 69; Electrolystor, 160; Transformer Transistors, 149
ampullae of Lorenzini, 88
analog bus system (AMIPLEX), 226; actuator bus, 231
analog data systems, 226, 236
analog input board, for personal computer, 15, 322
anemometer, 44
aquaplaning, 94
argon, 224
ASCII (American Standard Code for Information Interchange), 265
atmospheric railroad (Vacuum Railroad), 100

balls, multiple, supported on upward airstream (Juggling Airstreams), 34
"banana" effect, in CRTs, 280
bandgap, of light-sensitive semiconductor, 207
Barlow, Peter, 215
BASIC computer language, 180, 323
Bass, Thomas, 182
batteries, 301, 322
Becquerel, Edmond, 314
Beer-Lambert law, 209, 295
Bell, Alexander Graham, 160, 243
Bernoulli, Daniel, 32
Bernoulli's equation, 26
B field, 155
binary code, 236, 265
Black, Harold S., 158
black body radiation, 165
Blake, William, 198
Blanchard, Mike, 86
bolometer, 218
Bono, Edward de, xi
Boyle's Law, 55, 291
brass, stress corrosion cracking of, 307
bridge circuit, 216
Brook, Basil, 86
Brunel, Isambard Kingdom, 100
bubble: acoustic resonant frequency of, 51; in gas flow measurement, 190; -jet printers, 245; memories, 245; optical switches, 245; telegraph (Bubblegram), 245
bulk modulus, 258
buoyancy force, 137, 224

Callender's Cable & Construction Co., 88
capillary waves, 188
carbon fiber, 145
cathode ray tube (CRT), 279
cement kiln, 129
centripetal acceleration, 132
centripetal catapult (Rotapult), 140

chaotic effects: in droplet formation, 292; in oscillators, 2, 135
chaotic oscillation, 292, 293
chaotic pendulum clock (Chaotic Regularity), 4; correlation between successive swings in, 10
Chladni figures, 189
Ciufolini, Ignazio, 132
clocks, 2; quartz, as source of useful parts, 324
closed-circuit breathing apparatus, 306
closed loop, 207. *See also* feedback; open loop; servo motor
cochlea, 50
coherer, 66
color triangle, 281
combinatorial mathematics, 267
combinatorial Six-Wire Telegraph, 259
communication systems, 234
comparator, 230
compressed gas, energy in, 291
compressibility, 258
compression, of data, 270
"condenseritis," 307
constant-current telegraph, 237
contact, acoustic emission from (Singing Contacts), 65
contact noise, suppression by DC current, 66
convection, 224
conveyors, 122
Cooke, William F., 235
copper oxide: battery, 303; diode, 159
cosmic rays, 312
Coulomb, Charles, 94
Coulter, Wallace, 193; counter, 193
creep, in plastics, 212
CRT (cathode ray tube), 279
Cugnot, Nicolas, 92
current transmitter, 237

Daguerre, Louis, 300
daisy wire. *See* analog bus system (AMIPLEX)
databus. *See* analog bus system (AMIPLEX)
data compression, 271, 274
data glove, 211
datalogger, 15, 322
David and Goliath, 140
Davy, Sir Humphrey, 300
Day of the Triffids, The (Wyndham), 141
DC offsets on electrodes, 86
decoder, 266
delay effects, in human sensory-motor tasks, 257
Denny, Mark W., 88
dial telephone, 236
dice, 176; loaded, 176, 182
Dickens, Charles, 252
Diesel, Rudolf, 149
differentiator circuit, 242
diffraction, 311
diffusion equation, 256
digital data systems, 236
digital multimeter, 322
dipole electric field: 3D, 86; 2D, 87
disappearing filament pyrometer, 218
dispersion, in communication systems, 250
dondo (talking drum), 57
doublethink, 313
drag: Newton's approximation of, 38, 45; on sphere moving in air, 38
drift speed, of ions, 224
droplet formation, 292
drum: dondo, 57; pneumatic, 57
Duke, Thomas, 50
Dymo label-making system, 202
dynamic (sliding) friction, 94

eclipse equation, 210
Edison, Thomas, 160, 221, 234
Edison effect, 221
Einstein, Albert, 132, 170, 313, 318
elastomers, 199
electric field: dipole, 3D, 86; dipole, 2D, 87; effect of, on friction of ice, 113
electric motor, stop positions of, 179
electrodes, DC offsets on, 86
electrolysis, 160, 307; display system, 263
electrolystoronic oscillator, 167
electrolyte-contacted photovoltaic cell (Wet Solar Cell), 313
electrolytes, for batteries, 302
electrolytic amplifier (Electrolystor Amplifier), 160
electrolytic etching, 307
Electronic Elastic, 198
emitter follower, 84, 208
enabling technologies, xii
energy, in compressed gas, 291
Epstein, George, 270
equatorial mounting, for solar tracking, 216
escapement, for model control, 286
Evans, D.C.B., 111
eye, human: lens, 311; pupil, 311; receptors, 281

Faraday, Michael, 189, 300, 301, 318
feedback, 237; in amplifiers, 156, 167, 207; in human-controlled systems, 257
ferrites, 154
festoon lamp, 263
Feynman, Richard, 213, 225
field-follower boat, 79
figure of merit (FoM), 127

filament lamp, 151, 165, 213, 221; lifetime of, 222; radiation from, 165, 217
Fitzgerald, Lieutenant L., 302
Fleming, John, 221
flow cytometry, 193
flow meter, soap-film, 191
fluidized bed, 25
fluorescent colored inks, 283
flywheel, as energy store, 142
FoM (figure of merit), 127
fool's gold, photoelectric effect of, 319
four- to sixteen-line decoder, 266
Fox Talbot, William Henry, 300
freeze/thaw cycle oscillator (Glacial Oscillations), 13
freezing, 13
freezing point depression, 17
French, Tom, 199
frequency analysis, of letters in English, 270
frequency doubling effect, 68
friction: of aluminum and ice, 112; of copper and ice, 112; effect of electric field on, 113; and ice, 108; law of, 93, 125; nanocrack theory of, 94
fusée barrel, 5

gain factor of transistor, variation with temperature, 223
Galileo, 4
gas flow, measurement of, 190
gas flow meter using bubbles (Coulter's Bubbles), 190
gedanken (thought) experiment, 76
Gerde, Eric, 94
Gladstone, William, 301
glugging, musical, 51
Goliath of Gath, 140
Goodyear, Charles, 199
governor, centrifugal, 134
Graetzel, Michael, 317
Grahame, Kenneth, 260
gravity and inertia, 132
Gray, Elisha, 243
Great Western Railway, 100
ground effect, 174
gun, for satellite launch, 220

Hagfeldt, Anders, 320
Hallwachs, Wilhelm, 314
hammerhead shark, 79
hard magnetic material, 156
Henry, Joseph, 235
Hero's steam turbine, 120
hexagonal close packing of spheres, 163
H field, 155
high-pass frequency filter, 243
histogram, instant. *See* analog bus system (AMIPLEX)
Hooke, Robert, 131
hot-wire anemometer, 44
hourglass wheel (Everlasting Hourglass), 20
human senses, 184
Huxley, Aldous, 122
Huygens, Christiaan, 4
hydroclone, 136
hydroplaning, 94
hyperdimensional space, 105

ice, freezing process in, 17; friction and, 108
ideal gas law, 54
impulse, 77
information: content, 265; displays, 276; theory, 265
ink droplet variable light transmission window (Ink Sandwiches), 292
instant histogram. *See* analog bus system (AMIPLEX)
instrumentation, 184
integrator circuit, 243
Internet, boosting capacity of, 274
ionized gas, thermal convection in (Gravity Diode), 220
iron pyrites, photoelectric effect of, 319
irreversible processes, 213
isoprene, 199

jitter, 2
Jones, David R. H., 308

Lalande cell, 305
laminar flow, 26, 250, 255
Langley, Samuel P., 218
Lardner, Dionysius, 226
lateral thinking, xi
lava lamp, 138
Learning Resources Ltd., 291
LED information displays, 280
light-modulating tubular actuator (Light Tunnels), 204
lime kiln, 129
lithium, in batteries, 301
Lloyd's of London, 176
local action, in batteries, 301
low-pass frequency filter, 243
lubricant: high-viscosity, 24; water on ice as, 110

Macintosh, Charles, 199
magamp, 149
magnetic amplifiers (Transformer Transistors), 149
magnetic field strength (H field), 155

magnetic flux density (B field), 155
magnetic glass, 154
magnetic mine, 88
magnetization (M field), 155
Magnus, Heinrich G., 174
Magnus effect, 35, 174
Mamola, Karl, 173
Marconi, Guglielmo, 66, 221
Marder, Michael, 94
Maxwell, James Clerk, 87, 281
Maxwell's equations, 87
mayonnaise, optical opacity of, 202
mechanical random number generator (Motor Dice), 176
melting, 18
Minnaert, Marcel, 55
modular arithmetic, 180
Moivre, Abraham de, 176
momentum, law of conservation of, 38, 76, 143
monorail, Ballybunion to Listowel, 100
Monte Carlo analysis, 180, 323
Morse code, 265
Morse telegraph, 235
multimeter, digital, 322
multiple-valued logic, 270
multiplexed bus system. *See* analog bus system (AMIPLEX)
Murphy's law, 309
muscles, 288

Navier-Stokes equation, 32, 256
Newton, Isaac, 50
Newton's disks, 278
Newton's second law, 7, 59
Newton's third law, 76
Niepce, Joseph, 300
nucleation, 18
Nuclepore membrane filters, 312
nudge button, for Motor Dice, 177

Ohm's law, 319
op amp (operational amplifier), 318
open loop, 208. *See also* closed loop; feedback; servo motor
optical flex sensor, 204
optimization, 105
oscillators: chaotic effects and, 2, 135; in clocks, 2; electrolystoronic, 167; freeze/thaw, 13; high-speed, for random number generator, 180; and pendulum, 4

packing, of spheres on surface, 163
Panjandrum, 115–16
parallel-wire telegraph (Six-Wire Telegraph), 264; combinatorial, 266–70; *x-y* grid, 270–74
Pawle, Gerald, 116
pendulum oscillator, 4
perfect pitch, 184
permanent magnets, 156
permeability, 155
persistence of vision, 279
Petrenko, Victor, 113
Philips Electronics, 150
Philistines, 140
photoelectric effect, 170, 313, 318
photon, 316, 318
photovoltaic cell, 314
ping-pong balls, suspension in air stream, 34
pitch: perfect, 184; of propeller, 44
pitting corrosion, 309
Planck, Max, 313, 318
plug flow, of gas and liquid in tube, 249
pneumatic actuator using balloon/net principle (Balloon Biceps), 287
pneumatic controls, 287
Pneumatic Drum, 57
Pneumatic Morse, 252
Poiseuille's equation, 255
Poisson, Siméon-Denis, 32
Poisson's ratio, 201
probabilistic processes, 176
PROM (programmable read-only memory), 263
propellers, 77
pulsed jet boat propulsion (Giant Putt-Putt Boat), 73
pulse-jet engine, 120
pulse-length code telegraph (Servo Telegraph), 235
pulse oximeter, 204
pyrometer, disappearing filament, 218
Pythagoras' theorem, 86, 208, 290

QBasic, 180
Q factor (oscillator quality), 60
quantum mechanics, 170, 181, 313, 318
quartz, 198

radiation, from filament lamp, 165, 217
railroad: atmospheric, 101; friction of wheels on rail, 93; Vacuum Railroad, 100
Railway: Great Western, 100; South Devon, 101
ram-jet engine, 121
random number generators, 179
RC (resistor-capacitor) frequency filters, 242
reflection, from mirror, magnification of angle, 189
regelation, 110, 113
relativity, theory of, 132
resistivity, change with temperature. *See* temperature coefficient of resistance
resistor-capacitor (RC) frequency filters, 242

resonant frequency of bubbles (Musical Glugging), 51
reverse ice vehicle (Naggobot), 108
reversible processes, 213
revolution counter using waves (Coffee-Cup Revolution Counter), 185
Reynolds number, 32, 38, 250, 256
ripples, radial, 189. *See also* waves
rotarudder, 85
rotary kiln, 129
rotating disc displays (Moving Messages), 276
rotating liquid-filled bottle (Centripetal Chaos), 134
rotating/spinning tube (Waltzing Tube), 171
rotating tube conveyor (Tubal Travelator), 122
Royal Philips Electronics, 150
rubber: elastic properties of, 198; light transmission through, under strain (Electronic Elastic), 198
Rutherford, Lord, 63, 224

saccades, 280
sail-driven railroad car (Railroad Yacht), 41
sand: flow of, 27; and hourglass, 23; in Tubal Travelator, 128
sand-glass, 20
sand yacht, 43
satellite launch, using gun, 220
saturation: of amplifier, 167; of magnetic material, 153
Sayre glove, 211
Schrödinger's equation, 151
scintillation, of alpha particle on screen, 63
Scott, Robert Falcon, 108
seals, pressure-actuated, 106
selenium diode, 159
senses, human, 184
servo motor, 84, 207, 235; for model airplane, 240, 286
seven-segment display telegraph (Seven-Segment Telegraph), 259
Shakespeare, William, 264
shift register random generator, 180
Shute, Neville, 116
silicon, photoelectric effect in, 319
silver, photoelectric effect in, 319
skating equation, 112
sliding (dynamic) friction, 94
sling, 140
slippage, in air-lift pump, 246
soda bottle, as pressure vessel, 291
soda lime, as CO_2 absorber, 306
sodium-sulfur batteries, 302
sodium thiosulfate, exothermic crystallization of, 18
Soemmering, Thomas, 259, 263
soft focus, 311
soft magnetic material, 156
solar cell (Wet Solar Cell), 313
solar tracker (Reverse Electric Lamp Solar Tracker), 213
South Devon Railway, 101
space character, 260
speaking tube, 252
spinnaker, 44
spreadsheet software, 323; in simulation, 8, 180
square loop magnetic material, 153
stackfreed mechanisms, 5
static friction, 94
statistics, of dice throws, 179
Stefan-Boltzmann law, 165
Stokes, George, 32
stress corrosion cracking, 307
stun shot, 144
sunglasses (Unusually Cool Sunglasses), 307
superconductors, and magnetic propulsion of ships, 72
supergun (V-3), 221
suppliers, 321
surface tension, 250
Swan, Joseph, 221

talking drum (dondo), 57
tea leaves, migration of, 185
technologies, enabling, xii
teleautograph, 243
telegraph, 235; bubble, 245; constant-current, 237; Morse, 235; parallel-wire, 264; servo, 235; seven-segment display, 259; six-wire, 259; ticker-tape, 234; Wheatstone, 235
telewriter, 243
temperature coefficient of resistance, 216; of copper, 218; of tungsten, 216
temperature coefficient of transistor gain, 223
Tennyson, Alfred Lord, 115
thermal battery (Red-Hot Batteries), 301
thermal convection, 224; in ionized gas (Gravity Diode), 220
thermionic vacuum tube, 221, 225
thiosulfate, sodium, exothermic crystallization of, 18
three-cornered hat, 163
three- to eight-line decoder, 266
ticker-tape telegraph, 234
tip jet principle, 116
toboggan, aluminum runners for, 112
transformer, 150
transportation spectrum, 123
travelators, 122
Trevithick, Richard, 42

Twain, Mark, xii, 252
twisted-string actuator, 97

unary code, 235
uniselector, 236
units, 324

V-3 supergun, 221
vacuum tube, 225
variable light transmission window using ink droplets (Ink Sandwiches), 292
Verne, Jules, 220
Visual Basic, 180
Vitruvius' Odometer, 120
vortex rings, 76
vulcanization, 199

Wall of Death (ride), 126
water clocks, 20, 25
Watt, James, 134
waves: gravity, on water, 188; on membrane, 60; surface tension (capillary), 188
Wheatstone, Sir Charles, 216, 235, 265–66, 276
Wheatstone bridge circuit, 216
whirling arm vehicle (Boadicea's Autochariot), 115
wind-driven railroad car (Railroad Yacht), 41
window, variable light transmission, using ink droplets (Ink Sandwiches), 292
wind tunnels, 32
worm, motorized multi-segment (Electric Worms), 93
Wyndham, John, 141

x-y parallel-wired telegraph (Six-Wire Telegraph), 270, 275

Zimmerman, Thomas, 211